Biofuels:
Potential and Challenges

Biofuels:
Potential and Challenges

Editors

A.K. Pandey and A.K. Mandal

Tropical Forest Research Institute,
Jabalpur

SCIENTIFIC
PUBLISHERS (INDIA)
www.scientificpub.com

Published by:
Scientific Publishers (India)
5-A, New Pali Road, P.O. Box 91,
Jodhpur – 342 001 (India)

E-mail: info@scientificpub.com
www.scientificpub.com

Print:2021

ISBN: 978-81-7233-696-7

Typeset by Rajesh Ojha

Printed in India

PREFACE

In recent times serious concerns have been expressed both by developed as well as developing economies of the world over the increase in global warming, ozone layer depletion and number of other environmental problems occurring due to massive carbon footprints left by use of fossil fuels as a result of exponentially amplified demand in transportation and other purposes. The rising and volatile prices of petroleum products compelled entire world to search for new and more efficient technologies and alternative sources of energy. The huge energy demand has forced India to depend on other countries for oil and the country is spending huge amount due to high price of the crude oil in the international market. Biofuels, therefore, assume great importance and has become a high priority. The term biofuel is referred to as liquid or gaseous fuels that are predominantly produced from biomass. The biofuels that are currently the focus of attention are: (i) bioethanol (ii) biodiesel (iii) biomass derived fuels (iv) biogas and (v) oil bearing algae. The major strategic advantages of using biofuels are: (a) providing energy security (b) saving foreign exchange (c) promoting environmental security (d) meeting climate change commitments (e) promoting renewable energy sources and (f) generating rural employment opportunities.

The biofuel industry in India is poised to make important contributions to meet India's energy needs by supplying clean, environment friendly fuel. India commenced on its biofuel use journey in 2003. Basically, the Indian biofuel sector is driven by ethanol. The ethanol industry derives much benefit from improved agricultural practices in sugarcane cultivation, more efficient production processes and the use of alternate feedstocks including cellulosic material. On the other hand, the biodiesel industry is at the incubation stage and large-scale cultivation of Jatropha and other species and the infrastructure for oilseed collection and oil extraction is being established so that the industry can be placed on a rapid growth track. The biofuel sector is constantly developing in the country.

In order to deliberate the key issues by scientific and research community and industry to accelerate the growth of biofuel industry, Tropical Forest Research Institute, Jabalpur organized a National Conference on **"Biofuels: Potential and Challenges"** from 25–26 February, 2009. The conference has brought together researchers, policy makers, industries and all other stakeholders and had productive discussions on how best to meet India's growing biofuel needs.

Papers presented during the conference have been screened, edited and included in this proceedings. Following themes are covered:

- Biofuels: Global and Indian scenario
- Biofuel potential: Current and future prospects
- Biofuels and environment
- Biofuels: Challenges and sustainability issues
- Biofuels plantation development and management

We hope that the publication would provide a valuable source of knowledge to enhance the understanding of biofuels.

Biofuel cannot be the *saviour* alone, but it has a key role to play by both displacing fossil fuels and sequestering atmospheric carbon.

Editors

CONTENTS

Biofuels:
Global and Indian Scenario

Chapter - 1

BIOFUELS: AN OVERVIEW

A.K. Pandey and Pankaj Bhargava

Tropical Forest Research Institute, P.O. RFRC, Mandla Road, Jabalpur 482 021 (M.P.) E-mail: akpandey10@rediffmail.com

ABSTRACT

Biofuels are alternative to petro-based fuels, derived from biomass and being used to generate energy. At present they have become a focal point on energy solutions all over the world for energy security, trade balance, green house gas (GHG) emissions reduction, rural development and income generation. This article provides an overview on biofuels, comprising of their need, type/generations, potential feedstock options and various other related issues. It highlights the global and Indian scenario with respect to bioethanol and biodiesel production. Important researchable issues, challenges, constraints and possible drawbacks of biofuels are also discussed.

Keywords: Biofuels, energy security, potential feedstock, researchable issues, challenges.

INTRODUCTION

Energy is one of the major inputs for the economic development of any country. The energy sector assumes a critical importance in developing countries particularly due to ever-increasing energy needs; fluctuating petroleum prices; depleting fossil fuel reserves and increasing environmental concerns. Moreover, the need of energy is increasing continuously, because of increases in industrialization and population.

The basic sources of energy are fossil fuels, solar, wind, hydro and nuclear (Kulkarni and Dalai, 2006) and 90% of the commercially produced energy is from fossil fuels. Fossil fuels comprise of petroleum, natural gas and coal, are finite, exhaustible and non-renewable in nature. Therefore there is a growing demand for judicious use of energy and to explore better renewable alternatives. The burning of an enormous amount of fossil fuel (petroleum, natural gas and coal) has increased the CO_2 level in the atmosphere, causing global warming and climate change.

Consequently, there is a lot of concern among developing countries for evolving efficient energy crisis management plans. Much of the energy supply in the world comes from geo-politically volatile economies. With world energy consumption predicted to increase 54 % between 2001 and 2025 and global demand for petro-based energy sources to 40% by 2025 (NETL, 2005), considerable focus of world economies is being directed towards the search and development of renewable, sustainable and carbon neutral energy sources to meet future needs (EIA, 1999). The only way to safeguard the economy from above mentioned issues is to lower the dependence on petroleum in the economy. The most practical and least disruptive strategy to achieve this objective is to use alternative fuels. In order to enhance energy security, many countries, have been emphasizing production and use of renewable energy sources such as bioenergy, which is also emerging as an industry in the current economic environment.

The term "Bio" is a Greek prefix meaning "life" or a living organism and is a modifier or a qualifier that expands meanings to form new words, e.g., biology or biological. In the context of bioenergy, the prefix "bio" suggests that the energy is derived from recently living material excluding fossil fuels (oil, natural gas, coal) even though fossil fuels are derived from plants and animals. Besides alleviating the reliance of energy-driven economies on limited fossil fuel sources, bioenergy is continually to receive increasing attention from those concerned with promoting agricultural and environmental sustainability through the reduction of carbon emissions, which is an important component of climate change mitigation. Bioenergy is also considered by some to be a potentially significant contributor towards the economic development of rural areas and a means of reducing poverty via creation of employment and improving quality of living-supporting the achievement of important 'Millennium Development Goals' (FAO, 2005).

Biomass has been the most important energy source for humans since the discovery of fire, and today it is still the main source for almost half of the world's population. It is defined as the biodegradable fraction of products, waste and residues from organic non-fossil material of biological origin that is readily available in a renewable or recurring sustainable basis and also can be used as a energy source by means of thermal/chemical/biological conversion. Biomass being a renewable source of energy is seen as a long-lasting and sustainable solution to the energy crisis. They are described as renewable, representing their capability to be replenished mainly because of their inexhaustible supply, ability to re-grow or to be recycled. They derive energy from organic materials that contain chemical energy derived from sunlight. Examples of biomass are wood and its waste, straw, organic materials such as starch, oilseeds and animal fats, or cellulose, animal manure, sugar cane and many other agricultural products and byproducts. Biomass sequesters the atmospheric CO_2 through photosynthesis. If biomass is grown in a sustainable manner, its combustion will not have any impact on the CO_2 balance in the atmosphere, because the CO_2 emitted by the burning of biomass is offset by the CO_2 fixed by photosynthesis (Macedo, 1992 and Shay, 1993). Large-scale introduction of biomass energy could contribute to sustainable development on several fronts, environmentally, socially and economic (Turkenburg, 2000). However, it is important that energy from biomass be harnessed in a sustainable manner to provide an attractive alternative to current petro-based fuels. These alternative fuels derived from biomass are called as biofuels.

Biofuels are defined as organic primary and/or secondary fuels derived from biomass which can be used for the generation of thermal energy by combustion or by other technology. They comprise purpose-grown energy crops, as well as multipurpose plantations and byproducts (residues and wastes) (FAO 2000). They can be utilized with little change to current technologies and have significant potential to improve sustainability and reduce greenhouse gas emissions.

Why biofuels are needed?

Biofuels produced from microorganisms, plants, vegetable sources, grains, corns and organic wastes are superior fuel than fossil fuel from the environmental point of view. Moreover, the use of biofuels becomes compelling in view of the tapering of automotive vehicle emission standards and court interventions. Other benefits of biofuels usage are:

- Reduce dependence on crude oil imports.
- Provide energy security, especially for the rural areas as suitable renewable energy options for decentralized distributed generation (DDG) of electricity and for motive power applications (water pumping, milling, etc.).
- Create employment for the rural people through plantation, seed procurement, oil extraction.
- Provide nutrients to soil, checking soil erosion and land degradation through plantation of biofuel crops.
- Rehabilitate degraded lands and wastelands through greening.
- Address global concern related to carbon emissions as provided in the framework convention on climate change.

Currently biofuels are important agenda on energy discussions all over the world for energy security, trade balance, GHG emission reduction, rural development and income generation.

According to Inter-governmental Panel on Climate Change (2007), the record high oil prices as well as concern over potential threats to energy security and the global environment have made biofuels an urgent global priority. Biofuels can increasingly satisfy the energy needs in an environmentally benevolent and cost effective manner while reducing dependence on import of fossil fuels and thereby providing a higher degree of energy security. The transition from a fossil fuel-based economy to a biofuel-based economy depends on the improvement of biofuel yields via the successful and sustainable development of biofuel feedstock options and technologies.

Biofuel feedstock options

A specific plant or substance used for generating or producing bioenergy is called a feedstock. The feedstocks are usually converted into a more easily usable form, which may be solid, liquid or gaseous fuel. The search for suitable sources of biomass for generation of biofuels is actively going on in different parts of the world. Every country is looking for the suitability of biofuel feedstock depending upon its geographical position, energy needs and abundance of biomass. Since biofuels can be produced from a diverse set of crops, each country is adopting a strategy that exploits the comparative advantages it holds with respect to such crops. For example, sugarcane and maize are the main feedstock for ethanol in Brazil and USA, respectively together accounting for 90% of the global production of ethanol, while palm oil from Malaysia is the single largest source of biodiesel globally (Martinot, 2005).

Large amounts of biomass from forest and agricultural activities such as branches, tree tops, straw, corn stover and bagasse from sugarcane can be utilised as feedstock for bioenergy. Biofuel currently falls into two categories: ethanol, which is compatible with petrol engines, and biodiesel, which is compatible in diesel engines. Many plant species can provide suitable biofuel yields. Ethanol is currently made from two basic feedstocks: starch-based feedstocks, such as corn, grain, wheat, barley and grain sorghum; and sugar-based feedstocks, such as sugarcane, sugar beets, fruits, citrus molasses and cane (sweet) sorghum. Biodiesel has more diverse feedstocks than ethanol. It is derived from oilseeds, primarily rapeseed (canola), sunflower, soya and palm oil, as well as animal fat and vegetable waste products. While these feedstocks currently dominate biofuel production, feedstocks will likely expand into more environmentally beneficial cropping systems in the near future. For example, pilot plants are being developed for cellulosic ethanol production, which uses cellulose rather than plant starches. Potential feedstocks for cellulosic ethanol include dedicated energy crops such as grasses and trees, as well as residues from agriculture (wheat straw, corn stover, and sugarcane bagasse), industry (paper pulp, sawdust) and forestry. Scientists are also researching other liquid biofuel and feedstock options.

Classification of biofuels

Biomass can directly or indirectly be converted to biofuels which can be in solid, liquid or gaseous forms.

Liquid biofuels

Large quantities of liquid biofuels are presently used in many countries and the potential exists to greatly expand their use in the future. The two most common kinds of liquid biofuels are ethanol and biodiesel, but a range of other liquid fuels exist and are also being developed.

Ethanol is currently produced in large quantities by fermenting the sugar or starch portions of agricultural raw materials. The feedstocks used for ethanol production vary by region e.g. sugar cane in Brazil, grain and corn (maize) in North America, grain and sugar beets in France, etc. Because ethanol from sugar and starch directly competes with food production, people are working to commercialize technologies to produce ethanol from cellulose, which makes up the bulk of all plants and trees and is inedible.

Biodiesel is typically composed of methyl (or ethyl) esters of long chain fatty acids derived from vegetable oils or animal fats. It is produced chemically by transesterification of oils obtained from the pressing seeds of oil-seed plants both edible like rapeseed, soyabean and the fruits of oil palms and non-edible, like jatropha and karanja etc. Waste cooking oil can also be converted to biodiesel.

Other liquid biofuels are Straight Vegetable Oil (SVO), Biobutanol, Biokerosene and Pyrolysis oil. SVO or Pure Plant Oil (PPO) is a diesel type fuel. It occurs naturally in plant seeds such as sunflower, soyabean, rapeseed, jatropha, karanja and many others. Waste cooking oil can also be converted to SVO. SVO can be used in its pure form or mixed with diesel/biodiesel. Biobutanol (butanol or butyl alcohol) is an alcohol similar to ethanol but with a higher energy density produced by fermentation of biomass.

Solid biofuels

Solid biofuels include wood, manure or charcoal burned as fuel as well as more recent innovations like high-density clean burning pellets. Solid biomass can be burned for heat or to produce electricity either by itself or as part of a co-firing power plant. Wood can be utilized for bioenergy in the following forms: as firewood, wood charcoal (charcoal), wood-fired biomass boilers, wood gasification (wood gas) - especially waste wood, wood pellets and wood residues.

Charcoal is the dark grey residue consisting of impure carbon obtained by removing water and other volatile constituents from animal and vegetation substances. Charcoal is usually produced by slow pyrolysis, the heating of wood, sugar, bone char or other substances in the absence of oxygen. The resulting soft, brittle, lightweight, black, porous material resembles coal and is 50% to 95% carbon.

Wood and other forms of biomass (organic wastes) can be pressed into pellets (wood or biomass pellets).Due to their low moisture content, regular shape and high density, pellets can be burned very efficiently and are relatively easy to transport. They are often used for heating and electricity generation.

Gaseous biofuels

The most common kind of gaseous biofuel is biogas or biomethane, which is composed mostly of methane and carbon dioxide. It is produced from the anaerobic digestion or fermentation of biomass including manure, sewage sludge, municipal solid waste, biodegradable waste or any other feedstock. Biogas can either be burned to produce heat and electricity or purified to be used as a vehicle fuel, sometimes mixed with natural gas. Other type of gaseous biofuel is synthetic natural gas (SNG) or biopropane. It is generated by gasification or fermentation of biomass and additional methanation and cleaning. It is also called as syngas.

Another potentially important gaseous biofuel is bio-hydrogen. It is widely acknowledged that hydrogen is an attractive energy source to replace conventional fossil fuels, both from the environmental and economic perspective. When hydrogen is used as a fuel it generates no pollutants, instead it produces water which can be recycled to make more hydrogen. Bio-hydrogen is produced by using fermentative biological processes utilizing organic wastes as raw material.

Generation of biofuels

Biofuels can be grouped in 'generations' according to the type of technology they rely on and the biomass feedstocks they convert into fuel. Overview of different biofuels with their related feedstock and conversion processes is presented in Table 1.

First generation biofuels

First-generation biofuels are biofuels made from sugar, starch, vegetable oil, or animal fats using conventional technology (Bioenergy, 2010). The basic feedstocks for the production of first generation biofuels are often seeds or grains such as sunflower seeds, which are pressed to yield vegetable oil that can be converted to biodiesel, or wheat, which yields starch that is fermented into bioethanol.

Table 1. Overview of the first and second generation biofuels with their related feedstock and conversion processes

	Generic name	Chemical composition	Feedstocks	Technology
1st generation	Biodiesel	Methyl or ethyl esters of fatty acids (FAME); Hydrocarbons (products of cracking)	Oil crops (e.g. palm, soyabean, rapeseed, canola jatropha, karanja etc.), waste oil (e.g. frying oil), and animal fats	Cold/hot pressing, extraction & transesterification (Homogeneous, heterogeneous, and bio-catalysis); Hydrogenation (hydrocracking)
	Bioethanol	Ethanol	Sugars (glycosides) and starch from biowaste and woody biomass (sugarcane, sugar beet, cereals)	Hydrolysis & fermentation
	Vegetable oil	Straight Vegetable Oil (SVO) – triglycerides of fatty acids	Oil crops (e.g palm, soyabean, rapeseed, canola, jatropha, karanja etc.)	Cold/hot pressing, extraction, and purification
	Biogas	Methane, hydrogen and light hydrocarbons	Biomass (humid)	Anaerobic digestion
2nd generation	Cellulosic bioethanol	Ethanol	Lignin, cellulose and hemicellulose from bio-waste	Lignin, cellulose and hemicellulose from bio-waste
	Bio-SNG	Lignin, cellulose and hemicellulose from bio-waste	Lignocellulosic biomass	Pyrolysis, gasification, methanation
	Synthetic biofuels	Hydrocarbons (BTL/FT), methanol (biomethanol), mixed heavy alcohols, dimethyl ether (bio-DME)	Lignocellulosic biomass	Pyrolysis, gasification, synthesis
	Biohydrogen	Hydrogen	Lignocellulosic biomass	Pyrolysis, gasification, water gas shift reaction (WGSR)

Second generation biofuels

The biofuels obtained from non-food items are termed as second generation biofuels. These include waste biomass, the stalks of wheat, corn, wood, agricultural wastes and byproducts. Second generation (2G) biofuels use biomass to liquid technology (Oliver and David, 2009), including cellulosic biofuels (Somerville, 2008). Cellulosic ethanol (bioethanol) production uses non-food crops or inedible waste products and does not divert food away from the animal or human food chain. However, lignocellulose, the woody structural material of plants is abundant and potential biofuel feedstock. Scientists working in New Zealand have developed a technology to use industrial waste gases from steel mills as a feedstock for a microbial fermentation process to produce ethanol (Fisher and Lawrence, 2007; Voegele, 2009).

Third generation biofuels

Third generation biofuel, is a biofuel from algae. Algae are low-input, high-yielding feedstocks to produce biofuels. It is claimed that algae can produce up to 30 times more energy per acre than land crops such as soyabeans (Hartman, 2008), but these yields are to be realised.

One advantage of many biofuels over most other fuel types is that they are biodegradable and so relatively harmless to the environment if spilled. Second and third generation biofuels are also called advanced biofuels.

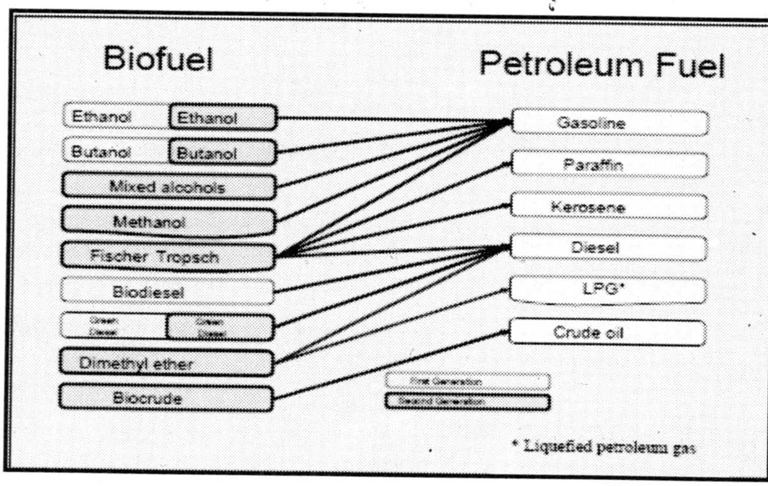

*Source : United Nations Conference on Trade and Development (2008)

Fig. 2. Substitutability of biofuels with common petro-based fuels

The substitutability of various biofuels for petro-based fuels is presented in Figure 2. Alcohol fuels can substitute for gasoline in spark-ignition engines, while biodiesel, green diesel (SVO) and DME (dimethyl ether) are suitable for use in compression ignition engines. The Fischer-Tropsch process can produce a variety of different hydrocarbon fuels, the primary one of which is a diesel-like fuel for compression ignition engines. While there is much attention on biofuels for the transport sector, the use of biofuels for cooking (Figure 3), has wide relevance globally, especially in rural areas of developing countries. In all cases, combustion of biofuels for cooking will yield emissions that are lower (or far lower) than emissions from cooking with solid fuels. Some 3 billion people in developing countries cook with solid fuels and suffer severe health damages from the resulting indoor air pollution (WHO, 2006; IEA, 2006). Thus, biofuels could also play a critical role in improving the health of billions of people. The comparison between the first and second generation biofuels on different aspects is presented in Table 2.

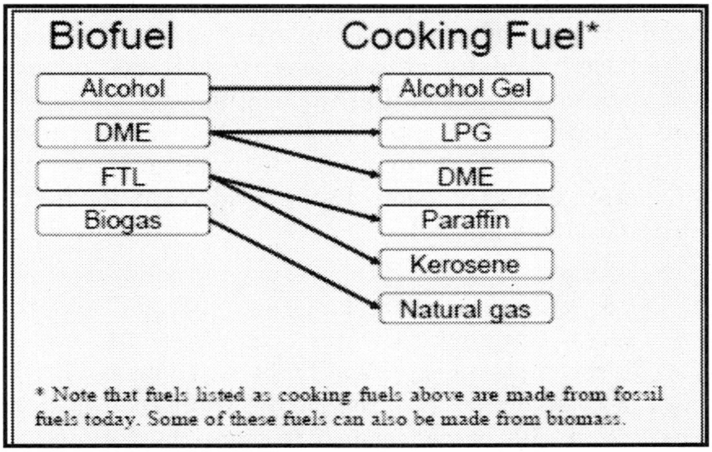

*Source: United Nations Conference on Trade and Development (2008)

Fig. 3. Substitutability of biofuels for clean fossil fuels used for cooking

Table 2. Comparison between First and Second generation biofuels

Parameter	1st Gen.	2nd Gen.
Biofuels readily usable in existing petroleum infrastructure	Yes	Yes
Proven commercial technology available today	Yes	No
Relatively simple conversion processes	Yes	No
Need of markets for byproducts of biofuel production	Yes	Yes/No
Capital investment per unit of production	Lower	Higher
Feedstock cost per unit of production	Higher	Lower
Total cost of production	High*	Lower
Minimum scale for optimum economics	Modest	Large
Land-use efficiency	Low	High
Food vs. Fuel competition	Yes	No
Feasibility of using marginal lands for feedstock production	Poor	Good
Ability to optimize feedstock choice for local conditions	Limited	High
Potential for net reduction in fossil fuel use	Modest*	High
Potential for net reduction in petroleum use	Good*	Better
Potential for net reduction in greenhouse gas emissions	Modest*	High

*Except for first-generation Brazilian sugar cane ethanol, that would get a more favourable mark.

Source:United Nations Conference on Trade and Development (2008)

Environmental and Social Issues

Biofuel feedstock production can have negative or positive environmental and social effects depending upon the local situation and factors. These include the feedstock type, the methods/technologies used to process or cultivate and harvest them and moreover, what would be the alternative land use particularly in case of energy crops. Biofuel feedstock production could lead to deforestation and raise issues of land tenure, water use and pollution, representing imperative and politically delicate issues in many countries. The capacity for long-term land use planning and enforcement is important to avoid or minimize the detrimental impacts of unsustainable expansion. Using food crops for biofuel production can also disrupt food markets.

Carbon emissions of biofuel feedstock production are also a controversial issue. Emissions from land use changes and from diesel combustion in farm equipment, water pumping, and production of fertilizer all erode the carbon benefits of biofuels. Different methodologies for estimating life cycle emissions produce different results. Most appear to arrive at estimates that show modest to significant emission benefits (Macedo *et al.*, 2008; Wang *et al.*, 2007; Armstrong *et al.*, 2002; Sheehan *et al.*, 1998). The change in greenhouse gas (GHG) emissions per mile travelled by replacing petro-based fuels with biofuels is depicted in Table 3. The net carbon loss of land conversion also needs to be considered and is one of the greatest sources of debate and uncertainty. Soil and plant biomass are the two largest biologically active stores of terrestrial carbon and hold about 2.7 times more carbon than the atmosphere. If land is cleared to allow for cultivation of food or energy crops, the carbon contained in the standing biomass and some of the carbon stored in the soil will be released in the atmosphere. A "carbon debt" is thus incurred when native ecosystems are converted to cropland. This carbon debt is the difference between the amount of carbon stored in standing biomass and soil before land clearing and that of the crop grown in its place. The actual carbon debt is thus highly dependent on the type of ecosystem that is being cleared and the crop that replaces it.

Table 3. Change in GHG emissions per mile traveled by replacing petro-based fuels with biofuels

Feedstock	Change from gasoline or diesel	Source
Sugar	-81%	Macedo *et al.*, 2008
Corn	-19%	Wang *et al.*, 2007
Cellulose	-86%	Wang *et al.*, 2007
Wheat	-47%	Armstrong *et al.*, 2002
Soya bean	-78%	Sheehan *et al.*, 1998

Global scenario

World biofuel potential depends on the biofuel feedstocks, their amount that can be grown on available land, biomass conversion technologies, biofuels production capacity and the economic competitiveness of biofuels compared with other options.With the development of motorized vehicles in the beginning of the 20th century, visionaries such as Henry Ford and Rudolf Diesel advocated for producing liquid transportation fuels from biological materials. But because public policy and extensive research and development supported cheap fossil fuels, these bio-based industries never came to existence. However, the oil shocks of the 1970s, once again brought both alcohol (ethanol) and combustible oil seed extracts (biodiesel) to the public attention. Brazil capitalized on its large sugarcane industry in 1980s and began producing sugar-based alcohol for automobile use. At the same time in the United States, corn-based alcohol was blended with gasoline to produce "gasohol" (10% ethanol) for cars. Political and industrial support for alternative fuels increased modestly until recently. Governments around the world have now created financial incentives for biofuel due to factors including industry lobbying, increased fossil fuel prices, climate change concerns and the need for energy independence.

Demand for biofuel is rising especially due to mandatory blending requirements adopted by large energy consumer countries. In order to contribute to energy security and to abide by the requirements of the Kyoto protocol, many have developed ambitious plans to further substitute biofuel for fossil fuel. However, the share of biofuels in overall global fuel consumption was marginal in 2006 (less than 1%) but the growth rate of biofuel production is enormous (IEA 2007b).

Ethanol, a predominant biofuel, is produced today from sugarcane and cereal grains mainly corn. Biodiesel is produced from oilseeds or palm oil. Table 4 & 5 lists the major ethanol and biodiesel producing countries in the world. The United States (US) is the largest producer and consumer of ethanol in the world, producing about 37% (4.86 billion gallons) of the world ethanol production (13.1 billion gallons). Brazil is the second largest producer with 36% (4.76 billion gallons) of world ethanol production. As seen from Figure 4, the world ethanol production has grown rapidly at a compound growth rate of 10 percent per annum since 1975, and at 23 percent per annum from 2001 through 2006. This acceleration may be attributed to push towards ethanol in the United States. Similarly, world biodiesel production has grown at a rate of 35 percent per annum since 1991; the majority of the boom comes from the biofuel initiative in the European Union countries. As depicted in Table 5, Germany is the leading producer of biodiesel (41% of world market share) with the production of 799 million gallons, followed by the US (20%), France (11%), Italy (7%) and other countries. Together, bioethanol and biodiesel account only for around 2% of the global annual consumption of 1200 billion liters of gasoline (in energy equivalents) (Dufey, 2006). However, the contribution of biofuels to energy supply is expected to expand rapidly.

The international trade in biofuels is still in an early stage and it is interesting to see the potential for trade in biofuels amongst the major producing countries. Brazil invested heavily in ethanol production during the energy crisis of 1970s and now has one of the world's most advanced production and distribution systems. The US has proposed 36 billion gallons of alternative fuel by 2022 which would replace about 15% of gasoline consumption in the country. The Europoean Union (EU) is also targeting a 10% share of biofuels in the transport fuel market by 2020.

For production of ethanol, US, China, France, Germany, Russia, and Canada mainly use corn as their main feedstock, whereas, Brazil and India use sugarcane, which is more energy efficient. In the US, about 90% of the ethanol is produced from corn (about 22% of total corn production in 2007) and in China, about 80% of the ethanol is corn-based, with the remainder produced from cassava and wheat (Konishi and Koizumi, 2007). For biodiesel production, all the remaining countries in Table 5 use rapeseed as their main feedstock, except for the US which uses soyabeans.

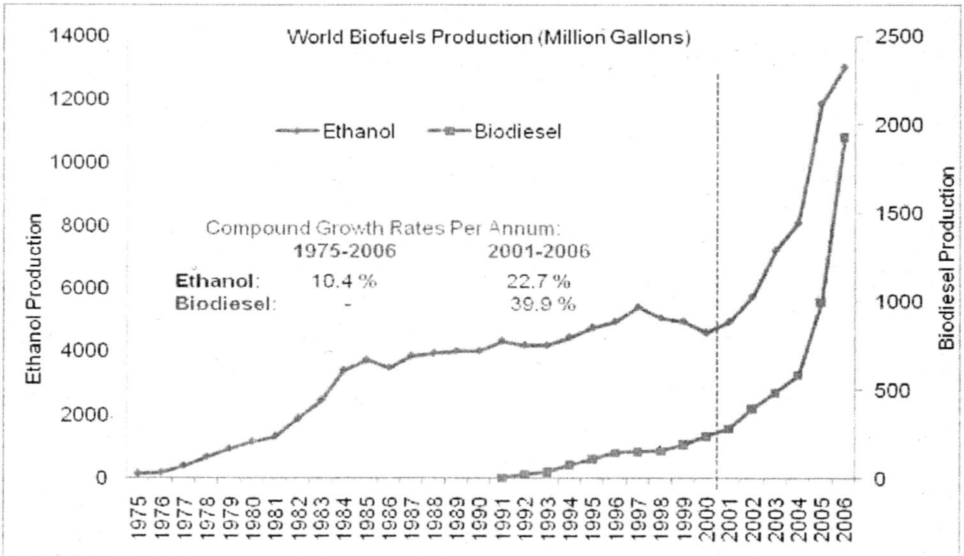

Fig. 4. World Production of Ethanol and Biodiesel (million gallons)
Source : Earth Policy Institute, 2006

Table 4. The major ethanol producers in the world

Country/Region	Ethanol production (Million gallons)	Share of total ethanol production (%)
United States	4,856	37.3
Brazil	4,763	36.5
China	1,083	8.3
India	486	3.7
France	251	1.9
Germany	202	1.5
Russia	171	1.3
Canada	153	1.2
Other countries	1,068	8.2

Data sources: Earth Policy Institute, 2006; Renewable Fuels Association, 2007; European Biodiesel Board, 2007.

Table 5. The major biodiesel producers in the world

Country/Region	Biodiesel production (Million gallons)	Share of total biodiesel production (%)
Germany	799	41.4
United States	385	20.0
France	223	11.6
Italy	134	7.0
United Kingdom	58	3.0
Austria	37	1.9
Poland	35	1.8
Czech Republic	32	1.7
Other countries	227	11.8

Data sources: Earth Policy Institute, 2006; Renewable Fuels Association, 2007; European Biodiesel Board, 2007.

Indian scenario

India is steadily gaining importance in the world energy scene. It is already fifth largest economy and a major energy importer. The Indian economy is expected to grow at a rate of over 6% per annum and the petroleum imports are projected to rise to 166 MT by 2019 and 622 MT by 2047 (TERI, 2002). Figure 5 provides the oil production and consumption forecast in India (IEA, 2002). The domestic supply of crude oil will satisfy only about 22% of the demand and the rest will have to be met from imported crude (Mandal, 2005). India is a big importer of crude oil and spends about Rs. 341887 crores of foreign exchange every year to meet 75% of its oil needs (GOI, 2009). There is a growing need for energy security as any disturbance in the supply of petroleum fuels or increase in petroleum prices can have negative impact on the growth of Indian economy.

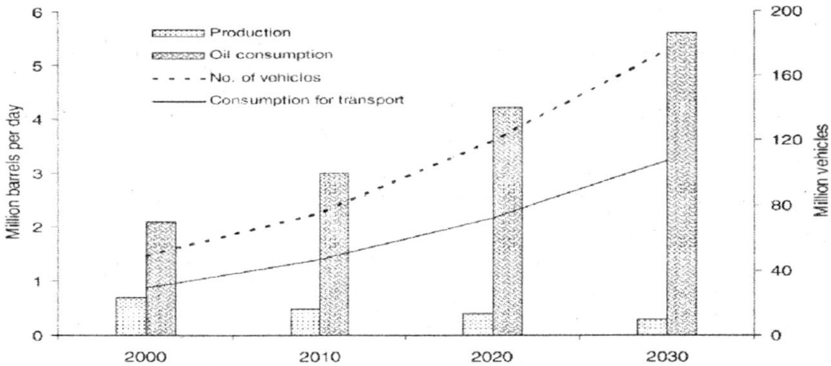

Fig. 5. Forecast of oil production, number of vehicles and oil consumption in India
Source: International Energy Agency (2002)

India has lot of forest residue, agricultural residue, wild grasses, non-fodder crops, etc., which may be used for the biofuel production programme. But a thoughtful selection of the biomass is required to avoid any ecological imbalance. The interest in biofuels has grown dramatically during the last few years. The chief rationale for biofuels in India is energy security. Better environmental performance, greening of wastelands and creation of new employment opportunities are seen as some of the other advantages of biofuels. The two biofuels that are currently the focus of attention are (i) bioethanol and (ii) biodiesel.

Biofuels have generated considerable interest among the government, research community, industry and general public. The development of biofuels in the India can be tracked by the key milestones outlined in following table.

Table 6. Milestones in biofuel development in India

Period	Milestone
1950	Use of biofuels- Ethanol (Power alcohols) in World war II
1975	Large number of committees and studies undertaken
1979-80	Trials on cars and other vehicles undertaken by Indian Oil Corporation (IOC) & Indian Institute of Petroleum (IIP) with 10 and 20% ethanol blend

1990-92	Trials undertaken in Delhi on around 92 cars
2001-02	Pilot project involving trial in 3 oil depots (2 Maharashtra & 1 UP) to cover around 350 petrol stations
2001	Statement of Minister of Petroleum in parliament in December for the rationale/benefits of ethanol/biofuels and intention to blend
2002 (March)	S.J.Chopra's committee report concluding that ethanol was the oxygenate for blending with petrol was placed in parliament
2002 (September)	Notification for blending of 5% ethanol in 9 states and 4 union territories from January 2003
2003 (March)	Autofuel policy in reference to biofuels
2003 (April)	Report of the Committee on Development of Biofuel by Planning Commission
2003	A national mission on biodiesel (NMB) was proposed to be constituted with the Ministry of Rural Development as the nodal ministry. Under the proposed demonstration phase of NMB, Jatropha plantations on 0.4 million ha. of wastelands were to be raised
2003	Reference to Tariff commission by Ministry of Petroleum and Natural Gas to determine price of ethanol
2005 (October)	Announcement of biodiesel purchase policy
2008 (September)	Announcement of National Biofuel Policy
2009 (December)	Announcement of revised National Biofuel Policy

*Source : Planning Commission, Government of India

National policy on biofuels

The Government of India approved a National Policy on Biofuels in December 2009, setting an indicative target to raise blending of biodiesel with diesel to 20% by 2017 and scrapping taxes and duties on biodiesel. Moreover, well funded government programmes for rural development are already undertaken to subsidize the establishment of biodiesel plantations on a large scale throughout India. While the federal policy has only recently been approved, several state governments took the lead and established their own biofuel policies, each setting its own priorities and employing particular policy mixes. Key features of the policy are:

- 20% blending of biofuels is targeted by 2017.
- A certification mechanism would be put in place for the blending exercise that would have to conform to Bureau of Indian Standards (BIS) specifications.
- The policy has also suggested removing all central taxes on biodiesel.

- Recommend minimum support prices for biofuel crops like Jatropha and other non-edible oilseeds with provisions of periodic revisions.
- Minimum Purchase Price (MPP) for the purchase of bioethanol by the Oil Marketing Companies (OMCs) would be based on the actual cost of production and import price of bioethanol. In case of biodiesel, the MPP should be linked to the prevailing retail diesel price.
- Promote biodiesel production from non-edible oilseeds in waste/degraded/marginal lands.
- Discourage plantations in fertile, irrigated premium farm land.
- Focus on domestic production of bio-diesel feed stock and not permit imports.
- Biodiesel and bioethanol may be brought under the ambit of "Declared Goods" by the Government to ensure unrestricted movement of bio-fuels within and outside the States.
- Set up of an inter-ministerial National Biofuel Coordination Committee under the Chairmanship of the Prime Minister and a Biofuel Steering Committee under the Chairmanship of the Cabinet Secretary for high level coordination and policy guidance or review on various aspects of biofuels development in India.
- In regard to research in biofuels, a Sub-Committee under the Steering Committee would be constituted led by the Department of Biotechnology, Ministries of Agriculture & Rural Development and coordinated by the Ministry of New and Renewable Energy.
- Major thrust to be given to Research, Development & Demonstration with focus on plantations, processing and production technologies including second generation cellulosic biofuels.

India's Biofuel Drivers

- The cornerstone of India's energy security strategy is to focus efforts toward energy self-reliance and developing renewable energy options like biofuels.
- Adoption of environmental friendly biofuels to meet improved vehicle emission norms.
- Developing an alternative usage for crops like sugarcane and its byproducts as feedstock for biofuels to support farm income.
- Improve utilization of wastelands and other unproductive land for cultivation of biofuel feedstock.

- Enhance rural employment and livelihood opportunities by promoting production and marketing of biofuel feedstocks.

Status of Biofuel Production in India

Ethanol production in India

India is the second largest sugar producer and fourth largest producer of ethanol in the world. Unlike Brazil, where ethanol is produced directly from sugar cane juice, and the United States, which uses corn for production, India produces ethanol from bagasse. Bagasse or molasses is the waste product after the extraction and refining of sugar from sugar cane. Ethanol production in India therefore has a marked advantage as its production could potentially leave sugar prices unaffected. Further, by blending petrol with 10% biofuel, 80 million litres of petrol could be saved annually in the country (The Hindu, 2006).

The first phase of the ethanol-blended petrol was launched in January 2003, but it took the industry about three years to stablilize. Major issue was that the ethanol imported from Brazil was available at a lower price than domestic ethanol. Even at the 5% blend level being implemented currently there is a shortfall of 225 million liters of ethanol for the oil companies whose current demand is put at 435 million liters (Sify Business, 2006).

India produced 1749 million liters of ethanol in 2004 mainly from sugar cane (Peterson, 2006) and has a very high potential to increase it further by using sweet sorghum, sugar beet and sugar cane juice as potential feedstock options. Another potential source of ethanol can be cellulosic energy crops and crop residues as well as other waste products high in cellulose.

Biodiesel

India consumes more than 250 million tonnes of fossil fuels every year. This comprises of approximately 40 million tonnes of diesel. The Planning Commission, Government of India launched “National Mission on Biodiesel” with a view to find a cheap and renewable liquid fuel based on vegetable oils (Shukla, 2005). The rural development ministry has been appointed as the nodal ministry for implementing the programme. This mission is being carried out in two phases – the first phase involving a demonstration stage for plantation of *Jatropha* on 0.4 million hectares and associated research activities for establishing the commercial viability of the fuel, and phase two for carrying out a self-sustaining expansion of the biodiesel programme. Biodiesel production has reached a decisive stage and the country has introduced 5% blend of biodiesel with conventional diesel in different states (Behl, 2006). Projected demand of petro-based fuels and corresponding blending requirements are represented in Table 7.

Table 7. Forecast of biofuel blending requirements

Projected demand for petrol and diesel								
Year	Petrol demand Mt	Ethanol blending requirements (metric tonnes)			Diesel demand Mt	Biodiesel blending requirements (metric tonnes)		
		@5%	@10%	@20%		@5%	@10%	@20%
2006-07	10.07	0.50	1.01	2.01	52.32	2.62	5.23	10.46
2011-12	12.85	0.64	1.29	2.57	66.91	3.35	6.69	13.38
2016-17	16.40	0.82	1.64	3.28	83.58	4.18	8.36	16.72

Source: Planning Commission, Government of India. Report of the Committee on Development of BioFuel

Though India is the fourth largest producer of edible oilseeds in the world, it produces only 60% of its total oilseed requirement and the rest is met through imports. Despite the low overall oilseed production, the country has a potential not only to become self-sufficient but to produce surplus oilseeds simply by following the improved low-input technologies of oilseeds production and by proper delineation of government policies favorable to oilseeds production. However, these low input technologies have demonstrated 14-100% increase in seed yield over the existing practices under different conditions (DOR, 2005).

Biodiesel is rapidly replacing diesel as a more efficient, cheap, and clean alternative. Plans are being chalked out to cultivate Jatropha plants on barren land to use its oil for biodiesel production. Now it is used for railway engines and it is recommended to raise plantations of these plants everywhere in unused areas through government sectors. Biodiesel is being used experimentally to run state transport corporation buses in Karnataka. University of Agriculture Sciences at Bangalore has identified many elite lines of *Jatropha curcas* and *Pongamia pinnata*. Large scale activities have been initiated quite recently. For example, large-scale plantations have been initiated in North-East India and Jharkhand through a Memorandum of Understanding signed between D1 Oils and Williamson Magor.

According to the National Oilseeds and Vegetable Oils Development Board of the Ministry of Agriculture there are about ten species of tree borne oilseeds (TBOs) with economic potential for biodiesel production, including *Jatropha curcas* (Ratanjot), *Pongamia pinnata* (Karanja), *Azadirachta indica* (Neem) and *Madhuca indica* (Mahua) (NOVOD, 2004). Proponents of biodiesel in the country focus almost exclusively on Jatropha and to a lesser extent on Karanja. Other species have not received much attention. The focus on Jatropha is justified mainly on the basis of two arguments: first of all, Jatropha is a shrub, i.e. it does not grow into a tree. Therefore, it is easier to harvest than large trees and has a much shorter gestation period. Since the time span between investment and return is shorter, more people will be willing to start cultivating this crop. Second, the seed collection period of Jatropha does not coincide with the rainy season in June-July, when most agricultural activities take place. This makes it possible for people to generate additional income in the slack agricultural season (Negi *et al.,* 2006). Karanja has become the second most important feedstock of the biodiesel sector for the reason that it is traditionally planted in several states and therefore well known to people. Being a multipurpose tree that is a source not only of oil but also of animal feed, manure, fire wood and substances with medicinal uses, farmers already integrate Karanja into their farming systems. Karanja is also systematically planted on public land or along roads and it is already a common practice to collect and sell the seeds.

The availability of TBOs can be enhanced considerably without any extra land and inputs if proper network for procurement from seed collectors is established. There is a considerable scope to enhance the collection of seeds from the existing trees by developing infrastructure facilities such as seed/produce procurement centers equipped with facilities for drying, decorticating, cleaning, grading, depulping, storing and oil extraction near the areas of collection of TBO. Establishment of biodiesel processing units near the procurement centers will further help in reducing the cost of transportation of the raw oil to the biodiesel processing plant. This should result in reasonable remuneration to the primary seed collector and also help in getting a quality product by reducing losses caused due to delayed and improper handling of the material at different stages in the existing trade of TBO in India. With the recent central government drive to produce biodiesel from TBO, many state governments have given very high priority to plantations of *Jatropha* for biodiesel production. Information from various sources indicates that area under *Jatropha* plantations in the country has gone up to 20,000-30,000 hectares. Governments of states like Chhattisgarh, Gujarat and Madhya Pradesh have drawn up plans to take up *Jatropha* plantations on massive scale.

The most important characteristic that distinguishes oil-bearing trees from other cash crops is the fact that they require very few nutrients to survive and therefore can also be grown on less fertile land. For example, Jatropha to survive only needs a minimum of around 600 mm of rainfall per year and temperatures that do not go below about 3°C. However, it is an input-responding plant, meaning that fertile land, fertiliser and pesticides as well as better irrigation will lead to much higher oilseed yields (Jongschaap *et al.,* 2007).

There are three ways of cultivating oil-bearing trees. First, they can be grown as boundary plantation, e.g. around fields or along roads, railway tracks and canals. Second, they can be planted in monoculture as block plantations. Third, oil-bearing trees can be cultivated through inter-cropping with other species, which is the best option when it is used for afforestation, but also possible when it is grown on fields. They are planted under the poverty alleviation programmes and National Employment Guarantee Scheme (NREGA) that deal with rural development and helping in employment generation. Boundary plantations of oil-bearing trees, especially of Jatropha and Karanja, are already common in India, even if the seeds are not used for SVO or biodiesel production. There remains a certain range up to which this kind of cultivation can be extended, but the amount of oilseeds produced will still remain marginal compared to the amount that could be reached through cultivation on regular plantations, either through monoculture or inter-cropping. These plantations can, in turn, be set up on three types of land: regular agricultural land, regular forestland as well as underutilised land (often called wasteland). Country has 63.85 million hectare of wastelands accounting for 20.17% of total geographical area (NRSA, 2005) of which 50% can be suitably used for growing TBO plantations like those of Jatropha and Karanja.

Biofuel strategy for India

India is pursuing a two-fold strategy for promotion of biofuels by: (a) providing budgetary support for research, development and demonstration of technologies; and (b) promoting private investment through fiscal incentives. Several ministries are involved in policy making, regulation, promotion and development of the biofuel sector in India. The role of different ministries involved in biofuel sector is presented in Table 8. In addition, various state governments have drafted policies and set up institutions for promoting biofuels in their respective states. Several states have also formed rules for allocation of wastelands to various entities for the plantation of biodiesel feedstock crops.

Table 8. Indian ministries involved in the biofuels sector

Ministry	Role
Ministry of New and Renewable Energy	Overall policymaking role for promoting development of biofuels. Also support research and technology development for production of biofuels.
Ministry of Petroleum and Natural Gas	Responsible for marketing biofuels as well as development and implementation of pricing and procurement policy
Ministry of Agriculture	Research and development for production of biofuel

	feedstock crops (sugarcane/sweet sorghum, etc. for ethanol; and Jatropha and other non-edible oilseed species for biodiesel)
Ministry of Rural Development	Responsible for promotion of Jatropha plantations on wastelands
Ministry of Science and Technology	Supports research in biofuels crops, especially in the area of biotechnology

A number of developmental activities are being taken up in the country for the production of biofuels which include 5% compulsory blend of ethanol in petrol and 5% biodiesel blend in diesel. Nevertheless a strategy to use both fuels (ethanol and biodiesel) as blends will be more beneficial. The prospective strategy consists of producing ethanol and biodiesel from crops to be grown on the same piece of land in different seasons. Since sugar cane is a long duration crop, sweet sorghum which is a 4-month crop is more suitable for such rotation (ICRISAT). According to the developed methodology, it is proposed that from a given piece of land two crops can be taken, viz. sweet sorghum during monsoon followed by a winter oilseed or a monsoon oilseed followed by sweet sorghum in winter. In addition to ethanol and oil, they will also yield food grain, meal suitable for animal feed and biomass residues which can also be converted to ethanol. These crops can give a reasonable output even under rainfed conditions with low external input. Though the crops can be grown under a rain fed situation, in most cases just 2-3 irrigations at critical stages can produce a significant increase in both net returns and bioenergy production. Four of the oilseed crops considered viz. rapeseed, mustard, safflower and linseed can be grown only in winter, while groundnut, sunflower and sesame can be grown in monsoon or winter and castor, niger, soyabean and cotton only in monsoon. Sweet sorghum can be grown in either monsoon or winter, but there is considerable decrease in its stalk yield accompanied by an increase in the yield of grain in the winter crop.

Biodiesel value chain in India

For better understanding and assessment of the developmental impacts of biodiesel production and consumption, some general aspects of the biodiesel value chain should be considered. For example, at the cultivation stage the type of land and type of plantation have important impacts on socio-economic and environmental effects. Different ways of processing the raw material imply different cost structures and require different technical capacities. Not all of them are suitable for the same conditions. Different end-products are consumed by different people at different levels – local or more distant – and have different developmental impacts. Moreover, the use of byproducts allows people to earn additional income. As not all crops generate the same byproducts, some crops may be more economically viable than others. The following figure (Figure 6) presents the simplified value chain of biodiesel comprising of three steps: cultivation, processing and consumption.

Impact of biofuel feedstock on food/feed/trade

India does not produce any ethanol from cereal grains (maize etc.), and thus, there has been no impact of the ethanol programme on the domestic market for food, feed and trade of cereal grains and byproducts. Similarly, as the biodiesel programme is based on the use of non-edible vegetable oil, biodiesel production should not have an impact on feed, food and trade of oilseeds, vegetable oils and other edible products. However, as the production of ethanol for fuel is basically from sugar molasses, it did not had a significant impact on the production, prices and trade of sugar for food and industrial use. Despite a decline in production of sugarcane/sugar and consequently sugar molasses, higher prices of alcohol vis-a-vis fixed ethanol prices have limited fuel ethanol production. If ethanol prices are allowed to be linked to sugar molasses prices, it may impact the availability of sugar molasses for use in cattle feed, and the use of alcohol for industrial purposes.

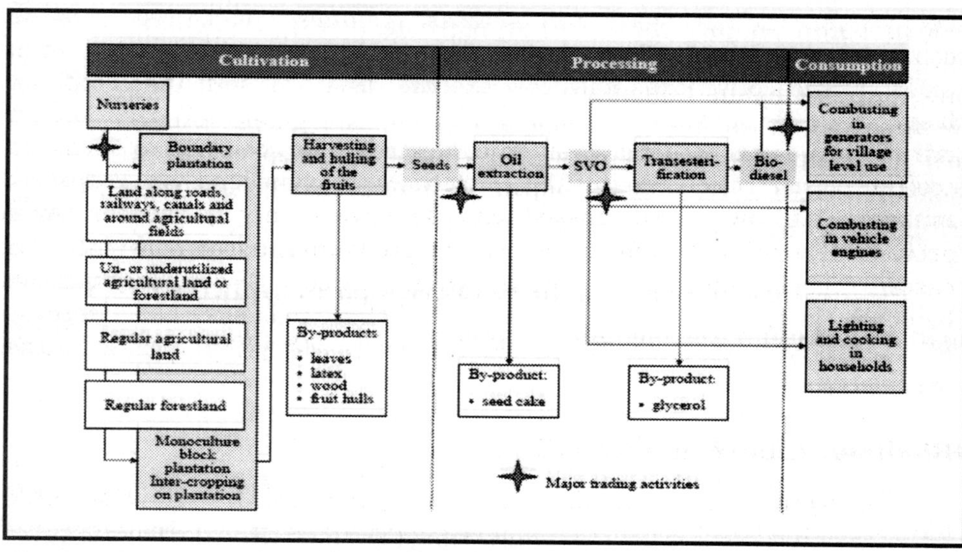

Fig. 6. Biodiesel value chain
Source : Altenburg *et al.*, 2009

Bioenergy use in other areas

In India, biomass-based fuels support over 80% of home energy use (mostly for cooking and heating) and consist of agricultural byproducts (crop residues, cow dung etc.) and gathered fuel wood. Biomass is also used as industrial fuel by small and cottage industries in the organized sector. Total biomass energy in the household sector and unorganized sector accounts for almost one-third of country's total primary energy consumption. India launched a National Project on Biogas Development in 1981-82 with the objective of utilizing cattle manure and human waste for fuel for rural households along with manure for agricultural fields. Currently, there are about 4.1 million family type biogas plants, against an estimated target of 12 million biogas plants in the country. However, evaluation studies show that less than half of the installed plants are operational.

Biomass resources like crop residues, agro-industrial waste, fuel wood etc., are also used for generation of electricity through biomass gasification. Some industries (sugar, paper pulp, rice mills etc) are using the industrial waste for cogeneration of electricity and heat energy to meet their total/partial requirement, and/or sale of excess power to distribution grids. Industrial co-generation did not receive a great deal of attention in the past as cheap electricity and fuel were abundantly available. With the increasing electricity costs and unreliable supplies of electricity from the public distribution grids, several industries are increasingly developing co-generation. Currently, most of the cogeneration activity is sugarcane bagasse based. However, there is significant potential in breweries, distilleries, rice mills, textile mills, fertilizer plans etc. for undertaking cogeneration.

The GOI has initiated several programs and schemes for promoting renewable energy sources including biomass based energy sources, details of which can be accessed from their website http://mnes.nic.in/. The potential and achievement of biomass-based commercial energy in India is represented in Table 9.

Table 9. Biomass-based commercial energy potential and achievement in India

S.N.	Source/System	Estimated potential	Achievement (2009)
1	Bio power (agric residues & plantations)	16,000 MW/1	683.3 MW
2	Sugarcane bagasse based cogeneration	3,500 MW	1033.7 MW
3	Non-bagasse based biomass co-generation/2	-	150.9 MW
4	Biomass gasifier	-	160.3 MW
5	Energy recovery from waste	2,700	89.9 MW
6	Family type biogas plants	12 million	4.1 million

Notes: /1 – Megawatt; /2 – Cogeneration by other agro-industries like rice mills, pulp, etc.

Source: Ministry of New and Renewable Energy, GOI.

Some key researchable issues for biofuels

- Development of biomass feedstocks with physical and chemical properties that facilitate processing, production and utility.
- Development of new micro-organisms that are high-temperature tolerant, ethanol-tolerant, and able to ferment variety of feedstock.
- Development of a wide variety of crops and technologies those are suited to the diverse socio-economic and environmental conditions.
- Utilization/value addition of byproducts generated during production of biofuels.
- Optimization of biofuel blend and performance through trial runs.
- Reduction of wastage by developing efficient waste utilization processes/methodologies.
- Sensitization of financial institutions for providing financial assistance.
- Establishment of demonstration plots for biofuel crops and development of farmer oriented programs to encourage them for cultivation of biofuel crops on marginal lands.

Challenges in promoting biofuels

- Institutional capacity and awareness: there is lack of awareness of the opportunities of small scale biofuel and capacity to develop programs.
- Lack of capacity at all levels – regional, national and local for policies, strategies and regulations development and implementation for promoting small scale biofuels.
- Low awareness on costs and benefits of small scale biofuel initiatives.
- Lack of credit providers and affordable financing for small farmers to purchase seeds, equipment and other inputs to establish biofuels business.
- Lack of supportive policies including tax incentives.
- Limited awareness and knowledge on which feedstock to adopt especially for higher yield.
- Avoid biofuel production using edible oil due to its competition with food requirements.
- Limitation for market development and understand market needs.
- Support farmers and entrepreneurs to meet the necessary market assessment costs.

Constraints

Despite the apparent success of biofuels production in the country, following factors potentially contribute towards reluctance to take a more aggressive approach towards development of biofuel sector.

- Lack of policy attention paid to the introduction and distribution of modern, efficient and clean bioenergy production system.
- Inattention of forestry and agricultural agencies towards the development, management and use of bioenergy resources.
- Lack of availability and proper selection of feedstock.
- Higher feedstock prices and high cost of production.
- Lack of infrastructure for biofuel production using diverse feedstocks.
- Development of proper storage techniques and their feasibility.
- Lack of value addition and proper utilization of byproducts.
- Lack of efficient marketing channels.

Possible drawbacks of biofuels

Despite the positive promises of changes presented by biofuel production, there are few possible drawbacks also:

- Possibility of less food production as there will be an over-aggressive scenario among developing countries to meet the demand for fuel.
- The over aggressiveness in producing biofuel will also result in rapid depletion of soil nutrients, induce and worsen erosion, degradation of water resources, displacement of habitats to cause disruption in biodiversity as well as threaten the traditional livelihoods of local communities.

Biofuels are needed to displace petro-based fuels, which contribute to global warming and are of limited availability. It is envisaged that the biofuel sector in India can be successful only through the implementation of R&D programmes in integrated manner involving the farmers' co-operative society, NGOs, automobile manufacturers, oil marketing companies, State Govt. and Central Govt. departments.

REFERENCES

Altenburg, T., Dietz, H., Hahl, M., Nikolidakis, N., Rosendahl, C. and Seelige, K. 2009. Biodiesel in India: Value chain organisation and policy options for rural development.German Development Institute (DIE) (43) ISBN 978-3-88985-379-0.

Behl, H.M. 2006. Bio-diesel from Jatropha: the road map. *Renewable Energy* (Akshay Urja) 2:12-18.

BioEnergyBroch_FINAL05.indd(PDF).2010.http://esa.un.org/unenergy/pdf/susdev.Biofuels.FAO.pdf. Retrieved 2010-07-14. Department, Rome,

http:// www.fao.org/docrep/008/j5135e/j5135e00.htm.

Directorate of Oilseeds Research (DOR). 2005. Frontline demonstrations in oilseeds: Annual report, 2004-05. Directorate of Oilseeds Research, Hyderabad. 44 p.

Dufey, A. 2006. Biofuels Production, Trade and Sustainable Development: Emerg ing Issues. International Institute for Environment & Development, London.

Energy Information Administration (EIA), 1998. Annual energy outlook 1999, with projections to 2020, in DOE/EIA-0383. 1998, Department of Energy, Washington, DC, USA.

Fisher and Lawrence M. 2007. Carbon gas is explored as a source of ethanol. *New York Times*. 24th April 2007.

Food and Agriculture Organization (FAO). 2003. *Bioenergy Info Notes: FAO and Bioenergy*, Forestry, Food and Agriculture Organization (FAO) 2005. Bioenergy and the Millennium Development Goals, Forestry Department, Rome, http://www.fao.org/docrep/008/j5135e/ j5135e01.htm.

Food and Agriculture Organization (FAO). 2000. The Energy and Agriculture Nexus", Environment and Natural Resource Working Paper 4, Annex1.Available at http://www.fao.org/docrep/003/X8054E/x8054e00.htm

Government of India (GOI). 2009. Basic statistics on Indian petroleum and natural gas 2008-09. Ministry of Petroleum and Natural Gas, Government of India. 17 p.

Hartman, E. 2008. A Promising Oil Alternative: Algae Energy. Washington Post. http://www.washingtonpost.com/wp-dyn/content/article/2008/01/03/AR2008010303907html. Retrieved 2008-01-15.

Inter-governmental Panel on Climate Change (IPCC). 2007. Climate Change 2007: Impacts, Adaptation and Vulnerability. Contribution of Working Group II to the Fourth Assessment Report of the Intergovernmental Panel on Climate Change, Cambridge University Press, Cambridge, UK, 976p.

International Energy Agency (IEA). 2007. Energy technology essentials – biofuel production, Paris; online: http://www.iea.org/Textbase/ techno/ essentials2.pdf.

International Energy Agency (IEA). 2004. Biofuels for transport: An International Perspective, OECD/IEA Paris, France.

International Energy Agency IEA, 2002. World Energy Outlook, 2nd edn. International Energy Agency, Paris.

International Energy Agency. 2006. Energy for cooking in developing countries. World Energy Outlook, Chapter 15. Paris.

Jongschaap, R.E.E., Corre, W.J., Brindaban, P.S. and Brandenburg, W.A. 2007. Claims and facts on *Jatropha curcas* L. Global *Jatropha curcas* evaluation, breeding and propagation programme, Wageningen: Plant Research International B.V. (Report 158).

Konishi, K. and T. Koizumi. 2007. Biofuels Policies in Asia. Presented at the Eighth Meeting of the Global Club of Directors of Agricultural Economics Research Institutes and Agency, Chicago, June 10-11. http://www.aeriasonline.org

Kulkarni, M.G. and Dalai, A.K. 2006. Waste cooking oil-an economical source for biodiesel: A review. *Industrial and Engineering Chemistry Research*, 45: 2901-2913.

Macedo, I.D.C. 1992. Energy from biomass and wastes. *Biomass Bioenergy*, 3: 77-80.

Mandal, R. 2005. Energy – alternate solutions for India's needs: biodiesel. Adviser for the Planning Commission, Government of India.

National Energy Technology Laboratory, 2005. Peaking of World Oil Production, Impacts, Mitigation and Risk Management.

National Oilseeds and Vegetable Oils Development Board (NOVOD) 2004. 1st R&D report on TBOs, New Delhi: Government of India; online: http://www. novodboard.com /1st% 20 R&D%20report.pdf.

National Remote Sensing Agency (NRSA). 2005. Wasteland Atlas of India, Ministry of Rural Development, Department of Land Resources, Government of India and National Remote Sensing Agency, Department of Space, Government of India, Hyderabad.

Negi, K., Komal, J.K. and Panjan, P. 2006. Promotion of biofuel in India: issues and prospects. Delhi Society for Promotion of Wastelands Development.

Oliver R.I. and David A.K. 2009. Quo Vadis Biofuels. Energy and Environnemental Science 2: 343.

Peskett, L., Slater, R., Stevens, C. and Dufey, A. 2007. 'Biofuels, Agriculture and Poverty Reduction' Paper produced for the DFID Renewable Natural Resources and Agriculture Team, ODI, London.

Peterson, J.B.D. 2006. Ethanol production from agricultural residues. *International Sugar Journal* 108:177-180.

Shay, E.G. 1993. Diesel fuel from vegetable oils: Status and Opportunities. *Biomass Bioenergy*, 4: 227-242.

Shukla, S.K. 2005. Experiences of Chattisgarh biofuel development authority. Biofuels India Vol. III (4):12-13.

Sify business. 2006. Sugar coops to supply ethanol to oil PSUs (http://sify. com/finance/fullstory.php?id=14120285).

Somerville, C. 2008. Development of Cellulosic Biofuels(PDF). U.S. Dept. of Agriculture. http://web.archive.org/web/20070927005653/http://www.usda.gov/ oce/forum/2007+Speeches/PDF+PPT/CSomerville.pdf. Retrieved 2008-01-15.

The Energy and Resources Institute (TERI) 2002. DISHA (Directions, Innovations and Strategies for Harnessing Action) for sustainable development, The Energy and Resources Institute, New Delhi, India.

The Hindu. 2006. Distillery Unit set to produce Ethanol from sweet Sorghum, The Hindu, article published on March 28, 2006.

Turkenburg, W.C. 2000. Renewable energy technologies. In: Goldemberg, J. (Ed). World Energy Assessment. United Nations Development Programme, New York, USA, pp: 219-272.

Voegele, E. 2009. Lanza Tech develops waste gas to ethanol technology. *Ethanol Producer Magazine*. 27th August 2009.

World Health Organization. 2006. Fuel for Life: Household Energy and Health. Geneva.

Chapter - 2

BIODIESEL SECURITY: IMPORTANCE OF MULTIPLE NON-EDIBLE OILSEEDS

N.K. Sahoo[1], Jagdamba Prasad[2], S.N. Naik[1]

[1]Centre for Rural Development and Technology,
Indian Institute of Technology, Delhi, Hauz Khas, New Delhi-110 016
[2]Department of Applied Science and Humanities, GLAITM,
17-Km Stone, NH-2, Choumuha, Mathura, UP
E-mail: sn_naik@hotmail.com

ABSTRACT

The biodiesel research has focused to a greater extent on the non-edible oil-seed plant, *Jatropha curcas*. To achieve the objective of national biofuel mission of blending petro-diesel with a planned 13 million tonnes of bio-diesel by 2013, the Planning Commission of India emphasized the production of biodiesel mainly from Jatropha oil and to a smaller extent from Karanja. However, at this point of time when there is a huge gap between demand and production of biodiesel, more and more emphasis is needed to be given on the exploitation of the potential of all other non-edible oil plants like Neem (*Azadirachta indica*), Karanja (*Pongamia pinnata*), Kusum (*Schleichera trijuga*), Mahua (*Madhuca indica*), Kaner (*Thevetia peruviana*), Nohar (*Mesua ferrea*), Moringa (*Moringa oleifera*), Tung (*Aleurites fordii*) etc. (specially, Neem and Karanja). While Jatropha is used as a live fence for protection of agricultural fields, Karanja and Neem are found abundantly in wild throughout India and can be grown on road sides and other similar places. Besides their health benefits, people have used Karanja and Neem plants for multiplicity of purposes, since ages. Although they have a gestation period of 5-7 years, their resource requirement is very less and have decades of productive life. Karanja has nitrogen fixing capacity in addition to 27-39% oil in kernel. Similarly, Neem kernel bears upto 52% oil by dry weight along with its pesticidal and other properties.

Keywords: Biodiesel, indigenous, neem, karanja

INTRODUCTION

Oil provides 95% of energy demand of transportation and agricultural sectors and the demands are continuously increasing. This ever increasing oil demands, depleting oil reserves and negative environmental impacts of petro-fuel use have created a renewed interest in fuels of biological origin (Azam *et al.*, 2005) particularly biodiesel, to ensure energy security for a sustainable development. The biodiesel are alkyl esters of triglycerides produced through transesterification of fresh and used vegetable oil of edible and non-edible nature as well as from animal fats. This processing also results in valuable byproducts like glycerine, biopesticide and bio-manure in the form of cake.

As raw material for biodiesel, rape and soyabean seed oils cover around 80% of feedstock requirement of USA and France. Other raw materials are Palm oil (Malaysia), Olive oil (Spain), Cotton seed oil (Greece), Jatropha oil (Nicaragua), beef tallow (Ireland), used frying oil (Australia), and other waste oil and fats (USA). Although edible oils have encouraging results for biodiesel conversion particularly in developed countries, they can't be dispensed for the same in developing countries like India.

At present, India is the second-biggest buyer of vegetable oils after China, the fact that over 72% of India's crude oil requirement is fetched from imports. Annual consumption of edible oil of India is about 10 million tonnes, half of which is fetched by imports. During the oil-year 2006-07 although non-edible oil imports have declined by 11 per cent, edible oil import has increased by 7 percent (Business Line, Nov. 16, 2007). Thus the demand for biodiesel feedstock rather can be fulfilled by different non-edible oils from seeds of cultivated as well as wild plants. In this regard, India being very rich in biodiversity has a great potential for producing bio-diesel from non-edible oil sources (Azam *et al.*, 2005). Contrary to this fact, the biodiesel research in India has focused mainly on *Jatropha curcas* and very little on other feedstocks. However the plantation of Jatropha is gaining momentum throughout the country but have to go a long way to achieve the goals of national biofuel mission.

By the year 2003, France was using diesel with 5% biodiesel blending (B5), USA with 20% blending (B20) where as even pure biodiesel (B100) is being used in Austria, Germany, Sweden like countries in unaltered vehicles(Kale and Vijayalakshmi, 2003). India is planning to ensure using B20 by 2013 (which seems, only a mere promise), unless the scientific community, investors and policy makers emphasize on utilization of the rich resource base of indigenous tree borne oilseeds (TBOs). Dependence on only one feedstock (*Jatropha curcas*) for biodiesel production may not fulfill the national requirements.

This article has emphasized on promotion of all possible TBOs, particularly *Pongamia pinnata* and *Azadirachta indica* Juss. for biodiesel production.

The national biofuel mission of India

The nationwide biofuel mission of India has been launched in July, 2007. To achieve the objective of blending petro-diesel with a planned 13 million tones of biodiesel by 2013, the Planning Commission of India emphasized the production of biodiesel mainly from Jatropha oil and to a smaller extent from Karanja (http://www.jatrophabiodiesel.org/indian programs.php). The objectives of National Mission on Biodiesel are: (1) to concentrate on demonstration projects to be implemented by the year 2006-07, with an investment of Rs. 1500 crore ($300 million) on 0.4 million hectares (phase-I), (2) Follow up of the Demonstration Projects, will consist of a self sustaining expansion (11.2 million ha) of the programme being started in the year 2007 leading to production of biodiesel required in the year 2011-12 (Phase II). The report has given maximum emphasis on Jatropha as a feedstock and Karanja, Neem and others a subsidiary position (Planning Commission report, 2003).

Jatropha curcas: biodiesel feedstock that has got highest attention

Native of South America this plant *J. curcas* has some favourable features. Jatropha is a tall bush or a small tree growing up to 6m height, suiting agroforestry plantation. The vegetation is not browsed and hence used as live fence around crop fields. The plant is suitable for plantation in wastelands, degraded lands, green capping of bunds, shallow soils, railway track sides.

The plant occurs all over India. It is drought tolerant, grows in arid to semi-arid conditions (annual rainfall 480-2380 mm). It develops without much care and irrigation and need hardly any application of pesticide. The plant has smaller gestation period of about 1.0 year with a productive lifespan of 40-50 years. Fruiting occurs once to twice a year (whole the year in permanently humid region). Leaves and seeds have insecticidal properties (Jatrophin from seeds is anti-cancerous; oil is purgative and also used for external skin applications; other products are dye, paint, soap, lubricant etc.). Oil content is around 46-58% in dry kernel and 30-40% in dry seed. The free fatty acid (FFA) value in oil is at a favourable range (approx. 2%) (Papitha and Paul Raj, 2007).

Uncertainties in Jatropha plantation

Jatropha is associated with low and uncertain yield (Tomar and Gour, 2004). Being a plant of arid to semi arid region, under wider irrigated condition it shows high vegetative growth and low seed production (Papitha and Paul Raj, 2007). During dry period the plants need irrigation. The survival rates for direct seedlings are low. The plants are potential of growing as weed in uncultivated land (Rama Rao, 2003). Vegetation is not of much use as plants neither have palatability nor have any fuel-wood value.

The plants bear toxicity in seed and vegetation and also allelopathic properties (http://www.jatrophabiodiesel.org/jatrophaplantation.php). Storage period for seeds is low. Occurrence of viral (leaf mosaic) and fungal (leaf curl, gummosis) disease and pest infestations (fruit borer) are being reported (Aswatha Narayana *et al.*, 2006; Wani *et al.*, 2006; Openshaw, 2000; Fu *et al.*, 2007) in different instances. Jatropha cannot be grown on lands with waterlogged condition or with slope exceeding 30°. Additionally, the plant is highly sensitive to frost. The ideal climatic requirement of jatropha is below 600 mm annual rainfall in moderate climatic conditions, 1200 mm in hot climatic zones and a less than 9 soil pH.

Table 1. A summary of various aspects of different biodiesel feedstocks of India (Azam *et al.*, 2005; Papitha and Paul Raj, 2007; Gupta and Guleria, 1982; Bringi and Mukerjee, 1987; Ermel, 1995; Damodaram and Hegde, 2002; Folkard *et al.*, 1994; Duke, 1983; Gupta, 1994)

Species	Distribution	Oil in kernel (%)	Gestation period (yr)	Fruiting life (yr)
Jatropha	At moderate climatic setups	46-58	<1	40-50
Karanja	Throughout India	25-39	4-7	40-50
Neem	Throughout India	47.6±5.5*	3-4	150-200
Mahua	Throughout sub-Himalayan India (with variation in species)	46-52	8-15	60
Kusum	Tropical settings of India	59-72	-	-
Silk cotton	Throughout India	20	-	50
Sal	14% of total forest area of India	15-20	15	>100
Undi	Along sea coasts	50-73	-	20-30
Nahor	Forests of N.E. India, Karnataka, Kerala	45	-	-
Tung	Northeast India	53-60	3	30
Kaner	Throughout India	62-67	-	-
Drumstick	Throughout India	40	1-2	-
Rubber	Humid tropical regions, India	40-50	-	-

Other important oil yielding plants of India

There are innumerable plants (explored and unexplored) with the potential of yielding various types of oils from different body parts. But plants yielding higher amount of oil are only commercially viable for bio-diesel production. A database prepared by BGIR/BSI (Botanical Garden of Indian Republic, Botanical Survey of India) has listed 287 plant species under oil yielding plants category and some more under other categories. Many of these plants though available abundantly in India have got only little attention for bio-diesel production *except Jatropha curcas*. Kusum (*Schleichera trijuga*), Pili Kaner (*Thevetia peruviana*), Nahor (*Mesua ferrea*), Mahua (*Madhuca indica*), Silk cotton (*Bombax ceiba*) Rubber (*Hevea brasiliensis*), Nag champa/ Undi/ Ball nut (*Calophyllum inophyllum*), Neem (*Azadirachta indica*), Karanja (*Pongamia pinnata*), Drumstick (*Moringa oleifera*) and Tung (*Aleurites fordii*) come under this category. There are studies reporting availability of more than 400 oil yielding plant species in India (Maheswari, 1989). Various properties of some selected plants have been presented in Table 1.

Two important native biodiesel feedstocks: Karanja and Neem

As Jatropha has its own salient features, Karanja and Neem being native to India have their positive aspects to be promoted as biofuel feedstocks besides their uses in other multiplicity of purposes.

Karanja (*Pongamia pinnata* Pierre)

Karanja is native of Western ghats and found throughout India up to 1200 m above sea level (ASL). The climatic requirement is semi-arid to semi-humid with medium to high rainfall (500-2500 mm). The plant is medium sized glabrous tree with crown height 5.5m or more. It is tolerant to water logging, salinity and alkalinity of soils as well as resistant to drought, high wind, moderate frost and medium to high rainfall. It posses nitrogen fixing capacity and is suitable for afforestration in watershed areas, on field boundaries, nala bank, on wasteland, on tank foreshore and dry parts of the country. It is recommended as a shade tree for pastures and windbreak for tea and avenue plantation. Large plantations on roadsides during last two decades have increased its seed yield. Seed matures during different months in different parts of the country covering all most the whole year. Seed yield increases with increasing canopy.

Similar to Jatropha its vegetation is not palatable by livestock but sometimes the leaves are used as scarcity fodder, especially in arid regions. It is used as green leaf mulch and as fuel wood due to high calorific value (4600 Kcal/kg). Oil is used in tanning agents, soap, candles, medicines, insecticidal and antibacterial preparations. Oil cake is blended with nitrogenous fertilizers for inhibiting nitrification and pest attack. Oil cake is a good source of manure, and is described as a "useful poultry feed". Karanja bark fibre is used for rope making.

Neem (*Azadirachta indica*)

Azadirachta is native to India and Burma, growing in tropical and semi-tropical regions has established itself in 30 countries worldwide in Asia, Africa and Central and South America. Adapted to a wide range of climates and annual rainfall (200 to 1200 mm) it can be found even at 1500 m ASL (Tewari, 1992). It is a fast growing tree, growing up to 15-20 m high or more on diverse soil types. Neem can't grow on poorly drained or waterlogged soil. It can dwell on dry, stony, shallow soils and even on soils having hard calcareous or clay pan, at a shallow depth. Neem plantation neutralizes acidic soils, increases fertility and water holding capacity of soil.

The plant flowers normally once a year but twice in some humid zones. Almost all parts of the plant have myriad of medicinal properties: insecticidal, anti helminthic, antileprotic, antipoisonous, antiheamorrhodial, antidermatonic, astringent, blood purifying, acrid, refrigerant, anti-oxidant etc. The tree is a renewable resource for organic farming as a provider of home grown agro-chemicals and nutrients which are biodegradable, non-toxic and effective. Neem cake with fertilizer and pest repellent property has made it a favoured agriculture input. Neem leaves have also been used to enrich the soil. Deoiled cake supplemented with chemical fertilizer synchronizes uptake.

Neem has the quality to purify air and the environment of noxious elements. Besides temperature moderation it prevents the occurrence of many diseases too. Being a hardy, multipurpose tree, it is ideal for afforestation, reforestation and for rehabilitation of degraded, semi-arid and arid lands. The tree also bears timber value.

Biodiesel feedstock potential of India

It is estimated that the existing potential of tree borne oilseeds (TBOs) in the country is 3-3.5 million tones but only 0.5-0.6 million tones are being collected. The Table 2 below is showing the pathetic situation of the utilization of biodiesel feedstock out of the huge potential available in the country.

Feedstock security in time and space

It is a known fact that different plants bears and matures fruit at different timing of the year. No plant singlehandedly can provide year round feedstock for oil production. In Table 3 some plants have been noted to provide fruits year round but that is possible by adding up fruit harvest timing of the plant species of different geographical regions. In the similar way, when fruit harvesting time of all the possible biodiesel feedstocks will be considered, it can ensure year round production of biodiesel. This possibility can be visualized from Table 3 where harvesting time of some of the possible biodiesel feedstocks have been listed.

Table 2. Yield potential and % utilization of some bio-diesel feedstocks (Gupta 1994; Subramanian, 2005)

Oil seed species	Seed production		
	Potential (tones/yr)	Actual % produced	Cultivation yield (tones/ha)
Jatropha (*Jatropha curcas* L.)	50,000	NA	0.4-12
Karanja (*Pongamia pinnata*)	135,000	6	6-9
Neem (*Azadirachta indica*)	400,000	25	2-5
Mahua (*Madhuca indica*)	490,000	17.3	-
Kusum (*Schleichera trijuga*)	200,000	15	-

Table 3. Fruit harvest timing of some potential biodiesel feedstocks of India (Azam *et al.*, 2005; Papitha and Paul Raj, 2007; Gupta and Guleria, 1982; Bringi and Mukerjee, 1987; Ermel, 1995; Damodaram and Hegde, 2002; Folkard *et al.*, 1994; Duke, 1983; Gupta, 1994; Subramanian, 2005)

	Jan	Feb	Mar	Apr	May	Jun	Jul	Aug	Sep	Oct	Nov	Dec
Jatropha	*	*				*	*	*				*
Undi	*	*	*									
Silk cotton				*	*							
Sal				*	*	*						
Nahor					*	*						
Kusum						*	*					
Neem						*	*	*				
Rubber							*	*	*			
Tung									*	*	*	
Mahua#				*	*	*	*	*	*	*	*	
Kaner#	*	*	*	*	*	*	*	*	*	*	*	*
Karanja#	*	*	*	*	*	*	*	*	*	*	*	*
Drumstick#	*	*	*	*	*	*	*	*	*	*	*	*

Fruit harvest timing of different regions of India have been summed up

Similar to time distribution (Table 3) of different biodiesel feed-stockds, the space distribution (Table 1) needs to be considered for ensuring biodiesel security. Different species of plants have occurrence limited to their specific geo-climatic requirements. This makes different species of plants to perform optimum in their respective region of natural occurrence. This is one of the reasons, why not any single species but all possible plant species should be promoted and explored to their maximum potential for getting maximum output. At the same time, suppose any species of plant establishes in different regions of the country, it may not provide the expected output. As evident from Table 1, except some, number of species of plants occurs throughout India but when all the plant species are considered together, the feedstock production covers totality of the wild land of the country. Therefore, in order to get maximum production of feedstocks and to ensure biodiesel security throughout the country, all native plants should be given priority. Moreover, promotion of any monoculture is in no way good for the biodiversity of ecosystem.

Benefits of multiple feedstock use

- **Ecology and biodiversity:** From ecological view point, the indigenous plants in their native lands can be relied upon for stability in production with minimal resource input and from inducing any ecological or environmental upheaval. Indigenous plants, being part of an ecological system act in harmony with both the organisms and environmental components (soil, water and air) of the system. Multiple plant species represent biodiversity thus ensures stability in the ecosystem and sustainability in feedstock production.
- **Continuity of feedstock:** Multiple feedstock utilization promises continuous flow of resource in the biodiesel production which otherwise will not be.
- **Easier processing and utilization:** This may open options for easier processing and utilization of biodiesel by blending optimally the oils or biodiesels having different positive aspects for best results. Research and technology developments will help to deal with the diversity of feedstocks.
- **Redistribution of load and benefits:** Diversifying the biodiesel feedstock ensures optimal utilization of all possible options which in turn will redistribute the load and benefit to all components of the system (resource base, production, service and consumption). It will re-distribute incomes throughout the country and utilization of the non-edible oil seed will provide income to the tribal and landless individuals.

- **Biodiesel-link to other multipurpose utilities:** Establishment of biodiesel-links for various underutilized multipurpose trees will facilitate the other utility options of the trees rather than competing with their existing utilities. For example, Neem oil can be utilized very well for bio-pesticide and other preparations by extracting Azadirachtin like active compounds, before going for biodiesel production. More biodiesel means more trees and hence more bio-derivatives for organic farming.

Management option for best utilization of available space

- Crop-field margins, bunds, railway track sides, arid lands etc. could be planted with Jatropha like shrubby vegetations, in some cases Karanja also.
- Nala bank, wasteland, tank foreshore etc. with water logging possibility could be planted with Karanja.
- Road sides could be planted with different big and hardy trees like Neem etc.
- Wastelands, watershed area where a vast expanse of land is available could be planted with mixtures of different habits of plants depending upon their characteristics, requirement and adaptability, the climatic and soil conditions.
- Similarly, intercropping with some selective agricultural crops could be done to enhance production and ensure sustainability of the system.

In all the cases emphasis should be given on planting a mixture of trees to manage the canopy space for best utilization of soil and solar resource.

Biodiesel production from multiple feedstock

Use of multiple feedstocks for biodiesel production will not be a difficult task as the general principle and procedures of bio-diesel production are the same for all feedstocks. Different ways to reduce viscosity include dilution, microemulsions, pyrolysis, catalytic cracking and transesterification. Out of these processes, transesterification is the preferred one because of its process simplicity, end products (esters) and commercially valuable byproduct (glycerol). After trans-esterification the bio-diesel need to be separated from alcohol-glycerol mixture by settling and then subject to water-washing followed by moisture removal to make it usable.

In the transesterification of different types of oils, triglycerides (fat/oil) react with alcohol where exchange of the alkoxy group of an ester compound by another alcohol takes place to form esters and glycerol (Miguel *et al.,* 2001). The reactions are often catalyzed by an acid or a base or both sequentially. The process consists of a sequence of reversible reactions (Fig. 1).

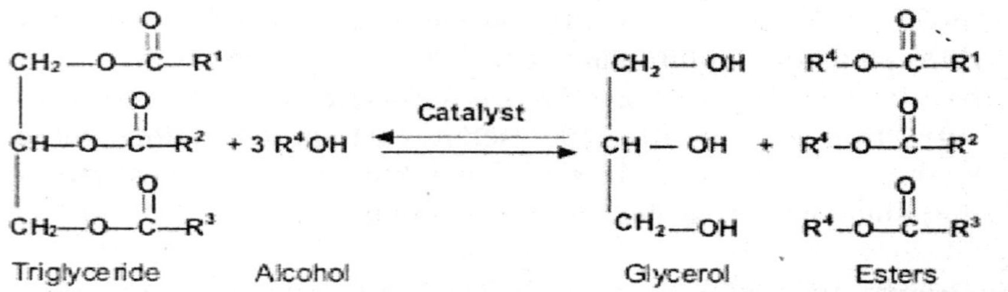

Fig. 1. General equation for transesterification of triglycerides

Where, R^1, R^2 and R^3 are long chain alkyl groups. R^4 of the short chain alcohols can be methyl, ethyl, propyl or butyl but methyl is in common use.

The overall process is normally a sequence of three consecutive steps (Fig. 2), of reversible nature. The steps consecutively are conversion of triglyceride to diglyceride to monoglyceride and finally to glycerol producing one ester at each step. Hence, from one triglyceride, three esters are formed. The byproduct, glycerol has numerous applications.

$$\text{Triglyceride} + R^1OH \rightleftharpoons \text{Diglyceride} + RCOOR^1$$

$$\text{Diglyceride} + R^1OH \rightleftharpoons \text{Monoglyceride} + RCOOR^1$$

$$\text{Monoglyceride} + R^1OH \rightleftharpoons \text{Glycerol} + RCOOR^1$$

Fig. 2. Three steps in transesterification of triglycerides, Where R^1 is alkyl group.

The most important variables (Ma and Hanna, 1999; Schuchardta *et al.*, 1998; Freedman *et. al.*, 1984; Meher *et. al.*, 2006; Agarwal, 2007; Marchetti *et al.*, 2007) affecting the yield of biodiesel from transesterification are: (a) Reaction temperature (b) Molar ratio of oil to alcohol and type of alcohol (c) Catalyst (d) Reaction time (e) Mixing intensity (RPM) (f) Moisture and Free Fatty Acid (FFA) (g) Co-solvent addition.

CONCLUSION

It is clear that though Jatropha shows its strength as a very good feedstock for biodiesel production, other feedstocks particularly Neem and Karanja are equally potential. The main strengths of these two feedstocks are that they are abundant in wild, native to India and occur throughout the country, and multipurpose in use. Besides these two, other feedstocks with occurrence, limited to specific geo-climatic setups are boon for the regions in particular and in turn for the country as a whole. There will be no need to promote or propagate them specially; rather people will propagate these plants and explore the potentials by their own, once the economic link of bio-diesel is established. The potential in oil yielding life span and quantities are remarkably higher for all the listed feedstocks. The landless individuals and marginal farmers have an additional income by working as labourers or selling seeds from wild. The tribals particularly will be more beneficial by selling the variety of feedstocks from wild. This strategy will ensure biodiesel security (ensured availability in time and space), take care of biodiversity, ecology and economy of the region and eventually a sustainable development.

REFERENCES

Agarwal, A.K. 2007. Biofuels (alcohols and biodiesel) applications as fuels for internal combustion engines. *Progress in Energy and Combustion Science* (33):233-271.

Aswatha Narayana, D.S., Shankarappa, K.S., Govindappa, M.R, Prameela, H.A, Gururaj, Rao, M.R. and Rangaswamy, K.T. 2006. Natural occurrence of jatropha mosaic virus disease in India. *Current Science* 91(5):584-586.

Azam, M.M., Waris, A. and Nahar, N.M. 2005. Prospects and potential of fatty acid methyl esters of some non-traditional seed oils for use as biodiesel in India. *Biomass and Bioenergy* (29):293-302.

Bringi, N.V. and Mukerjee, S.K. 1987. Karanja seed (*Pongamia glabra*) oil. In: Non-traditional Oilseeds and Oils of India (Ed. Bringi, N.V.), Oxford IBH Publishing Co. Pvt. Ltd., New Delhi, pp. 143-166.

Damodaram, T. and Hegde, D.M. 2002. Oilseed situation. Directorate of Oilseeds Research, Hyderabad, India.

Duke, J.A. 1983. Handbook of Energy Crops (unpublished). Available online at http://www.hort.purdue.edu/newcrop/duke_energy/dukeindex.html,

Ermel, K. 1995. Azadirachtin of neem seed kernels from different regions of the world. In: The Neem Tree (*Azadirachta indica* A. Juss.) and other Meliaceous Plants, (Ed. Schmutterer, H.), Weinheim (Germany)/New York: VCH, pp. 89-92.

Folkard, G., Sutherland, J. and Shaw, R. 1994. Water Clarification using *Moringa oleifera* Seed Coagulant. Loughborough University, United Kingdom.

Folkard, G.K. and Sutherland, J.P. 1996. *Moringa oleifera* - a tree and a litany of potential. *Agroforestry Today* 8(3):5-8.

Freedman, B., Pryde, E.H. and Mounts, T.L. 1984. Variables affecting the yields of fatty esters from transesterified vegetable oils. *JAOCS* 61:1638–1643.

Fu, G., Huang, S.L., Wei, J.G., Yuan, G.Q., Ren, J.G., Yan, W.H. and Cen, Z.L. 2007. First record of *Jatropha podagrica* gummosis caused by *Botryodiplodia theobromae* in China. *Australasian Plant Disease Notes* 2:75-76.

Gupta, B.N. 1994. India. In: Durst, P.B. Ulrich Kashio, W.M. (Eds.) Non-Wood Forest Products in Asia. Rapa Publication 1994/28, Bangkok.

Gupta, T. and Guleria, A. 1982. Non-wood forest products in India. Oxford and IBH, New Delhi, www.fao.org http://www.jatrophabiodiesel.org/indianPrograms.php

http://www.jatrophabiodiesel.org/ jatrophaPlantation.php

Kale, V. and Vijayalakshmi, P. 2003. Biodiesel- a boon. In: Tree borne oil seeds as a source of energy for decentralised planning (Ed. Radhakrishna, P.), *Renewable Energy Science Series XII*, Ministry of Non-Conventional Energy Sources, pp. 141-149.

Ma, F. and Hanna, M.A. 1999. Biodiesel production: a review. *Bioresour. Technol.* 70:1-15.

Maheswari, J.K. 1989. Processing and utilization of perennial vegetation in the arid-zone of India. In: Proceedings of the FAO Expert Consultation on the Role of Forestry in Combating Desertification, Saltillo, Mexico 24-28 June 1985, FAO Conservation guide 21, Food and agriculture organization of the United Nations Rome, http://www.fao.org/docrep/T0115E/T0115E0g.htm

Marchetti, J.M., Miguel, V.U. and Errazu, A.F. 2007. Possible methods for biodiesel production. *Renewable and Sustainable Energy Reviews* 11:1300-1311.

Meher, L.C., Vidya Sagar, D. and Naik, S.N. 2006. Technical aspects of biodiesel production by transesterification — a review. *Renewable and Sustainable Energy Reviews* 10:248-268.

Miguel, V., Trubiano, G., Perez, G., Borio, D.O. and Errazu, A.F. 2001. Kinetic analysis of enzymatic esterification of fatty acids and ethanol. *Stud. Surf. Sci. Catal.* 133:619-624.

NewswireToday- /newswire/ - Delhi, New Delhi, India, 05/07/2007 http://www.newswiretoday.com/news/17698/)

Openshaw, K. 2000. A review of *Jatropha curcas*: an oil plant of unfulfilled promise. *Biomass and Bioenergy* 19:1-15.

Papitha, P. and Paul Raj, S. 2007. Tree borne oil seeds. In: *Proceedings of National Workshop on Biodiesel*, Madurai Kamraj University 17-18 October, 2007:55-78.

Planning Commission. 2003. Report of the Committee on the Development of Biofuels. Planning Commission, New Delhi, India.

Rama Rao, D., Nanda, S.K. and Sastry, K. 2003. Strategies for augmenting potential of vegetable oils as bio-diesels. In: Tree Borne Oil Seeds as a Source of Energy for Decentralised Planning (Ed. Radhakrishna, P.), *Renewable Energy Science Series XII*, Ministry of Non-Conventional Energy Sources, pp. 48-56.

Schuchardta, U., Serchelia, R. and Vargas, R.M. 1998. Transesterification of vegetable oils: a review. *J. Brazil. Chem. Soc.*, 9:199-210.

Subramanian, K.A., Singal, S.K., Saxena, M. and Singhal, S. 2005. Utilization of liquid biofuels in automotive diesel engines: An Indian perspective. *Biomass and Bioenergy* (29):65-72.

Tewari, D.N. 1992. Monograph on Neem (*Azadirachta indica* A. Juss.). Dehra Dun, India, International Book Distributors, pp 297.

Tomar, V.S., and Gour, V.K. 2004. Cultivation and management of *Jatropha curcas* plantations. In: Proceedings of National Conference on Bio-diesel for IC Engine Technologies and Strategies for Rural Application (Eds. De, D., Mehta, C.R., Pandey, K.C.), Cenrtal Institute of Agricultural Engineering, Bhopal, India, December 3-4, 2004:26-46.

Wani, S.P., Osman, M., D' Silva, E. and Sreedevi, T.K. 2006. Improved livelihoods and environmental protection through biodiesel plantation in Asia. *Asian Biotechnology and Development Review* 8 (2):11-12.

Chapter - 3

JATROPHA: AN ENVIRONMENT FRIENDLY SOURCE FOR FUEL AND AN OPPORTUNITY FOR EMPLOYMENT GENERATION IN RURAL AND URBAN INDIA

R.L. Srivastava, N.K. Bohra and D.K. Mishra

Silviculture Division, Arid Forest Research Institute, Krishi Mandi, New Pali Road, Jodhpur-342 005

ABSTRACT

Energy use is directly proportional to the economic activity once a country goes from "developing" to "developed". The importance of energy, is paramount. India ranks sixth globally, accounting for 3.5 percent of the world commercial energy demand. This is primarily met by fossil fuel sources viz., coal (52%), oil and gas (44%), hydropower (2%), nuclear (1.7%) and the rest from non-conventional sources. In USA and Europe biodiesel from edible oil sources (soyabean, rapeseed) is used. In India, since edible oil demand is higher than its domestic production, there is no possibility of diverting this oil for production of biodiesel. There are many tree species, which bear seeds rich in non-edible oil. Species like *Pongamia pinnata, Calophyllum inophyllum, Jatropha curcas, Euphorbia antisyphilitica, E. tithymaloides, E. caducifolia, E. royleana, E. neerifolia*, *Calotropis gigantea* and *C. procera* have been found promising as petro crops. Among these species recently there has been thrust on biodiesel potential of *J. curcas*. It is a fast growing and long-lived plant, easy to propagate, can grow and survive with minimum inputs in marginal land and not browsed by animals and seeds not eaten away by birds. Time taken for net yield is between 2 to 5 years based on soil and rainfall conditions. Besides environmental benefits, biofuels are likely to emerge as an alternative transport fuel and help in generating employment both in urban and rural areas. This paper describes the use of biofuels, its advantages and role in generating employment in India.

Keywords: *Jatropha curcas*, biodiesel, wastelands, employment generation.

INTRODUCTION

India is emerging very fast as a major force in the global economy. During the present phase of accelerated development, it is inevitable that energy consumption increases many folds as compared to the present per capita energy consumption which is very low in comparison to the other developed countries. India's total energy requirement has doubled over the last three decades, from 147.05 million tones of oil to 437.69. However, the scenario is bound to change very fast and the gap between demand and supply of crude oil will exert a recurrent pressure on the import bills. On the other hand, non-renewable fossil fuels are getting reduced at a faster rate than ever before pushing the resources at the brink of exhaustion. There has been a renewed interest in biofuels in the past two years as government looks for alternatives to increasingly expensive oil-based fuels.

Depleting reserves of fossil fuel and increasing effects of pollution from these fuels demand eco-friendly alternatives, which can supplement or replace fossil fuels. In recent years, research has been directed to explore 'biofuels' – plant based fuel sources as a supplement or substitute of fossil fuel. Biofuels are renewable and environmentally safe. The most common types of biofuels that are being developed and used at present are bioethanol and biodiesel. Since these biofuels are derived from cellulosic biomass such as herbaceous and woody plants, agricultural and forestry residues, a prefix 'bio' is added before their names. At present biodiesel is reported to substitute about 10% of the world diesel consumption (Choo and Ma, 2000). Low cost labour, high unemployment, high cost of delivered petroleum products, and unfavourable balance of payments are some of the serious concerns which make biofuels an attractive option for India. In view of a strong agricultural and forestry base, both these biofuels – ethanol and biodiesel – are likely to emerge as an alternative transportation fuel, beneficial for the environment and economy of India. Ethanol has been explored in many countries as a potential fuel to reduce harmful emissions. Brazil has been producing ethanol from sugarcane molasses and juice for more than forty years. Ethanol, thus produced is being used as transport fuel as a blend (24%) in petrol (gasoline) and in the neat form (alcohol) in Brazil (Jagadeesan, 2001). Similarly, in US, the use of ethanol produced by corn maize as an oxygenate in Reformulated Gasoline (RFG) is well established. Other countries, which have experience of using ethanol as a transport fuel, include France, Indonesia, the Philippines, Guatemala, Costa Rica, Argentina, The Republic of South Africa, Kenya, Thailand and Sudan.

Biodiesel

Biodiesel is a fuel produced by a chemical process between vegetable (virgin or used) oil or animal fats and alcohol in the presence of a catalyst.

Biodiesel development worldwide is well documented (Korbitz, 1998). Biodiesel has acquired respectability among diesel fuel producers, diesel engine manufacturers and end users. In the last 15 years, 28 countries have embarked on biodiesel projects and biodiesel production. Till date a total of 85 production plants of biodiesel have been identified, with capacities ranging from pilot plant to industrial scales up to 1,20,000 t/year (Choo and Ma, 2000). Biodiesel is currently used commercially in France, Germany, Italy, Australia, Sweeden, the Czech Republic, Slovakia and USA.

Biodiesel (fatty acid alkyl esters) is a cleaner-burning diesel replacement fuel made from natural, renewable resources such as fresh and used vegetable oils and animal fats. Vegetable oils/fats are triglycerides of fatty acids, which on transesterification are converted into their alkyl esters to produce biodiesel (Sonntag, 1982). Esters of vegetable oils and fats are used because they have low viscosity and better fuel properties than pure vegetable oils and fats. Thus biodiesel is a mixture of methyl or ethyl esters of C_{16} to C_{18} fatty acids. Just like petroleum diesel, biodiesel operates in combustion-ignition engines (Choo *et al.*, 1997). Blend of upto 20% biodiesel (mixed with petroleum diesel fuels) can be used in nearly all diesel engines. These low level blends (20% and less) don't require any modifications and can provide the same payload capacity and range as diesel with following characteristics:

- Biodiesel is environment friendly: it is renewable, "more biodegradable than sugar and less toxic than table salt" (US National Biodiesel Board)
- Biodiesel can be used in any diesel engine
- Fuel economy is the same as conventional diesel fuel
- Biodiesel is a much better lubricant than conventional diesel fuel and extends engine life - a German truck won an entry in the Guinness Book of Records by travelling more than 1.25 million km (780,000 miles) on biodiesel with its original engine
- Biodiesel has a high cetane rating, which improves engine performance: 20% biodiesel added to conventional diesel fuel improves the cetane rating 3 points, making it a premium fuel
- Biodiesel can be mixed with ordinary petroleum diesel fuel in any proportion, with no need for a mixing additive.
- Even a small amount of biodiesel means cleaner emissions and better engine lubrication: 1% biodiesel will increase lubricity by 65% .

The current industry recommendation in the USA is that biodiesel be used within six months or reanalyzed after six months to ensure that the stored fuel meets ASTM specifications. A longer shelf life is possible depending upon the fuel composition and the use of storage enhancing additives such as antioxidants, generally standard conditions used for storage and handling of diesel can be used for biodiesel. It could be stored in tanks made of aluminum or steel under clean, dry and dark environment. To avoid growth of algae in fuel, water contamination need to be minimized and if necessary, some biocide should be added.

Several studies (e.g. in Mali, Nicaragua, India and Zimbabwe) have indicated that the *J. curcas* plant shows potential as an oil crop for biodiesel (Foidl and Elder, 1997; Henning, 1998). The Jatropha plant is Latin American in origin and is closely related to the Castor plant. It is a large shrub/small tree able to thrive in a number of climatic zones in arid and semi-arid tropical regions of the world. An easy to establish plant, drought resistant, can grow in areas of rainfall of 400-1200 mm. In addition it is valued for crop protection, prevents wind/water erosion, is not browsed by animals, reaches maximum productivity by 5th year, and has a 50-year life span. Planting is usually done at a density of about 2500 plants per ha.

Realizing the convenience of raising plantations of *J. curcas* as a pure crop with or without irrigation, projections for cultivation on wasteland has been made and it is estimated that cultivation of *J. curcas* on 2.67 million hectare would suffice to meet the target (5%) replacement by biodiesel (Table 1). The target is achievable in view of availability of about 175 million ha wastelands.

As shown in Table 1, the cultivation of Jatropha on about 5.34 million ha of wastelands would make the country self sufficient in meeting its biodiesel needs. However, raising of mixed plantations over monoculture should be preferred in view of ecological sustainability and multifeed technology to sustain the supply of feedstock.

Table 1. Projections for raising plantations of *Jatropha curcas* on wastelands

S.N.	Item	Yield/ requirement
1	Average fruit yield per plant (kg) after 5th year	2.0
2	Average seed yield per plant (kg); 50-60% of fruits	1-1.2
3	Number of plants per hectare (2x2m spacing)	2500
4	Average oil yield (litres) per hectare (30% of oil on seed basis)	750
5	Biodiesel yield (litres) per ha (95% efficiency)	712.50
6	Annual biodiesel requirement for 5% replacement (lakh tones)	20.0
7	10% replacement	40.0
8	Wasteland required for cultivation of Jatropha (million hectare) to meet 5% replacement	2.67
9	10% replacement	5.34

Other species of forest origin seeds which contain fatty oils in significant quantities and having adaptability of growing on wastelands need to be identified and suitable site-specific ecologically sustainable and economically viable plantation models with peoples' participation for wasteland, uncropped marginal lands/degraded forest lands may be developed for the production of biodiesel in the country. No doubt, the gestation period in case of oil seed trees will be longer, the benefits once start accruing will be available for a much longer period without much efforts and investment. For the purpose, besides launching a nation wide campaign to educate the masses, particularly the rural masses, under peoples' participation programme and wasteland development programme, improved and high yielding plant material could be supplied in rural areas free of cost with other essential inputs and technical know-how for planting them on community wastelands, and degraded forest lands. The villagers in return for raising and maintaining these plantations may be given exclusive rights for collection of the produce at nominal royalty. Moreover, to promote the use of biodiesel in the country, policy should be made to provide the subsidies initially for cultivation of energy plants and production of biodiesel from them.

Socio-economic and environmental considerations of using biodiesel in the country

Local production of bio energy is projected to have a broad range of positive economic, social and environmental implications. Upgrading eroded and deforested land, creation of employment and income is part of the argument. The national program wants to stop soil and forest degradation and its environmental implications, generate employment for the poor, in particular for women, reduce climatic change and improve energy security.

While the country is short of petroleum reserve, it has large Arabic land as well as good climatic conditions, potential to produce biomass to be processed into bio-fuels. Instrument to promote non-edible oils is hoped to be buy-back arrangements with oil companies to be put in place and mandatory use of bio-diesel blends. The Jatropha programme is to be combined with other programs of the Ministry of Rural Development to attract growers, entrepreneurs and financial institutions so that a "self sustaining programme of expansion takes off on its own, with the Government playing mainly the role of a facilitator. Hence, for the expansion phase, the Government will need "to give only marginal financial support". The rural community will have the first right of access to the oil for its own use. Responsibility for availability of sufficient processing units will be with the Ministry of Petroleum. Studies have revealed that "direct and indirect impact of bio-diesel e.g. employment generation, balance of trade, emission benefits etc. are substantial and need to be accounted for" while considering the duty structure on bio-diesel and HSD.

Economics of production of biodiesel is a determining factor in making biofuel programme successful. Production economics though dependent upon several factors is essentially and basically guided by the sustainable availability of the feed stock/raw material. Raw materials being exploited commercially by the biodiesel producing countries constitute the edible fatty oils derived from rapeseed, soybean, palm, sunflower, coconut, linseed etc, (Reed, 1993). Use of such fatty oils to produce biodiesel in India will not be feasible in view of a big gap in demand and supply of such oils in the country as the gap is presently met by huge imports. Increased pressure to augment production of edible oils has also put limitations on the use of these oils for production of biodiesel. Under Indian conditions only such plant sources can be considered for biofuel, which produce non-edible oil in appreciable quantity and can be grown in large scale on non-cropped marginal and wastelands. Seeds of more than 100 species of forest plants have been identified as source of fatty oils with an estimated potential of 11.3 lakh tones (Gulati *et al.*, 1984; Dayal *et al.*, 1999; Juyal *et al.*, 1995; Rajak, 1995). However, the realized potential of such oils is 6.6 lakh tonnes (Anon., 1994). It has been reported (Foidl and Elder, 1997; Srinivasa, 2001) that fatty oils derived from *Jatropha curcas* (Ratanjyot) and *Pongamia pinnata* (Karanj) are excellent feedstock for biodiesel production. A World Bank study reports that developing countries will not be able to leave themselves out of poverty without increased use of energy (Anon., 1999). Rural electrification has long been recognized as needed to improve conditions in rural areas and help stop migration of people to already over crowded cities. According to the Central Electricity Board in India, about 15% of villages were not electrified by March 1998 (Menon, 1999).

Biodiesel can be made through a simple chemical process technology. It could be made accessible to rural communities. It provides opportunity for immediately addressing benefits of energy needs particularly of remote areas where access to conventional sources of energy is not always possible and affordable. More reliable electricity can be produced from mini-grid system than central grid extension. On the basis of the study conducted (Rosenblum, 2000) energy benefits of biodiesel in the country are seen for decentralized micro-grid electricity generation at village level as well as a replacement for diesel fuel in small scale application such as irrigation pump sets, etc.

There is potential to generate an additional 12.58 crore new well-paid jobs in agribusiness and textile export. Our oil companies should create a bioenergy clusters in each village. The cluster will use locally available feedstock to produce biogas, biodiesel and ethanol. Rural people for cooking, running agricultural pumps, energising cold storages and powering transport vehicles, can use biogas. It can also produce electricity, bio-fertilisers and bio-pesticides. Waste heat from the generating set can boil drinking water for health and hygiene. Biodiesel and ethanol can be sold to oil companies for further processing and blending with diesel and petrol. Carbon credits under Kyoto Protocol will improve the clusters economic viability.

Ready energy at reasonable cost will induce the farmer to opt for mechanization and improve productivity. A rural woman can add to her family income by taking up a job in a local agribusiness unit. India has 108 million hectares of non–irrigated farmland where farmer's earnings are rather low. On a part of this land they can grow sweet sorghum, corn or similar energy-rich plants or grass and sell to the cluster at a good profit. Such feedstock can produce enough ethanol to meet out petrol needs.

Domestic production and utilization of biodiesel will result in significant economic benefits to both the urban and rural sectors, and the balance of trade. The cost and employment generated by nursery raising, site preparation, planting, maintenance and seed collection activities of Jatropha raising on wasteland are shown in Table 2, 3 and 4 respectively.

Table 2. Cost of nursery raising (2500 seedlings per hectare of plantation)

S.No.	Item	In Rs.
1.	Cost of polythene bags 10 kg (Rs. 60/Kg)	600.00
2.	Cost of collection of seeds (3 kg)	45.00
3.	Cost of soil, sand and FYM	1250.00
4.	Cost of filling of 3000 polybags excluding cost of soil-fertilizer mixture and sowing two seeds in each (200 polybags /MD) i.e.15 MD @ Rs. 89.00/MD	1335.00
5.	Maintenance for 4 months total 30 MD @ Rs. 89.00/MD	2670.00
	Total	5900.00

Table 3. Cost of planting and maintenance of Jatropha plantation (per ha)

S. N.	Particulars	Year of plantation			Total	Employment generation in MD		
		1	2	3		1	2	3
1.	Site preparation							
i	Site clearing (7 MD)	623.00	-	-	623.00	7	-	-
ii	One ploughing (Rs. 1000.00)	1000.00	-	-	1000.00		-	-
iii	Alignment & stacking (3 MD)	267.00	-	-	267.00	3	-	-

iv	Digging pits (2,500 No.) of 45 cm @40 pits/MD- 62.5 MD	5563.00	-	-	5563.00	62.5	-	-
v	Cost of FYM/Compost: 1 kg/pit @Rs. 400/MT	1600.00	-	-	1600.00	-	-	-
vi	Cost of pesticides	1250.00	-	-	1250.00	-	-	-
vii	Cost of biofertilizer including application 1.00/pit	2500.00			2500.00			
2.	Transportation and planting							
i	Cost of transportation @ Rs.0.50 / seedling	1250.00	-	-	1250.00	7	-	-
ii	Planting cost including application of pesticide (100 pits/MD), 25 MD	2225.00	-	-	2225.00	25	-	-
iii	Mortality replacements 500 pits/MD: 5 MD	445.00	-	-	445.00	5	-	-
3	Inter-culture operations							
i	Weeding & soil work: twice in 1st year & once in subsequent year @ 30 MD/weeding.	5340.00	2670.00		8010.00	60	30	
ii	*Irrigation: three irrigation in 1st year & one irrigation in 2nd year @ Rs. 500 per irrigation	1500.00	500.00	-	2000	14	5	-
4	Tending operation: Cut back at a height of about 30 to 45 cm in the 1st year. Pruning side branches in 2nd year: 500 plants/MD; 5 MD	445.00	445.00	-	890.00	5	5	-
	Total	24,008.00	3,615.00	.	27,623.00	189	40	40

Basis of calculation: Spacing: 2m x 2m; No. of plants per ha : 2500; Mortality replacements : 20%; wage rate : Rs. 89 per manday (MD) vide Govt of India /Min. of Labour/ notification no. S.O. 09(E) & 113(E) dated 3rd and 26th Jan. 2002.

* Irrigation is applicable in arid and semi arid areas where it is essential for initial establishment phase.

Note

- In hilly terrain, the expenditure of Rs. 6,000.00 should be incurred for digging/preparation of contour trenches and other soil conservation structures in place of ploughing and digging pits.

- In case of termite infected soils, termiticide to be mixed with soil mixture before filling the pits.
- The above estimate has been prepared on the basis of Rs. 89.00 per manday, which may vary as per prevailing rates of daily wages in different states.

Table 4. Employment generation in raising plantations of *Jatropha curcas* on wastelands

S.No.	Item	Mandays generated (Millions)
1	Wasteland required for replacing 5% diesel (million ha)	2.67
2	Raising of nursery, 2500 plants/ ha; 45man days/ha	120.15
3	Plantation activity; 2500 plants/ha; 189man days/ha	504.63
4	Maintenance of plantation; 2500 plants/ha; 40mandays	106.8
5	Seed collection and other activities; 66 man days/ha	176.22
6	Total employment generated for raising 2.67 million wasteland with Jatropha till 2nd year (274man days)	731.58
7	Continuous employment till 30 years for seed handling and maintenance activities (66man days)	176.22

Thus the cost of raising one hectare *Jatropha* plantation and its maintenance upto 2nd year at considering espacement 2 m x 2 m including nursery cost is estimated at Rs. 33,523.00 only. The activity may be viewed mainly as rural employment generation programme. During the first year itself 234 mandays per hectare would be generated with 40 mandays per hectare during the second year onwards. From 4th year onwards, as seed yield gets stabilized, 66 mandays are supposed to be generated per year per hectare which may extend up to 30 to 40 years of fruiting span of plantations. Implementation of similar projects in Mali and other countries have shown that proposed activities have high financial viability taking in to account end uses of products and by-products of the different components.

In addition to generate the employment opportunities, there are other financial benefits of raising these plantations. The oil cake left as residue after extraction of oil from the seeds is rich in nitrogen and are excellent organic fertilizers. Organic biofertilizers in form of these cakes will not only be available to the farmers right at their doorstep but also help to reduce huge import bill and subsidies make on chemical fertilizers.

High unemployment, low cost of labour, high cost of delivered petroleum products and unfavourable balance of payments make biodiesel production an economic and environmentally lucrative venture. Domestic production of biodiesel starting from large scale plantation, collection of seeds, their processing into fatty oils and their conversion into bio-diesel will generate more employment as it is three to six times more labour intensive per unit of production than fossil fuels. Further, utilization of the by products (e.g. glycerine) and residual products (e.g. seed cake, a good source of organic fertilizer) left after fatty oil extraction would offset the production cost of biodiesel.

Carbon trading potential

India can tap the $52-billion global market for carbon trading by encouraging production and use of biofuels. Biofuels, apart from enhancing energy security, ensuring employment and development and mitigating environmental pollution, can be instrumental in carbon trading if certain criteria of the clean development mechanism (CDM) of the Kyoto Protocol of the United Nations Framework Convention on Climate Change (UNFCCC) are fulfilled.

According to a recent study by Point Carbon, the potential of global carbon market over the next several years is around $52 billion per annum. Growing at a rate of 4% per year, India, the sixth largest producer of greenhouse gases (GHGs), contributing almost 3% of the world's total emissions (including CH_4 from waste generated by cattle) is seen as one of the most attractive destinations for CDM linked investments. Estimates put the cumulative foreign direct investment (FDI) on account of such projects at about $2 billion, growing at the rate of $200 per year.

In light of increased evidences of climate change effects and their mitigation methodologies, several carbon market and investment mechanisms are slowly evolving. Though the carbon market dynamics are not transparent, the scenario indicates a huge potential in future. Till January 2004, the total volume traded in project-based transactions is 78 million tonne of carbon dioxide emissions (CO_2e). The buyer side included Japan with 41 per cent, The Netherlands and CFB with 23% each. According to estimates, if the CDM captures at least 35% of the global market, the estimated value to the concerned countries would be $18 billion.

Reductions in GHG emissions can be accumulated and traded as carbon credits. The CDM facilitates selling of these reductions in terms of certified emission reductions (CERs), a unit which equals to one tonne of CO_2 emissions.

The present market price of carbon credits is around $5 per CER, which translates into additional revenue of Rs 690 per tonne of bio-diesel consumed. This additional revenue from sale of carbon credits can be used to raise plantations of trees having oil-bearing seeds and materials or meet unforeseen expenses during stabilization period of bio-diesel technology. Large-scale plantation of trees having oil-bearing seeds and materials, like Jatropha and Pongamia species will fix carbon by photosynthesis via the carbon cycle. Thus the consumption of bio-diesel as an energy source either in stationary or mobile combustion leads to "no net-addition of CO_2 to the atmosphere."

Apart from making India self sufficient in meeting energy needs, reducing the import bill and generating employment, environmental benefits of using biodiesel in the country are seen significant. Biodiesel is considered "Climate neutral" because all of the CO_2 released during consumption is sequestered out of the atmosphere during crop growth. Combustion of one litre of diesel fuel results in the emission of about 2.6 kg of CO_2, a gas responsible for 'Green House Effect' (Anon., 1997). Combustion of biodiesel has been reported in a number of sources to have lowest emissions compared with petroleum diesel as presented in Table 5. Environmental benefits from using biodiesel depend upon how much of it is blended into regular diesel fuel.

Table 5. Average biodiesel emissions compared with conventional diesel

S.No.	Emission type	Percent reduction/ increase in emissions	
		Biodiesel* (100%)	Biodiesel** (5%)
1	CO_2	-78.0	-3.90
2	CO	-44.0	-2.25
3	Total unburned hydrocarbons	-68.0	-3.50
4	Ozone depleting potential of certain hydrocarbons	-50.0	-2.50
5	Sulfates	-100.0	-5.00
6	Particulate matter	-40.0	-2.0
7	Polycyclic aromatic hydrocarbons	-80.0	-4.0
8	Nox	+6	+0.25

* National Biodiesel Board, USA (Rosenblum, 2000); ** Projected from B_{100} results

Thus the development of biodiesel will help the country not only in balancing the huge import bill on account of petroleum products but also in generation of employment opportunities and meeting the environmental obligations such as reductions of GHGs, exhaust emissions, carbon sequestration etc. Under the Kyoto Protocol regulations, Euro-III norms will be implemented in the entire country by 2010. In order to meet these emission requirements, drastic reduction in sulphur content (<350 ppm) and higher cetane number (>51) will be required. Biodiesel meets all these important specifications and would help in improving the lubricity of low sulphur diesel (Laucey, 1998).

In the Indian context, biodiesel would not only serve to reduce import of petro-diesel but also facilitate largescale employment and accelerated rural development. Further large wastelands could be utilized for the cultivation of non-edible oil producing trees/shrubs for production of biodiesel. However, current literature on environmental impacts is limited, and there is a need to carry out extensive studies from all possible feedstocks and generate comprehensive data on environmental impacts including pollutant emission, global warming potential, and energy efficiency.

REFERENCES

Anonymous 1994. Strategies for development of non-traditional oil seeds. National Oilseeds and Vegetable Oil Development Board, Gurgaon, India. p. 10.

Anonymous 1997. The United Nations: Energy and Environment Basic, second edition. Food and Agriculture Organization of the United Nations, Bangkok.

Anonynous. 1999. Fuel for Thought: Environmental strategies for the energy sector. A World Bank Group Sector Strategy Paper: New York.

Choo, Y.M. and Ma, A.N. 2000. Plant Power. *Chemistry and Industry*: 534.

Choo, Y.M. and Ma, A.N. and Ong, A.S.H. 1997. Lipids: Industrial Application and Technology (Gunstone, F.D. and Padley, F.B., eds). Marcell Dekker Inc., New York. pp. 771-785.

Dayal, R., Jain, P.P. and Soni, P.L. 1999. Non-Wood Forest Products-A Chemical Approach. Proc. Intl. Workshop on Forestry Research in Conservation of Natural Forests (Negi, R.S. *et al.*, eds.) UNDP-ICFRE, Dehra Dun. pp. 42-47.

Foidl, N. and Elder, P.1997. Agro-industrial exploitation of *Jatropha curcas*. Biofuels and industrial products from *Jatropha curcas* (Gubitz, G.M. and Mathur, K.C., eds.). Dvb-Verlag, Graz, Austria.

Gulati, N.K., Suri, R.K.and Jain, P.P. 1984. Ways to attain self sufficiency in oils and fats. Proc. Nat. Sem. on Chemistry, Industry and Citizen (Suri, R.K. and Mathur, K.C., eds). Dehra Dun, Dec. pp. 59-72.

Henning, R. 1998. Use of *Jatropha curcas*: A household perspective and its contribution to rural employment creation. Paper presented at the Regional Workshop on the Potential of *Jatropha curcas* in Rural Development & Environmental Protection, Zimbabwe, May.

Jagadeesan, T.R. 2001. Flexible fuel vehicles with ethanol blends. *The Ethanol Alert* 1: 6-7.

Juyal, S.P., Singh, R.V. and Singh, J.P. 1995. Tree borne oil seeds of A.P., Bihar, M.P. and Orissa. *Proc.* Workshop on Strategies for development of tree born oil seeds of Niger in tribal areas. National Oil Seeds and Vegetable Oil Development Board, Gurgaon, India. pp. 3-25.

Korbitz, W. 1998. Proc. 1998 PORIM International Biofuel and Lubricant Conference. Palm Oil Research Institute of Malaysia. pp. 3-22.

Laucey, P. 1998. Fuel Lubricity Reviewed. *SAE* Paper No. 982567, International Fall Fuels and Lubricants Meeting and Exposition, Oct. 19-22. San Francisco, California.

Menon, M. 1999. Rural power scene still dim, says CEA. *Indian Express*, Bombay, January 10.

Rajak, R.L. 1995. Tree borne oil seeds, a vital source to supplement vegetable oil resources. Proc. Workshop on Strategies for development of tree borne oil seeds of Niger in tribal areas. National Oil Seeds and Vegetable Oil Development Board, Gurgaon, India. pp. 1-2.

Reed, T.B. 1993. Proc. First Biomass Conference of the Americas. 2: 797.

Rosenblum, J.L. 2000. Feasibility of biodiesel for rural electrification in India. Tellus Institute, Carnegie Mellon University: Draft report. pp. 1-15.

Sonntag, N.O.V. (1982). Bailey's Industrial Oils and Fat Products, 4th ed., Vol. 2 (Swern, D. ed). Wiley, New York. pp. 97-173.

Srinivasa, U. 2001. A valuable substitute for diesel in rural India. *Current Science* 80: 1483-1484.

Chapter - 4

INFORMATION ON INDIAN JATROPHA SPECIES FROM INTERNET AND LITERATURE

R. Indumathy and P. Ratha Krishnan

National Centre for Agroforestry, Gwalior Road, Jhansi 284 003
E-mail: indumathymca@yahoo.com

ABSTRACT

In India, more than 100 species of forest plants – Tree Borne Oil seeds (TBOs) have been identified as source of fatty oils. Jatropha is one such genus which includes some exotic species and very little known but eagerly looking to meet all our future energy requirements. A decade back itself few native Jatropha species were utilized in castor improvement programme and inter-specific hybridization has been attempted between different species of Jatropha with a limited success. Even though every one is talking and aware about *J. curcas*, the curious question before us is the wealth of Jatropha species in India. Even though twelve Jatropha species were notified by several Indian floras, the research has been confined with nine species only. This article will elaborate the total twelve Jatropha species available in India viz., *Jatropha curcas, J. gossipifolia, J. glandulifera, J. multifida, J. tanjorensis, J. podagrica, J. integerrima, J. pandurifolia, J. villosa, J. nana, J. heynei* and *J. maheswari* along with its botanical description which was collected from different floras and internet.

Keywords: Jatropha species, exotics, indigenous, botany and interspecific hybridization

INTRODUCTION

Each country is adopting a strategy that exploits the comparative advantages over a diverse set of crops to use as biofuels. Jatropha is potential to meet our future energy needs due to its early yield and oil content. Other than oil, several different toxic principles including lectin (curcin), phorbol esters, saponins, protease inhibitors and phytates neither the seeds nor the press cake nor the oil of *Jatropha curcas* are also available. The selection of possible species for interspecific hybridization / non toxic *J. curcas* is the possible options to overcome this problem. So, it is imperative to know the available Jatropha wealth of India.

Genus : Jatropha

Jatropha belongs to the tribe Joannesieae of Crotonoideae in the Euphorbiaceae family and contains 176 known species distributed in the tropical, subtropical Africa and America. They are herbs, shrubs or trees, monoecious (rarely dioecious), exudates watery to white latex. This genus possesses poisonous substance in the sap/seed. Indumentum simple hairs and sometimes glandular hairs, leaves alternate, often digitately lobed. Flowers are terminal cymes with a single pistillate flower at the end of the primary axis. Sepals-5, free, imbricate; petals-5, mainly free; staminate disc annular or 5 free glands, stamens 6-10, in two whorls; pistillate foliaceous annular, 5–lobed; fruits capsular to tardily dehiscent and sub-drupaceous (Anon, 1959, 1998).

The following are the Jatropha species (Fig. 1) available in India and its botanical descriptions:

1. *Jatropha curcas* L.: Named by Linnaeus, Greek word means "Jatros"- doctor, "trophe" - food / nutrition which implies medicinal uses. A large shrub, native to tropical America and introduced to Asia and Africa by the Portuguese (Heller, 1996). This is a diploid with 2n = 22 chromosomes. Leaves 3-5 lobed, cordiform, stipules deciduous. The inflorescence is complex, monoecious with protandry. The first branching is racemose and subsequent branching are cymes. The inflorescence is a cyathium which appears as a single flower. Generally flowered between September to January and second flowering in June also reported. It is propagated by seeds or by cuttings. This 50 years of rotation species can yield 2-4 tones of seed / ha with 30 – 42% of oil content after three years of planting (Raina and Gaikwad, 1987).

2. *Jatropha gossipifolia* L.: Perennial ornamental shrub upto 3 m tall, native of Brazil. The palmately opposite leaves are 3-5 lobed, deciduous in winter. Stipules ciliate, glandular margin of leaves, petiole and leaf blade covered with glandular hairs. Flowers unisexual, monoecious, purple red with yellow centers and produced in clusters on branched stalk in the upper leaf axils. It is intolerant to shade with the yield potential of 500 kg/ha of seeds with 28-30% oil content.

3. *Jatropha glandulifera* Roxb.: An evergreen shrub of 2 m height, distribution confined to the black cotton soils of Deccan and Carnatic. It is known as Adalai in Tamil. This species contains much clear pale-yellow juice, tender parts purplish, trunk short, stout and have smooth papery bark. Seeds ellipsoid, black coloured with 20-27% of brownish yellow oil. Flowering and fruiting is throughout the year and propagated by seeds and by cuttings (Singh, 1988).

Fig. 1. Well known Jatropha species in India

4. *Jatropha multifida* L.: A glabrous shrub, native to South America. It is cultivated in gardens, parks for its ornamental foliage and flower and known as coral plant. It is known as "Malai amanakku" in Tamil. Inflorescence tightly corymbiform, upto 28 cm long, flowers deep red, in axillary long peduncled cymes. Seeds obovoid, brown in colour contain 32-40% of oil. Flowering and fruiting appears during rainy season and it is mainly propagated by seeds.

5. *Jatropha tanjorensis*: It is a natural hybrid between *Jatropha curcas* and *Jatropha gossypifolia* mainly confined to Metteurdam and Tanjore region of Tamil Nadu (Prabakaran and Sujatha, 1999). It is vigorous in growth but sterile in nature, growing upto 3 m, tender parts brownish in colour. Propagated by cuttings.

6. ***Jatropha podagrica***: Small shrub, upto1 m height, native of Panama. "Podagrica" is a Latin word meaning "Gouty" refers to the swollen base of the plant. Orange red or scarlet flowers on terminal long stalked cymes, unisexual, monoecious, male flowers are more than female flowers. Seed yield is 400 kg / ha with the oil content upto 54% (Raina and Gaikwad, 1987). Flowering and fruiting throughout year and propagated by division or seed.

7. ***Jatropha integerrima***: An erect ornamental shrub, native of West Indies, grows upto 6 m tall. Inflorescence terminal, peduncle upto 10 cm long. Scarlet or pinkish unisexual flower, monoecious. Seeds are small, slender, ellipsoid to ovoid. Propagated by cuttings.

8. ***Jatropha pandurifolia***: Shrubs, upto 3m tall, with slender, graceful branches, native to Cuba. "Pandurifolia" is Latin for "with fiddle shaped leaves". Inflorescence terminal cyme, dioecious, calyx purplish red in colour. Fruits a capsule, purplish green in colour (Bor and Raizada, 1944). Flowering throughout the year and propagated by cuttings or by seed .

9. ***Jatropha villosa***: Undershrub, shoots rusty – villow. Indigenous to India, known as "Thanakku" in Tamil. Flowers pale yellow. Var. Villosa confined to Coimbatore, Dharmapuri, Salem and Nilgiris where as Var. Ramanadensis to Ramanathapuram district of Tamil Nadu.

10. ***Jatropha nana***: A small, sparingly branched shrub, 30 – 45 cm high found in stony and waste places near Poona and Mumbai, being apparently endemic to the Deccan, dwarf of African type (Hooker, 1885). Root woody as thick as finger, stem round. Flowering occurs in May – July. Fruits: capsule 1 cm long, obovoid – oblong, flattered at the top, slightly 6- lobed.

11. ***Jatropha heynei***: A small (15-30 cm tall), low glabrous shrub, branching (upto 6) from a tuberous rootstock (weigh about 1 kg). Distribution confined to peninsula (Bala Krishnan, 1962). Flowering and fruiting season Aug -Sep and the fruit is capsule.

12. ***Jatropha maheswari***: An under shrub, narrow small and lanceolate leaves. Distributed in Kanyakumari and Tirunelveli districts of Tamil Nadu (Paramathma *et al.*, 2006).

CONCLUSION

Moving towards biofuels for a secure and clean energy, understanding of our Jatropha wealth and its potential is important. This information on available Jatropha species of India can be used to explore the potential for possible yield improvement strategy through inter-specific hybridization, grafting, and genetic transformation.

REFERENCES

Anonymous 1998. The United States Germplasm Resources Information Network (GRIN) database. *www.ars-grin.gov*

Anonymous 1959. The Wealth of India, Raw Materials,Vol V, CSIR, New Delhi.pp 93–297.

Bala Krishnan 1962. Bulletin of Botanical Survey of India, 1962, 3, 40.

Bor, N.L. and Raizada, M.B. 1944. Some beautiful Indian climbers and shrubs. *J. of the Bombay Natural History Society* XLIV(4): 499- 505.

Heller, J. 1996. Physic nut–*Jatropha curcas* L.: Promoting the conservation and use of underutilized and neglected crops. International Plant Genetic Resources Institute, Rome, Italy, 54p.

Hooker, J.D. 1885. Flora of British India (Vol. V). International Book Distributors, Dehradun, 910p.

Paramathma, M., Venkatachalam, P., Sampathrajan, A., Balakrishnan, A., Jude Sudhakar, R., Parthiban, K.T., Subramanian, P. and Kulanthaisamy, S. 2006. Cultivation of Jatropha and biodiesel production. Biofuel series – 2, Publications of Centre of Excellence in Biofuels, TNAU, Coimbatore, 42p.

Prabakaran, A.J. and Sujatha, M. 1999. *Jatropha tanjorensis* Ellis & Saroja, a natural interspecific hybrid occurring in Tamil Nadu, India. *Genetic Resource and Crop Evolution* 46, 213 – 218.

Raina, K.A. and Gaikwad, B.R. 1987. Chemobotany of Jatropha species in India and further characterization of 'curcas oil'. *The Journal of the Oil Technologists Association of India* XIX (4):81 – 85.

Singh, N.P. 1988. Flora of Eastern Karnataka. Mittal Publications, Delhi, Vol II,783 p.

Chapter - 5

JATROPHA CURCAS OIL AS A PROMISING BIOFUEL OIL FOR FUTURE

Bhumika Chauhan* Rashmi and Anita Tomar***

Department of Silviculture* and Chemistry**, Forest Research Institute, Dehradun. E- mail:-bhumidun@gmail.com

ABSTRACT

Jatropha curcas L., a non-edible oil seed, being hardy in nature, survive in harsh soil and climatic conditions and is a suitable species for wastelands and marginal farmlands. *Jatropha curcas* plant shows promise for use as an oil crop for biodiesel production. *Jatropha curcas* seeds contain about 25-44% of oil, which is mainly used for the production of fuel and after transesterification as biodiesel. Physical and chemical properties of curcas oil provide its potential to be used for petroleum substituting fuel. In recent years, research is focused to further assess its application as a fuel substitute. As a crop, Jatropha represents a renewable resource with a high energy content and oil yield.

For the present study seeds were obtained from fruits collected from Uttarakhand and used for experiment by decoating, sun drying and grinding. For recovering the oil from the seeds solvent extraction using petroleum ether method is used, by same method a golden yellow coloured curcas oil was obtained and characterized for determining the pH, refractive index, specific density, moisture percent, saponification value, protein percent, free fatty acids (FFA), unsaponifiable matter, iodine value and acid value according to the standard method of analysis. Oil yield of 38% was obtained. The results shows pH (9), refractive index (1.474), specific density (1.10), unsaponifiable matter (2.7%), moisture (0.2507%), saponification value (28.66%), protein (3.7%), acid value (1.9), iodine number (5.55), ester value (12.34) and presence of reasonable amount of free fatty acids.

The aim of this study was to characterize the potential of *Jatropha curcas* seed oil in the oil and petroleum industries.

Keywords: Renewable resource, solvent extraction, free fatty acids (FFA)

INTRODUCTION

Jatropha curcas L. is a drought resistant shrub/tree belonging to the family Euphorbiaceae cultivated in Central and South America and was distributed by Portuguese seafarers in Southeast Asia, India and Africa. Propagated by cuttings, is widely planted as a hedge to protect fields from browsing animals. The plant and its seeds are non edible to animal and humans; toxicity of seeds is mainly due to the presence of curcine and deterpine. The existing distribution of *J. curcas* shows that introduction has been most successful in drier regions of the tropics. It grows on well-drained soils with good aeration and is well adapted to marginal soils with low nutrient content. Seeds and oil yield can be attributed to a variety of factors such as climate, intensity of management, site quality, size of plant and genetics. The huge future requirement of oilseeds can be met by raising large-scale plantations of species valued for their oil yielding seeds. The potential use of extracted oil from *Jatropha curcas* as transesterified oil (biodiesel) or as a blend with diesel was studied by many oil and petroleum companies. Biodiesel is gaining worldwide acceptance as an environment friendly solution to energy problem. Biodiesel results in substantial reduction of unburnt hydrocarbons, carbon monoxide, and particulate matter. Indian Oil Corporation Ltd. (IOCL) reported that maintenance cost of vehicles running on biodiesel has reduced substantially. The biodiesel has no sulphur, no aromatics and has about 10% built in oxygen that helps it to burn freely. Higher cetane number improves the combustion.

Biodiesel also has to compete with diesel in prices even though it is environmentally clean and technically more efficient. However, many countries, like India, have large tracts of wastelands and tropical climate suitable for cultivating a variety of plants that yield non-edible oil. Cultivation for oil in degraded, waste, abandoned and unused lands will provide sustainability, employment generation, and much needed oil to replace fossil fuels. Biodiesel, the renewable liquid fuel from biological raw material is a good substitute for petroleum diesel. The country has a ray of hope since there is a potential to grow a variety of plants in different habitats. India has vast areas that are wastelands and are not being utilized for cultivation since these are unfertile, dry, sodic, saline, or alkaline. *J. curcas* is the first choice as it can grow in saline to alkaline soils, in arid to semi-arid conditions, low slopes of hilly areas, degraded and unused soils.

It is an accepted option for energy security, reduction in imports, rural employment, and improving agriculture economy. The calorific value and cetane number of *J. curcas* oil are comparable to diesel, but the density and viscosity are much higher. Since the density of oil is high, the engine performance, emissions and combustion parameters can be reached by optimizing the injector opening pressure, injection time, injection rate and enhancing the swirl level of the operated engine. High viscosity of *J. curcas* oil is not advantageous for the compression ignition engine. Pramanik (2003) has studied the performance of the engine using blends and Jatropha oil in a single cylinder (compression ignition) engine and made comparison with the performance obtained with diesel. Adequate thermal efficiencies of the engine were obtained with blends containing up to 50% volume of Jatropha oil. 40-50% of Jatropha oil can be replaced without any engine modification and preheating of the blends. In the oil industry, different processes must be done before oil extraction occurs. These different proceses are as follows: (a) dehulling, separating hull from nut, (b) de shelling, separating shell from kernel, (c) drying and (d) oil extraction. Physical, mechanical and chemical properties of seed and kernel are needed for the design of equipment to handle, transport, process, store and assess the product quality.

The aim of this study was to investigate the physical and chemical properties of *Jatropha curcas* L. seeds. The considered parameters were oil yield percent, refractive index, specific density, unsaponifiable matter, moisture percent, saponification value, protein percent, acid value, iodine number, ester value, moisture content, oil density calorific value, iodine value and water content of oil.

MATERIALS AND METHOD

Collection of seeds

For the study, *J. curcas* fruits were collected from Betalghat, Nainital, Uttarakhand. The place of origin was characterized by an annual precipitation of 1000-1200 mm and a temperature range between 15-35°C. *J. curcas* plantation was 11-12 years old and its oil % and seed yields were high. Seeds were harvested manually during November-December 2007 (harvest season) and stored in jute bags and transported to the Forest Research Institute, Dehradun. Fruits were kept in shade for drying; seeds were inspected and manually cleaned. Seeds were obtained by breaking down the fruit coats manually. Seeds were labeled according to their collection place. Kernel were obtained by breaking down the seeds and used for the solvent extraction method.

Extraction and isolation of fatty oil from seeds

Kernels were used to determine the physical, mechanical and chemical properties of *J. curcas*. Measurements of physical and chemical properties were performed in the Chemistry Division, Forest Research Institute, Dehradun.

The kernels from the seeds were removed and crushed to obtain a coarse powder. The powdered seeds were extracted with petroleum ether

(60-800°C) by using soxhlet apparatus. Removal of the solvent under reduced pressure gave pale yellow coloured fatty oil. The yield of fatty oil was determined on moisture free basis (Table 1). Volume of oil was recorded and expressed as oil content (%) as follow:

$$\text{Oil content}\,(\%) = \frac{\text{Oil weight}}{\text{Sample weight}} \times 100$$

The physico-chemical properties of the *J. curcas* seed oil were determined by using standard methods of analysis (AOAC) (4) (Table 2).

Table 1. The yield of fatty oil on moisture free basis

Place of Collection	Oil content (%)
Betalghat, Nainital	34.48%

Table 2. The physico-chemical properties of the fatty oils

S.No	Physical property	Value
1	Oil yield percent	38
2	Refractive index	1.474
3	Specific density	1.10
4	Unsaponifiable matter	2.7
5	Moisture percent	0.2507
6	Saponification value	28.66
7	Protein percent	3.7
8	Acid value (3.3)	1.9
9	Iodine number (16.24)	5.55
10	Ester value	12.34

Results

After solvent extraction 34.48% moisture free oil yield is obtained from the seeds collected from the Betalghat, Nainital. The physico-chemical analysis revealed the following results with reference to the particular parameters that were taken into consideration i.e. pH (9), refractive index (1.474), specific density (1.10), unsaponifiable matter (2.7%), moisture (0.2507%), saponification value (28.66%), protein (3.7%), acid value (1.9), iodine number (5.55), ester value (12.34) and presence of reasonable amount of free fatty acids.

Discussion

The extraction and physico-chemical properties of the oil will reasonably enhance the profit status of most of the oil industries. Biofuels offer great potential for global climate change mitigation, local pollution prevention, sustainable development and enhancing energy security. Though the steps are initiated towards development of biofuels in India, much more is needed to be done for effective utilization of biofuels including strict enforcement of emission norms as a matter of environment protection, technology development, guidance and assistance for biofuel projects and establishment of proper institutional framework for capacity building, wastelands allocation, plantation rising technology development, standardization and quality control for biofuel projects. Development of complete nursery package for this species will help in production of quality planting material which can withstand different soil and climatic conditions.

REFERENCES

Anonymous 1973. Official Standard and Recommended Methods, Second ed, *Society of Analytical Chemists, AOAC*, London, 675.

Chauhan, B., Rashmi, Srivastava, R.K., Kumar, D. and Tomar, A. 2006. "*Jatropha curcas:* Studies on Isolation, Estimation of Seed Fatty Oil and Storage Trials". Poster at *National Conference on Natural Products & Biodiversity: Chemistry and Utilization* from 2-3 November, 2006, Forest Research Institute, Dehradun.

Lyonia 2006. Seed oil content variation in *Jatropha curcas* Linn. in different altitudinal ranges and site conditions in H.P. India. *Lyonia* 11 (2): 31-34.

Pramanik, K. 2003. Properties and use of *Jatropha curcas* oil and diesel fuel blends in compression ignition engine. *Renewable energy* 28 (2):239-248.

Singhal, S. 2004. Biodiesel specifications for India, directions for development. International Conference on Biofuels, Perspective and Prospects held on September 16-17, New Delhi, India. 278-279.

Tomar, A., Srivastava, R.K., Kumar, D., Rashmi and Chauhan, B. 2006. Jatropha biodiesel: Promising clean alternative energy. Poster at *National Conference on Natural Products & Biodiversity: Chemistry and Utilization* from 2-3 November, 2006, Forest Research Institute, Dehradun.

Chapter - 6

POTENTIAL OF JATROPHA PLANTATIONS IN RAJASTHAN

Prakash Chandra Chaplot

Directorate of Extension Education, Maharana Pratap University of Agriculture & Technology, Udaipur-313 001
E-mail.: chaplotpc@yahoo.co.in

ABSTRACT

India's energy challenge can be met by TBO's focused primarily on *Jatropha curcas* for use as renewable alternate fuel to petrodiesel. It is well suited for cultivation in harsh tropical and subtropical climate. Survey data show that out of total wasteland (10.6 mha) available in Rajasthan, the area suitable for Jatropha plantation is only 5.6 mha. Thus to explore possibilities of sustainable and economic cultivation of Jatropha, there is need to work on systematic genetic improvement to produce quality seeds, planting geometry, adaptability to different edaphic and agro-climatic conditions, compatibility or suitability of its cultivation with agricultural/ silvicultural crops, response to fertilizer, gestation period, yield, marketing facilities, etc. Amongst various facets of researchable issues related to agro-techniques of this plant, experiments conducted on clay loam soils of southern Rajasthan, planting geometry of 3 x 3 m was found better for improved growth and yield of Jatropha. Regional survey conducted to identify elite plant material revealed that the oil content in the seeds ranged from 19 to 42%. The present status of this plant in Rajasthan indicates that research is required to generate database for laying guideline for enhancement of growth and productivity through suitable agro-techniques under prevailing agro-climate.

Keywords: Jatropha plantation, wasteland, Rajasthan

INTRODUCTION

India stands sixth in the world oil consumption, which has forced the country to depend on oil producing countries. India imports nearly 70% of its annual crude oil requirement, which is roughly 111 million tones, which has a considerable impact on the country's foreign exchange reserves. Considering the fact that India produces a meager 25% of its crude oil consumption, increased emphasis should be on biofuels to meet a significant fraction of the country's energy needs. Considering the rising price of fossil fuels and being a part of the clean environment policies, India started its biofuel journey in 2003 but announced biofuel policy on September 11, 2008. National Biofuel Policy drafted by the Ministry of New and Renewable Energy Sources assures that biofuel programme would not compete with food security and the fertile farm lands would not be diverted for plantation of biofuel crops. The policy deals with a numbers of issues like minimum support prices, subsidies for growers, marketing of seeds, subsidies and fiscal concession for industry, R & D, mandatory blending of autofuel with biofuel, quality norms, testing and certification. An indicative target of 20% by 2017 for the blending of biofuels-bioethanol and bio-diesel has been proposed in the policy.

Vegetable oils are renewable, non-toxic, bio-degradable and have low emission profiles. The raw materials being used commercially by the bio-diesel consuming countries constitute the edible fatty oils derived from rapeseed, soyabean, palm, sunflower, mustard, coconut and linseed etc. Use of such vegetable oils to produce biodiesel in India is not feasible in view of a big gap in demand and supply. Increased pressure to augment production of edible oil also put limitation on the use of these oils for production of biodiesel. Under Indian conditions, plants, which produce non-edible oils in appreciable quantities and suitable for large scale cultivation on non-cropped marginal and wastelands can be considered for biodiesel production. In this context, the best suited plant fulfilling the above criteria is *Jatropha curcas*. It is a large shrub or small tree, able to thrive under varied climatic condition with the rainfall of 250-1000 mm. It is traditionally used as biofence protecting cropland from freely ranging cattles. Being drought tolerant, it can be used to reclaim degraded or wastelands (Dubey *et al.*, 2005).

Rajasthan background

Rajasthan is situated in the north-western part of India. The climate varies from arid to sub-humid. On an average, winter temperatures range from 8° to 28°C and summer temperatures range from 25° to 46°C. It is characterized by low and erratic distribution of rainfall ranges from 150-1000mm most of which falls from July to September. The Aravalli ranges divide the state approximately into two parts. The area lying to the west is the arid region and the east can be termed as semi-arid region. The forest covered an area of about 26.6 lakh ha in state. The density of forest varies from dense to barren and highly degraded with or even without root stock. The major sources of irrigation are tube wells and canals. About 70.8% of the total land is irrigated by wells and tube wells followed by 27.6% by canal. The total geographical area of the state is 342.7 lakh ha comprising 173.9 lakh ha of net sown and 216.6 lakh ha total cropped area. The irrigated land accounted for only 17.13% of the agricultural land by all sources. The net sown area more than once is 42.2 lakh ha. The state has 48.7 lakh ha culturable wasteland, 25 lakh ha barren and uncultivated land and 99.4 lakh ha culturable fallow land. The land under non-agricultural uses is only 17.6 lakh ha.

Natural biofuel plantations

Many tree species bear fruits having seeds rich in oil which is found suitable for conversion into biodiesel. Out of these Neem, Karanj, Mahua and Jatropha are being considered for plantations on lands classified as waste and degraded. Of all the Tree Borne Oilseeds, Jatropha and Karanj stands at the top. Both tree species are found naturally growing in the south and south western part covering Udaipur, Rajsamand, Chittorgarh, Bhilwara, Dungarpur, Banswara, Sirohi and foot hills of Pali districts. Karanj is mostly found under semi-arid to humid and sub humid ecosystems on soil types ranging from stony to clayey around field boundaries, wastelands, nala bank stabilization, river valleys and tank foreshore. Mahua is most loving tree of tribals in Aravalli range of south Rajasthan. It is found on bunds of the field. The tribals collect flowers for edible and seed for oil purpose. Jatropha is grown extensively by the farmers in south and south west Rajasthan as a live fence on arable and non-arable lands particularly for protection of field against damage by livestocks, as hedge for protection of kachha house from rain showers and green caping of bunds and base of hills as it is upalatable to cattle and goals. Thus, Jatropha has attained the status of Addivasyon Ki Laxmi as the tribals earn handsome money simply by collecting its fruits for seed and selling to the local merchant.

Although no authentic data is available on revenue records regarding area under Jatropha, Karanja and Mahua in Rajasthan, but it is estimated that about 13-15 thousand ha area is under Jatropha in the form of fence and wasteland plantations. From this acreage about 10 thousand tones of Jatropha seed reaches to the market annually. The price of its seed fluctuates from Rs. 10-15 per kg.

Potential of Jatropha

Jatropha is a small tree or shrub growing under diverse agro-climatic conditions and even on gravel, sand and native soils under tropics and sub-tropics. Its water requirement is very low and can withstand long periods of drought by shading leaves to reduce transpiration loss (Anon., 2001). The only advantage of Jatropha is the ease of production even under harsh climate and early realization of seeds after plantation compared to other oil bearing trees (Table 1). Its oil can be used as a biodiesel that is environmentally safe and a cost effective renewable source of non-conventional energy.

India has tremendous potential as a producer of Jatropha. Out of 130 million ha of wasteland, 40 million ha is suited for cultivation. Such land has little agricultural value at present but is suitable for the cultivation of Jatropha. Jatropha was considered to be primary feedstock for biodiesel production due to its high oil content, ability to tolerate a wide range of climates and its productive life spans of over 40 years. The states having maximum acreage of wasteland are Rajasthan, MP, Chattisgarh and Gujarat. A few productive states have set up Biofuel Board and have announced policies in their respective states and a minimum support price for Jatropha to provide a fair price to the tribal farmers.

It is estimated that 66.90 MMT of petroleum diesel will be needed by 2011-12, which will require 13.38 MMT of biodiesel considering a maximum blending of 20%. For producing 13.38 MMT of biodiesel, an area of 11.19 million ha is needed under Jatropha plantation, which will yield revenue of Rs. 20,000 crore and employment to 12 million people (GOI, 2008). There is hardly any doubt that India has the potential to become not only a major consumer of bio-diesel but also one of the World's leading producers of bio-diesel.

Table 1. Suitability of Jatropha with other oil bearing trees

Characteristics	Jatropha	Karanja	Mahua
Ecosystem	Arid to semi-arid	Semi-arid to sub-humid	Sub-humid
Rainfall	Low to medium	Medium to high	Medium to high
Plant suitability	Wasteland, degraded live fence, shallow soils	Field boundary, nala bank, tank fore shore, wasteland	Field boundary, wasteland
Plants/ acre	1000	200	80
Gestation period	Short, 3 years	Long, 5 years	Long, 8 years
Yield/ plant (kg)	1.0 – 1.5	10 – 100	40 – 50
Oil content (%)	27 – 38	31 – 42	35 – 38
Protein (%)	38	35	16

Scope of Jatropha plantations in Rajasthan

In Rajasthan, over 105 lakh ha area is under wastelands. The area under gullied and ravenous land is 5.0 lakh ha, with and without scrub 27.1 lakh ha, degraded forest 12.5 lakh ha, degraded grazing/ pasture 12.2 lakh ha and barren, rocky/stony 4.8 lakh ha (NRSA, 2005). Substantial acreage of wasteland available is sandy or hilly regions, most industries are desperately looking to the state governments to give them state owned land at throwaway prices or to allocate large tracks of forest land for biodiesel plantations. A large portion of barren and uncultivated land reported in the state can be put under Jatropha plantations. A good portion of culturable wasteland, land put under non-agricultural uses and vast stretches of public land, railway track, roads and canals can be utilized for biofuel plantation. The survey data show that out of total wasteland available, the area suitable for Jatropha plantation is only 56 lakh ha. The biofuel authority of Rajasthan prepared a road map for development of biofuel plantation. Even if 5 lakh ha is brought under TBO in block plantations every year, the entire wasteland area can come under permanent vegetation yielding edible/ non-edible oils in about 9 years. Nearly 21 lakh ha wasteland of the state is in 11 districts where such programmes are being planned. Likewise, the area under agricultural crop is around 132 lakh ha. Since animals do not browse Jatropha, the crops can be protected from stray animals by forming a thick biofence, this would give valuable biodiesel yielding seed also.

Constraints of Jatropha cultivation

1. Lack of location specific recommended varieties and agro-techniques
2. Low and uncertain yield. Actual yield under field condition is only 0.5-1.5 kg per plant
3. No systematic attempts were made to develop Jatropha based agro-forestry models
4. Poor technical and extension support on cultivation and processing method due to lack of systematic research
5. Regularized marketing facilities to save the poor villagers from the clutches of middleman is not created
6. No buy back facility is extended by the Government
7. The fluctuations in yield due to climatic hazards
8. The infestation of crop due to insect, pest and diseases

Present need

Jatropha plantations can play an important role in greening a large area of the barren lands in the Mewar belt and other potential pocket through suitable Jatropha crop combination. The non-edible vegetable oil of Jatropha has potential of providing a viable alternative to diesel oil. It can also be utilized in various industrial products. The massive plantation can also play an important role in generating large-scale employment and additional income. The major considerations for promotion of Jatropha in the region are given in next page:

1. Development of improved and high yielding varieties alongwith agro-technologies suited to diverse agro-climatic condition
2. The collection and marketing of Jatropha seeds can be organized with support from appropriate marketing agencies at village/ tehsil level so that tribals get remunerative prices for their collections. This will also put check on premature harvest of seed
3. Identification of suitable crop combination with Jatropha.
4. Organization of awareness programme at village level
5. Strengthening of research and FLD for back stopping
6. Establishment of oil expeller at village level and biodiesel plants of different capacity on community basis at block level
7. Strengthen the marketing facility by way of forming the SHGs
8. Incentives/subsidy by State Biofuel Authority to TBO growers
9. Development of technology for boosting production from naturally growing plants

REFERENCES

Anonymous 2001. The Wealth of India, Vol. V, CSIR, New Delhi, pp 293-295.

Dubey, K., Khan, M.K., Srivastava, A. and Singh, V.K. 2005. Oil from wasteland- The Jatropha project in India. National conference on management of land resources and land use towards sustainable development. Institute of Environment Development Studies. Bundelkhand University, Jhansi, U.P.

GOI 2008. Government of India, Planning Commission. Report of committee on development on biofuels, Govt. of India, New Delhi.

NRSA 2005. National Remote Sensing Agency, Wasteland Atlas of India, Ministry of Rural Development, Department of Land Resources, Government of India and National Remote Sensing Agency, Department of Space, Government of India, Hyderabad.

Chapter - 7

JATROPHA- "THE WONDER PLANT"

Pradeep Kumar Singh,
SRIDA, Barela, Jabalpur

ABSTRACT

Jatropha is a potential bio-energy crop and can be used to alleviate soil degradation, desertification and deforestation. *Jatropha curcas* is a drought-resistant perennial, grows well in marginal/poor soil and thrive under extremely harsh climatic conditions and can withstand moisture stress. Jatropha produces seeds with around 37% oil content and can be potentially exploited to produce biodiesel. The plant also provides medicinal, environmental and ecological benefits. Seed production ranges from about 2 tons per hectare per year to over 12.5 t/ha/year, after five years of growth. Jatropha can help to increase rural incomes, self-sustainability and alleviate poverty.

Keywords: *Jatropha curcas*, multipurpose crop

INTRODUCTION

Jatropha is a valuable multipurpose crop to alleviate soil degradation, desertification and deforestation, which can be used for bio-energy to replace petro-diesel, for soap production and climate protection, and hence deserves specific attention. *Jatropha curcas* is a drought-resistant perennial, grows well in marginal/poor soil. It is easy to establish, grows relatively quickly and producing seeds for 50 years. Jatropha produces seeds with around 37% oil content. The oil can be combusted as fuel without being refined. It burns with clear smoke-free flame, tested successfully as fuel for simple diesel engine. The by-products are, press cake, a good organic fertilizer and glycerine. The oil also has insecticidal properties. Medically it is used for diseases like cancer, piles, snakebite, paralysis, dropsy etc. Jatropha grows wild in many areas of India and even thrives on infertile soil. A good crop can be obtained with little effort. Depending on soil quality and rainfall, oil can be extracted from the Jatropha nuts after two to five years.

J. curcas plant can survive and thrive under extremely harsh climatic conditions and can withstand moisture stress. Hence, it is considered suitable for planting on millions of hectares of wastelands in India, as it can produce bio-diesel in a decentralized manner. The plantation technology is simple and cost effective, cattle do not browse this plant, there is no damage by birds, it can be grown under rainfed conditions and once established it can continue to give seed production for a period of about 40-50 years. The leaves shed during the winter months, form mulch around the base of the plant. The organic matter from shed leaves enhance earth-worm activity in the soil around the root-zone of the plants, improving the fertility of the soil. It is considered as an ideal agro-forestry crop for generating source of self-employment on a large scale in rural sector and thus could be used as an effective means for poverty alleviation.

Uses of Jatropha

Medicinal plant: The latex of plant contains an alkaloid known as jatrophine, which is believed to have anti-cancerous properties. It is also used as an external application for skin diseases and rheumatism and for sores on domestic livestock. In addition, the tender twigs of the plant are used for cleaning teeth, while the juice of the leaf is used as an external application for piles. The roots are reported to be used as an antidote for snake-bites.

Raw material for dye: The bark of plant yields a dark blue dye which is used for colouring cloth, fishing nets and lines.

Soil enrichment: The de-oiled cake is rich in nitrogen, phosphorous and potassium and can be used as organic manure.

Leaves : Jatropha leaves are used as food for the tusser silkworm.

Insecticide/ pesticide: The seeds are considered anthelimintic in Brazil, and the leaves are used for fumigating houses against bed-bugs.

Alternative to diesel: It is significant to point out that, the non-edible vegetable oil of *Jatropha curcas* has the requisite potential of providing a promising and commercially viable alternative to diesel oil since it has desirable physicochemical and performance characteristics comparable to diesel. Cars could be run with this biodiesel without requiring much change in design.

Crop yield: It appears very difficult to estimate unequivocally the yield of a plant that is able to grow in very different conditions. Yield is a function of water, nutrients, heat and age of plant and other. Many different methods of establishment, farming and harvesting are possible. Yield can be enhanced with right balance of cost, yield, labour and finally cost per MT.

Seed production ranges from about 2 tons per hectare per year to over 12.5t/ha/year, after five years of growth. Although not clearly specified, this range in production may be attributable to low and high rainfall areas. From the experience in India and elsewhere, a plant density of 1,100 per hectare (spacing of 3 x 3 meters) has been found to be optimal, although in rainfed areas on poor soils a higher plant density of 1,666 has been felt to be more desirable. In such plantations Jatropha gives about 2 kgs of seed per tree. If planted in hedges, the reported productivity of Jatropha is from 0.8 kg. to 1.0 kg. of seed per meter of live fence. Oil content varies from 28% to 30% and with 94% extraction; one hectare of plantation will give 1.6 MT of oil if the soil is average, 0.75 MT if the soil is lateritic.

Economy: Jatropha can help to increase rural incomes, self-sustainability and alleviate poverty. Plantation of Jatropha alone is not economically attractive, as there is little income from it for first 2 to 3 years. Since Jatropha plant is initially small in height, castor can be intercropped with it to get income and oil for first 2 to 3 years.

With all these advantages – Jatropha plant is considered to be a "Wonder Plant"

75-76

Biofuel Potential: Current and future prospects

Chapter - 1

POTENTIAL AND PROSPECTS OF BIOFUELS FROM LIGNOCELLULOSIC FORESTRY RESIDUES

Sanjay Naithani[1], Alok Kumar Dubey[1], Gyanesh Joshi[1] and P.K. Gupta[2]

[1] Cellulose and Paper Division

[2] Chemistry Division, Forest Research Institute, Dehra Dun 248 006

ABSTRACT

Biofuels are drawing increasing attention worldwide as substitutes for petroleum-derived transportation fuels to help address energy cost, energy security and global warming concerns associated with liquid fossil fuels. The term biofuels is used here to mean any liquid fuel made from plant material or biomass that can be used as a substitute for petroleum-derived fuel. It could also bring environmental benefits, including lower green-house-gas emissions, because the biomass raw materials for producing biofuels are renewable. There are several types of biofuels and many different ways of producing them. The bulk of biofuels produced around the world such as bioethanol and biodiesel are common fuel for today. More recently, new technologies have emerged to enable the conversion of lignocellulosic biomass (wood, pulp and paper mill discards, forestry and agricultural wastes and residues, etc.) into bioethanol. Ethanol is usually produced from hydrolysis and fermentation of lignocellulosic biomass while biodiesel is produced mainly from oil-seed crops, including rapeseed and sunflowers through process of trans-esterification. Each fuel has its own unique characteristics, advantages and drawbacks. Hydrolysis of ligno-cellulosic biomass can open the way towards low cost and efficient production of ethanol from these abundant types of biomass. The conversion is more difficult than for sugar and starch because from lignocellulosic materials, first sugars need to be produced via hydrolysis. Lignocellulosic biomass requires pretreatment by mechanical and physical actions to clean and size the biomass, and destroy its cell structure to make it more accessible to further chemical or biological treatment. Also, the lignin part of the biomass is removed, and the hemicellulose is hydrolysed (saccharified) to pentose sugars such as xylose, arabinose and hexose sugar (Glucose, Mannose & Galactose). The cellulose can then be hydrolysed to hexose like Glucose. The sugars are fermented to ethanol by yeasts like *S. cerevisiae*, *Pichia stiptis* or bacteria like *Zymomonas moblis*, *E. coli* etc., which is to be purified and dehydrated. As only the cellulose and hemicellulose can be used in the process, the lignin is used for power production.

Keywords: Biofuels, lignocellulosic biomass

INTRODUCTION

The increased concern for the security of the oil supply and the negative impact of fossil fuels on the environment, particularly greenhouse gas emissions, has put pressure on society to find renewable fuel alternatives. These alternatives come from biomass. The large role of biomass is expected to play in future energy supply can be explained by the fact that biomass fuels can substitute more or less directly fossil fuels in the existing energy supply infrastructure. Intermittent renewable such as wind and solar energy are more challenging to the ways we distribute and consume energy, especially in the transport sector. The term biofuels is used here to mean any liquid fuel made from plant material or biomass that can be used as a substitute for petroleum-derived fuel. It could also bring environmental benefits, including lower greenhouse-gas emissions, because the biomass (raw materials) for producing biofuels is renewable. Biomass for biofuel is derived from living organisms, including plants, animals and their by-products. For example manure, garden waste and crop residues are all sources of biomass. It is a renewable energy source based on the carbon cycle, unlike other natural resources such as petroleum, coal, and nuclear fuels.

Biofuel is defined as solid, liquid or gaseous fuel obtained from relatively recently lifeless or living biological material and is different from fossil fuels, which are derived from long dead biological material. Also, various plants and plant-derived materials are used for biofuel manufacturing. There are two common strategies of producing liquid and gaseous fuels. One is to grow crops high in sugar (sugar cane, sugar beet, and sweet sorghum (ICRISAT)) or starch (corn/maize), and then use yeast fermentation to produce ethyl alcohol (ethanol). The second is to grow plants that contain high amounts of vegetable oil, such as oil palm, soybean, algae, Jatropha or Pongamia. When these oils are heated, their viscosity is reduced, and they can be burnt directly in a diesel engine, or they can be chemically processed to produce fuels such as biodiesel. Wood and its byproducts can also be converted into biofuels such as woodgas, methanol or ethanol fuel (Yunqiao *et al.*, 2007). It is also possible to make cellulosic ethanol from non-edible plant parts, but this can be difficult to accomplish economically (Inderwildi and King, 2009).

Liquid fuels for transportation

Most transportation fuels are liquids, because vehicles usually require high energy density, as occurs in liquids and can be provided most inexpensively by an internal combustion engine. These engines require clean burning fuels, in order to keep the engine clean and minimize air pollution. The fuels that are easier to burn cleanly are typically liquids and gases. Thus liquids (and gases that can be stored in liquid form) meet the requirements of being both portable and clean burning. Also, liquids and gases can be pumped, which means handling is easily mechanized, and thus less laborious.

Biofuels classification

Biofuels potential may be divided into three parts-

First generation biofuels: 'First-generation biofuels' are biofuels made from sugar, starch, vegetable oil, or animal fats using conventional technology. The basic feedstocks for the production of first generation biofuels are often seeds or grains such as wheat, which yields starch that is fermented into bioethanol, or sunflower seeds, which are pressed to yield vegetable oil that can be used in biodiesel. These feedstocks could instead enter the animal or human food chain, and as the global population has risen their use in producing biofuels has been criticised for diverting food away from the human food chain, leading to food shortages and price rises.

Second generation biofuels: Supporters of biofuels claim that a more viable solution is to increase political and industrial support for, and rapidity of, second-generation biofuel implementation from non food crops, including cellulosic biofuels. Second-generation biofuel production processes can use a variety of non food crops. These include waste biomass, the stalks of wheat, corn, wood, and special-energy-or-biomass crops (e.g. Miscanthus). Second generation biofuels use biomass to liquid technology (Inderwildi and King, 2009), including cellulosic biofuels from non food crops (Eurek Alert, 2009). Many second generation biofuels are under development such as biohydrogen, biomethanol, DMF, Bio-DME, Fischer-Tropsch diesel, biohydrogen diesel, mixed alcohols and wood diesel.

Third generation biofuels: Algae fuel, also called oilgae or third generation biofuel, is a biofuel from algae. Algae are low-input, high-yield feedstocks to produce biofuels. It produces 30 times more energy per acre than land crops such as soybeans. With the higher prices of fossil fuels (petroleum), there is much interest in algaculture (farming algae). The United States Department of Energy estimates that if algae fuel replaced all the petroleum fuel in the United States, it would require 15,000 square miles (38,849 square kilometers), which is roughly the size of Maryland. Algae, such as *Botryococcus braunii* and *Chlorella vulgaris,* are relatively easy to grow, but the algal oil is hard to extract.

Issues with biofuel production and use

There are various current issues with biofuel production and use, which are presently being discussed in the popular media and scientific journals. These include the effect of moderating oil prices, the "food vs fuel" debate, carbon emissions levels, sustainable biofuel production, deforestation and soil erosion, impact on water resources, human rights issues, poverty reduction potential, biofuel prices, energy balance and efficiency, and centralized versus decentralized production models.

Environment

The carbon in biofuels was extracted from atmospheric carbon dioxide by growing plants. Therefore, there is a usually low level of carbon emissions to produce and transport the fuels through use of fossil fuel energy. Therefore, releasing the carbon that the plants contain through combustion or other treatment does not increase the average amount of carbon dioxide in the atmosphere i.e. biofuels are carbon neutral. This is their attraction for energy conversion and storage.

Bioethanol as a biofuel from lignocellulosic biomass

As early as in 1925, Henry Ford stated the fuel of future will come from agricultural waste, weeds, sawdust, forestry residues. Today Henry Ford's futuristic vision can be easily understood (Levine, 1996).

Lignocellulose is a generic term for plant matter and "wastes or biomass," that are generated through forestry and agricultural practices, paper-pulp discards like mixed waste papers, timber industries and many agro-industries. Much of the lignocellulosic waste is often disposed of by biomass burning, which is not restricted to developing countries alone, but is considered a global phenomenon. However, such materials can potentially be converted into different value-added products including biofuels (Malherbe and Cloete, 2003).

Advantages of lignocellulose-based liquid biofuels

1. Biofuel sources are geographically more evenly distributed than the fossil fuels. Thus, the sources of energy will, to a larger extent, be domestic and provide security of supply.
2. Lignocellulosic raw materials minimize the potential conflict between land use for food (and feed) production and energy feedstock production. The raw material is less expensive than conventional agricultural feedstock and can be produced with lower input of fertilizers, pesticides, and energy.
3. Biofuels from lignocellulose generate low net greenhouse gas emissions, reducing environmental impacts, particularly climate change.

4. Biofuels might also provide employment in rural areas.

Technology of ethanol production from lignocellulosic biomass

The technology of conversion of lignocellulose-to-ethanol has following (Fig. 1) major unit operation i.e. Pretreatment, Hydrolysis (Acid or Enzyme), Detoxification, Fermentation, Distillation and Product recovery.

Lignocellulosic Biomass

↓

Pretreatment

↓

Hydrolysis (Acid or Enzyme)

↓

Fermentation

↓

Product Recovery

Fig. 1. Schematic diagram of production of ethanol from biomass

CONCLUSION

Today, there exists a significant demand in industrialized countries for biofuels, driven largely by regulatory mandates for blending of biofuels into petroleum fuels. This demand is likely to grow considerably in the years ahead, driven by increasingly energy security concerns. The limitations of first-generation biofuels in terms of direct food vs. fuel conflict, cost competitiveness, and greenhouse gas emissions reductions are not likely to be substantially different in developing countries than in industrialized countries. Sugar cane ethanol stands out among first-generation biofuels as suffering fewer limitations. The current debate on the sustainability of biofuels is largely focusing on the biomass like lignocellulosic biomass. Lignocellulosic biomass residues available from agricultural and forest processing constitute a potential source for production of chemicals such as ethanol, reducing sugars and furfural, using enzyme or acid-catalysed hydrolysis. To ensure that a low-cost energy feedstock is available, researchers are examining dedicated energy crops, wood and grass species that have been selected to produce high yields. The choice of the best technology for lignocellulose to bioethanol conversion should be decided on the basis of overall economics (lowest cost), environmental (lowest pollutants) and energy (higher efficiencies) issues.

ACKNOWLEDGEMENT

The authors are thankful to Director, Forest Research Institute, Dehradun (Uttarakhand) for permitting to carry out the research work on bioethanol project.

REFERENCES

ICRISAT: Sweet sorghum balances food and fuel needs.

Inderwildi, O.R. and King, D.A. 2009. Quo Vadis Biofuels. *Energy & Environmental Science* 2: 343.

Levine, J.S. 1996. Biomass burning and global change. In: Remote Sensing and Inventory Development and Biomass Burning in Africa (Vol.1). Levine J.S. (eds). The MIT Press Cambridge Massachusetts, USA pp 35.

Malherbe, S. and Cloete, T.E. 2003. Lignocellulosic biodegradation fundamentals and application: A review. *Environ. Sc. Bioethanol* 1:105-114.

Somerville, Chris. 2007. Development of Cellulosic Biofuels. U.S. Dept. of Agriculture.http://www.usda.gov/oce/forum/2007%20Speeches/PDF%20PPT/CSomerville.pdf. Retrieved on 2008-01-15.

Yunqiao, P., Zhang, D., Singh, P.M. and Ragauskas, A.J. 2007. The new forestry biofuels sector. *Biofuels, Bioproducts and Biorefining* 2(1):58-73.

Chapter - 2

PRODUCTION OF BIODIESEL FROM HIGH FREE FATTY ACID NEEM (*AZADIRACHTA INDICA*) OIL

Subhalaxmi Pradhan, N.K. Sahoo, P.K. Rout and S.N. Naik

Centre for Rural Development and Technology, Indian Institute of Technology, New Delhi-110016, India

E-mail: snn@rdat.iitd.ac.in, sn_naik@hotmail.com

ABSTRACT

Biodiesel has become a reality of the present world. Different tree borne non-edible oil seeds such as Jatropha, Karanja, Mahua, Simaruba and Neem are potential source for biodiesel production. Out of these plants, inspite of being native to India and with multiple utility Neem has not been emphasized much from the view point of biodiesel production. Biodiesel can also be prepared from Neem oil and the feasibility has been tested in this study. Whole Neem fruit (without depulping) and seed (depulped fruit) were extracted to determine the oil content and different physico chemical properties of the oil. The high FFA Neem oil was tried out for biodiesel (Fatty acid methyl ester of Neem oil) production following dual step transesterification process. The first step is the acid catalysed esterification to produce methyl ester by lowering the acid value and next step is alkali-catalysed transesterification. In first step the high FFAs of the oil was reduced with methanol and H_2SO_4 as catalyst for two hours at a reaction temperature of 55°C. In second step the product obtained from first step was transesterified with methanol and KOH as catalyst with a reaction temperature of 60°C for two hours. The yield of biodiesel after reaction is 95%. The product after transesterification reaction was monitored by GC and NMR and fuel properties of biodiesel were measured.

Keywords: Biodiesel, free fatty acid, transesterification, Neem oil

INTRODUCTION

Majority of energy consumed worldwide are from petrochemical sources, coal and natural gases which are finite and will be exhausted in near future. The present energy crisis and pollution problem caused due to the combustion of these fossil fuels are the incentives for the search of alternative fuels from renewable resources with less environmental impact. One of the alternatives to fossil fuels is biodiesel which is biodegradable, renewable, non-toxic and has low pollutant emissions. Biodiesel is the monoalkyl ester of long chain fatty acids derived from transesterification of vegetable oils and animal fat with short chain monohydric alcohols in presence of catalyst. The oilseed crops used for biodiesel production are sunflower, soybean, rapeseed, linseed, cottonseed, canola etc. (Peterson, 1986) majority of which are edible in nature. Use of edible oils for biodiesel production is unaffordable as the demand of edible oil has increased tremendously. Non-edible oils from various plants like Neem (*Azadirachta indica*), Jatropha (*Jatropha curcas*), Karanja (*Pongamia pinnata*), Mahua (*Madhuca indica*), Simarouba (*Simarouba glauca*) etc. are feed stocks available for biodiesel production (Meher *et al.*, 2006). Among these, Neem is available plenty throughout India and native to the country.

Neem *(Azadirachta indica)* is distributed in various countries like Asia, Africa, Central and South America. It grows in almost all types of soils including clay, alkaline, dry and stony soils. A mature tree has about 150-200 years of life span and can produce 30-50 kg of fruits per year. Hence about 5,40,000 tonnes of seeds are produced yielding 1,07,000 tonnes of oil and 4,25,000 tonnes of cake (Ilavarasi *et al.*, 2008). The oil content of the seed varies from 20-32.61% and that of kernel has an average oil yield of 47.6% in India (Kaura *et al.*, 1998). The oil is bitter in taste, yellow to dark in colour and has strong odour. Neem oil has been widely employed in manufacturing soaps, skin care products, waxes, lubricants, epoxy compounds, and as a fuel for lighting.

The main constraints of using vegetable oils directly as fuel are higher molecular weight, higher viscosities, poor cold- flow properties, deposit formation due to poor combustion and poor volatilites (Kale, 2005). The non edible oils contain large amount of free fatty acids (FFA) and lipid associates for which it is not directly adapted as fuel in diesel engines. The solution to viscosity problem has been attained by four different processes: dilution, micro-emulsification, pyrolysis and transesterfication (Srivastava *et al.*, 2000). Transesterification is the convenient process among the above techniques to produce clean and environmentally safe fuel. Transesterification is a process of reacting triglycerides with alcohol (methanol and ethanol) to obtain fatty acid alkyl esters and glycerol and the process is important for biodiesel production. The factors affecting tranesterification reaction are type and amount of catalyst, alcohol to oil molar ratio, reaction temperature, rate of stirring and reaction time, FFA content and water content in oil. When the FFA content of oil is high, transesterification reaction is difficult because of formation of soap with alkaline catalyst. The high FFA content oil thus needs acid catalysed pretreatment to lower the acid value by esterification of FFA, followed by alkali catalyzed transesterification for preparation of biodiesel (Canakei *et al.*, 1999, Canakei *et al.*, 2001). The present study deals with preparation of Neem oil methyl ester (NOME) from high FFA Neem oil following two step procedures; acid catalysed pretreatment followed by alkali catalysed reaction.

MATERIALS AND METHODS

Procurement of sample

Neem fruits were collected from IIT campus, New Delhi, India for the study. A fraction of the collected fruits were dried directly in hot air oven at 80°C till they become totally moisture free. Rest of the fruit was depulped by decomposing the pulp by soaking in water for 3-5 days to get the seeds. The seeds were sun dried and then dried in hot air oven till free from moisture. Both the oven dried fruit and seed were stored for further analysis.

Oil content and characterisation of Neem oil

Both oven dried fruit and seed were mildly grinded separately, for 1 minute in a domestic grinder and separated by sieving on a BSS 8 mesh (2 mm) test sieves. For oil content determination weighed amount of grinded fruit and seed were extracted separately in soxhlet apparatus using petroleum ether (boiling point 60-80°C). The extract was concentrated in rotavapour; the residual oil was cooled and weighed. The physicochemical properties (density, viscocity, acid value, saponification value and unsaponification matter) of the oil were determined. Then the oil was stored for biodiesel preparation.

GC-FID and GC-MS analysis

The fatty acid composition was determined by gas chromatographic analysis of the fatty acid methyl esters as follows (Hammond, 1993). 10mg of oil was refluxed for 90 minutes in a round bottom flask with 10 ml of transmethylating agent on a water bath. Then 10ml of petroleum ether (B.P- 40-60°C) and 10ml of water was added after cooling. The aqueous layer was pipette out and washing was repeated with second 10 ml of water. The petroleum ether layer was treated with anhydrous sodium sulphate (Na_2SO_4) to remove moisture and clean ether layer was evaporated in rotavapour. Then it was dissolved in minimum amount of petroleum ether. GC analysis was carried out against standard fatty acid methyl esters (sigma make) on a Varian CP-3800 Gas Chromatograph equipped with a flame ionization detector (FID) and a 30 m x 0.25 mm WCOT column coated with 0.25 mm film thickness of polyethylene glycol supplied by J & W (Carbowax column). Helium was used as the carrier gas at a flow rate of 1.0 mL/ min at a column pressure of 22 KPa. 0.2μl of sample was injected into the injection port of the GC using a split ratio of 50:1. Compound separation was achieved following a linear temperature program of 160°C (1 min), 160°C-240°C (4°C/min), 240°C (23 min), so the total run time was 45 minute. Percentage composition was calculated using peak normalization method assuming equal detector response. The peaks were identified by co-eluation of standard methyl ester samples procured from Sigma-Aldrich and analysed in the same GC conditions. The GC/MS analysis (Fig. 1) was carried out on a Varian Saturn 2200 GC/MS fitted with the same column and temperature programmed as above. MS parameters: ionization voltage (EI) 70 eV, peak width 2 sec, mass range 40-500 amu and detector voltage 1.5 volts. Peak identification was confirmed by comparison of the mass spectra with mass spectra available on NIST-1 and NIST-II libraries.

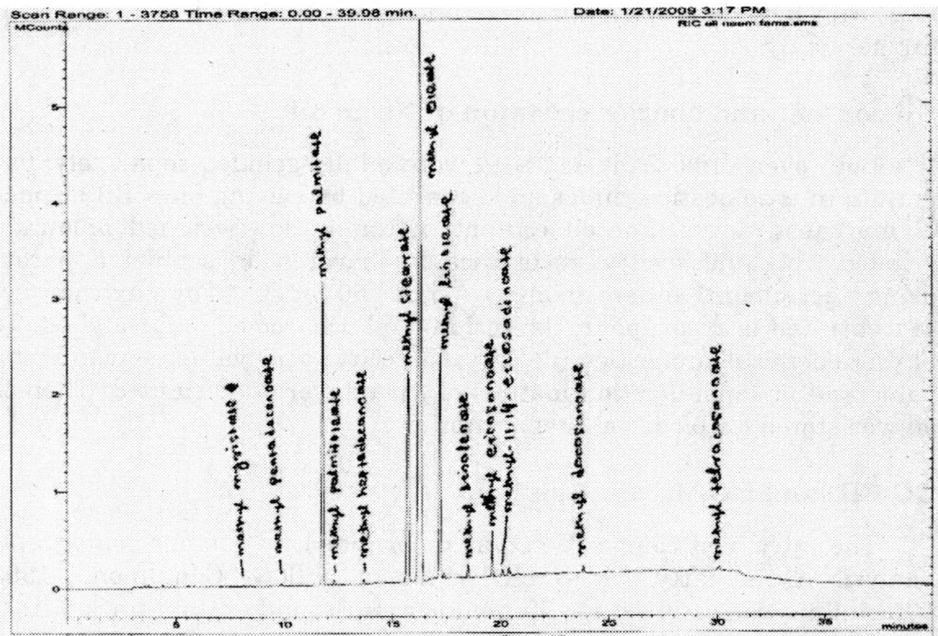

Fig. 1. GC/MS of fatty acid methyl esters from Neem Oil

Transesterification of Neem oil

Transesterification or alcoholysis is the displacement of one alcohol from an ester by another alcohol a process similar to hydrolysis; except that an alcohol is used instead of water. This process has been widely used to reduce the high viscosity of triglycerides. Neem oil methyl ester (NOME) was prepared from seed oil. A two step procedure (Acid catalyzed reaction and Alkali catalyzed reaction) was followed due to high acid value (25.24) of the oil (Meher *et al.*, 2006). Acid catalyzed pre-treatment was conducted at 60°C with methanol (MeOH)/ oil at a molar ratio of 6:1

and an acid (H_2SO_4) concentration of 0.5% w/w of oil. After every 10 minutes sample was taken out of the reactor and titration was carried out. The reaction was continued till the acid value was lowered and remained constant. The water and acid catalyst was removed before alkali-catalysed transesterification. The reaction for preparation of NOME was carried out with MeOH/oil at a molar ratio of 6:1 and catalyst (KOH) concentration of 1% w/w of oil at 65°C and stirring rate of 600ppm. The reaction mixture was separated in a separating funnel. The upper biodiesel was washed with hot distilled water (70-80°C) until the resulting water from washing remains neutral (phenolphthalein do not produce pink colour). The moisture was removed from biodiesel by drying at 90-100°C under vacuum and passing the ester layer over anhydrous sodium sulphate. The properties of the biodiesel flash point, cloud point, pour point viscosity and acid value of the methyl ester were determined.

The biodiesel after transesterification reaction was analysed in ^1H-NMR and GC–FID to know the percentage conversion of the reaction. Nuclear Magnetic Resonance Spectroscopy (NMR) of NOME was carried out using proton (^1H) techniques in DPX 300 MHz Brucker instrument. From the proton spectrograms of NOME and absorption resonance frequencies (δ) were recorded using $CDCl_3$ as solvent. The difference in the δ values specifies the particular number of protons in the form of peak splittings. The percentage conversion of biodiesel (NOME) was calculated from the ^1H-NMR results as given below in Eq-1.

$$\text{Percentage conversion of methyl ester (C)} = 100 \times \left(\frac{2 \times A_{ME}}{3 \times A_{CH_2}} \right) \quad (1)$$

Where, A_{ME} = Integration value of the protons of methoxy group of methyl esters

A_{CH_2} = Integration value of methylene protons

RESULTS AND DISCUSSION

The oil content of seed was found to be 25. 5% and that of whole fruit was 10.9%. The oil content of seed is more than that of fruit as the pulp contains no oil. The pulp and seed fraction in fruit are 54.2% and 45.8%. The physicochemical properties of both the oils are given in Table 1. The acid value of the seed oil (3.6) is less than that of the fruit oil (5.1). Seed oil has high saponification value (189.2) than that of fruit oil (188.2). This higher saponification value of seed oil is indicative of being a better biodiesel feedstock than that of whole-fruit oil. This is supported by the presence of higher unsaponifiable matter in fruit oil (2.09) than seed oil (1.9). Similarly, fruit oil is more viscous than that of the seed oil. The fruit oil after degumming has similar properties to that of seed oil. After one month storage the acid value of the seed oil increased to 25.24.

Table 1. Physico-chemical properties of oil

Properties	Neem seed oil	Neem fruit oil
Acid value (mgKOH/g)	3.6	5.1
Saponification value (mgKOH/g)	189.2	188.2
Unsaponifiable matter (%, w/w)	1.9	2.09
Density (gm/cc)	0.92	0.94
Viscosity @ 40°C (cSt)	25.4	39

Table 2 shows the fatty acid composition of the neem seed oil. The major compounds are oleic acid (53.8%) and linoleic acid (10.9%) along with the saturated fatty acids viz. palmitic acid (16.9%) and stearic acid (14.1%). The total percentage of unsaturated fatty acid is comparable with Jatropha and Pongamia oil (Akintayo, 2004, Naik *et al.,* 2008). So Neem oil may be considered as a potential feed-stock for biodiesel production.

Table 2. Fatty acid composition of Neem seed oil

Retention time	Fatty acid	Amount (%)	Identification
11.4	Myristic acid (C14:0)	0.1	Co-eluation, MS
13.4	Pentadecanoic acid (15:0)	<0.1	MS
15.5	Palmitic acid (C16:0)	16.9	Co-eluation, MS
16.5	Palmitoleic acid (C16:1)	0.2	Co-eluation, MS
18.3	Heptadecanoic acid (C16:2)	<0.1	MS
20.2	Stearic acid (C18:0)	14.1	Co-eluation, MS
20.8	Oleic acid (C18:1)	53.8	Co-eluation, MS
21.6	Linoleic acid (C18:2)	10.9	Co-eluation, MS
23.3	Linolenic acid (C18:3)	<0.1	Co-eluation, MS
24.4	Arachidic acid (C20:0)	1.4	Co-eluation, MS
25.5	Ecosenoic acid (C20:1)	0.4	MS
31.7	Docosanoic acid (22:0)	<0.1	MS
42.6	Tetracosanoic acid (24:0)	<0.1	MS
	Saturated fatty acids	32.6	
	Unsaturated fatty acids	65.4	

Due to high acid value of the Neem oil, it cannot be directly used for biodiesel production by base catalysis. The acid catalysed pre-treatment reduced acid value of oil to an extent and then the oil is suitable for alkali-catalysed reaction. Acid value of the oil decreased sharply as given in Fig. 2. After 90 minutes its value was below 4 and hence good for biodiesel synthesis. Then the alkali used for alkali-catalysed reaction compensates the acidity of pre-treated oil and remaining part act as transesterification catalyst.

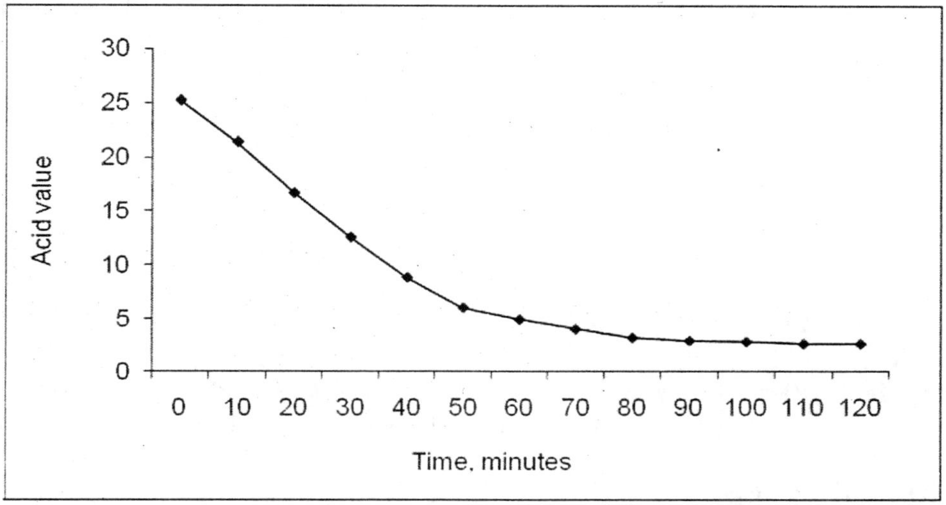

Fig. 2. Reduction of acid value against time during acid esterification of high FFA Neem oil

Trasesterification reaction for preparation of NOME was effectively carried out, which resulted in the desired products biodiesel and glycerol layer. The yield of methyl ester after the reaction was 95%. The percentage conversion was 95.2% as calculated from ^{1}H-NMR spectrum (Fig. 3) of biodiesel. The properties of biodiesel are given in Table 3 and it is comparable to the limits prescribed by American standards of biodiesel.

Table 3 Fuel properties of NOME

Properties	NOME	ASTM standards for Biodiesel
Density (gm / cc)	0.91	0.86-0.89
Flash Point (°C)	155	> 130
Pour Point (°C)	8.5	-
Cloud Point (°C)	15.5	-
Viscosity @ 40°C (cSt)	5.15	1.9-6.0
Acid value (mgKOH/g)	0.5	< 0.8

CONCLUSION

Biodiesel can directly be prepared from Neem seed oil but in contrary the whole fruit Neem oil needs degumming. High FFA containing Neem oil needs two step treatments, acid (1% v/v H_2SO_4) esterification followed by alkali catalyzed transesterification. This process gives a yield of 95% biodiesel. The fuel properties of the biodiesel are comparable to the diesel and within the limits prescribed by American standards of biodiesel.

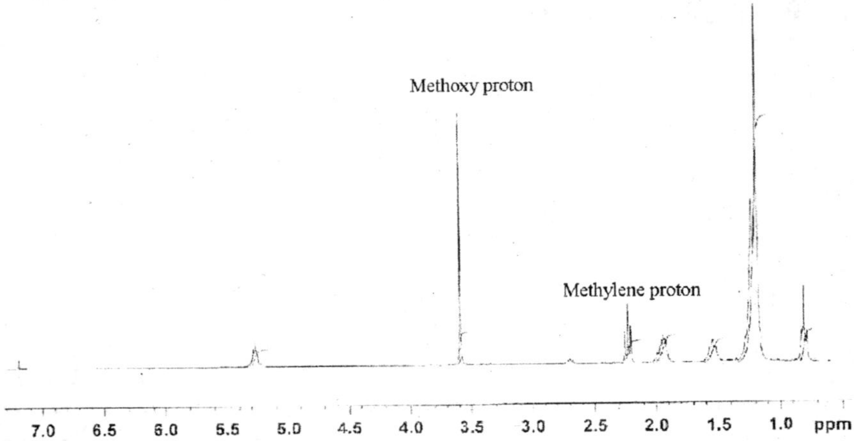

Fig. 3. ^{1}H-NMR spectrum of Neem oil methyl ester

REFERENCES

Akintayo, E.T. 2004. Characterisation and composition of *Parkia bioglobbosa* and *Jatropha curcas* oils and cakes. *Bioresource Technol.* 92: 307-310.

Canakei, M. and Gerpen, J.V. 1999. Biodiesel production via acid catalysis. *Trans. ASAE.* 42 (5): 1203-1220.

Canakei, M. and Gerpen, J.V. 2001. Biodiesel production from oils and fats with high free fatty acids. *Trans. ASAE* 44, 6: 1429-1436.

Hammond, E.W. 1993. Chromatography for the analysis of lipids. pp 81-85, CRC press.

Kale, V. 2005. Biodiesel gaining popularity in India. *INFORM* 16 (7): 466-468.

Kaura, S.K., Gupta, S.K. and Chowdhury, J.B. 1998. Morphological and oil content variation in seeds of *Azadirachta indica* A. Juss. (Neem) from northern and western provenances of India. *Plant Foods for Human Nutrition 52*: 293–298.

Meher, L.C., Vidya Sagar, D. and Naik, S.N. 2006. Optimisation of alkali-catalysed transesterification of *Pongamia pinnata* oil for production of biodiesel. *Bioresource Technol*. 97 (12): 1392-1397.

Meher, L.C., Vidya Sagar, D. and Naik, S.N. 2006. Technical aspects of biodiesel production by transesterfication – A Review. *Ren. Sust. Energy Rev.* 10, 3: pp. 248- 268.

Naik M, Meher L.C., Naik, S.N. and Das, L.M. 2008.Production of biodiesel from high free fatty acid Karanja (*Pongamia pinnata*) oil. *Biomass and Bioenergy* 32 (4): 354-357.

Peterson, C.L. 1986. Vegetable oil as diesel fuel: Status and research properties. *Trans. ASAE* 29 (5): 1412-1422.

Selva, I.P., Rao, L.N., Lyer, P.V.R., Ravichandran, K. and Rajendran, N. 2008. Biodiesel from high FFA saturated non-edible oil by using multi-step transesterification process. In: *Renewable energy and environment for sustainable development* 763-771.

Srivastava, A. and Prasad, R. 2000. Triglycerides-based diesel fuels. *Ren. Sust. Energy. Rev.* 4(2): 111-133.

Chapter - 3

INTER-POPULATION VARIABILITY OF *JATROPHA CURCAS* L. ACCESSIONS CHARACTERIZED BY RAPD MARKERS

Mamta Sahu[1], Mohd. Akram[2] and Munish Kumar[3]

[1] Department of Biochemistry; [2] Department of Plant Pathology and [3] Department of Forestry, C.S. Azad. University of Agriculture & Technology, Kanpur, 208002. E-mail : mamtacsa@gmail.com

ABSTRACT

Jatropha curcas L. grows well throughout India. The oil of Jatropha could be a renewable resource of energy and a viable alternative to petroleum products. Jatropha species are essentially cross pollinated, which offers the breeder ample scope to improve the yield potential. The determination of genetic relationships between accessions is important for Jatropha improvement because till date there is very less work done in this direction in India. Present study was under taken to assess the genetic relationships between different Jatropha accessions collected from different parts of Uttar Pradesh using RAPD markers. Twenty *J. curcas* accessions were analyzed using 8 random primers (SBSA series of Genetech Co., Ltd., China) of which 6 produced reproducible polymorphic banding patterns. A total of 59 bands were scored out of which 58.89% were polymorphic. The number of bands generated per primer varied from 3 to 12 and a minimum of 3 bands with SBSA-06 primer while maximum of 12 bands were observed with primer SBSA-04. Both SBSA-02 and SBSA-03 primers amplified unique bands of 900 and 500 bp in CSJ-38. Jaccord's genetic similarity co-efficient varied from 1.00 to 0.425. The maximum possibility of similarity coefficient 0.425 was observed in CSJ-38 with all 19 accessions of *J. curcas*. The results of RAPD analysis of *J. curcas* accessions revealed that the genetic diversity is very low in the present accessions which are being cultivated in Uttar Pradesh. This suggests that there is urgent need to widen the genetic base of the *J. curcas* so that the higher yield and oil may be obtained. In the present study genetically different accession CSJ-38 has been found superior in seed length and breadth. CSJ-38 accession could be utilized for widening the genetic base as well for improving other characteristics.

Keywords: *Jatropha curcas*, RAPD, genetic diversity, accessions

INTRODUCTION

Jatropha (*Jatropha curcas* L.) belonging to family Euphorbiaceae, which is named as Physic nut, Jangali arandi, Safed arandi, Jamalgota and Ratanjyot in different territories of India. It is also known as "Purging nut" in Mexico and Central America. The name *J. curcas* is derived from two Greek words, Jatros meaning Doctor and Trophe meaning Nutrition and used for medicinal purposes. Jatropha is native of South America and has a long history of its propagation by Portuguese into Africa and Asia. It grows well throughout India including Andhra Pradesh, Gujarat, Rajasthan, Karnataka, Chattisgarh, Uttarakhand, Tamil Nadu, Maharastra, Orissa and North Eastern States. These are some of the promising states where it occurs in the vicinity of villages and town as semi wild bush or shrub and also as hedge. It is a bushy plant with numerous side branches that arise from its main stem and it attains a height of 3-4 meters. Leaves are 10-15 cm long and 7-12 cm broad with pointed edges. The ripe fruits are about 2-5 cm large and yellow in color. The seeds resemble with castor seed in shape and are about 1.9 – 2.0 cm long and shape is either ovoid or oblong and is covered in dull brownish black capsules (Takeda, 1982 and Banerji *et al.*, 1985).

Jatropha seeds contain 46-58% of oil on kernel weight and 30-40% on seed weight. It shows promise for use as an oil crop for bio-diesel. The oil is renewable resource and a safe source of energy and a viable alternative to diesel, kerosene, LPG, furnace oil, coal and fuel wood (Takeda, 1982). Jatropha species are essentially cross pollinated, which offers the breeder ample scope to undertake screening and selection of seed sources for the desired traits (Ginwal *et al.*, 2005).

DNA marker based finger printing can distinguish species rapidly using small amounts of DNA and therefore can assist to obtain reliable information on their phylogenetic relationship. DNA markers are not typically influenced by environmental conditions and therefore can be used to help describe patterns of genetic variation among plant populations and to identify duplicated accessions within germplasm collections. RAPD is rapid method not requiring prior information regarding the genome of the study plant, and has been widely used to ascertain the genetic diversity in several plants (Belaj *et al.*, 2001 and Deshwal *et al.*, 2005). Use of such techniques for germplasm characterization may facilitate the conservation and utilization of plant genetic resources permitting the identification of unique genotypes or sources of genetically diverse genotypes (Kapteyn and Simon, 2002 and Welsh and McClelland, 1990). The determination of genetic relationships between accessions is important for Jatropha improvement because till date very less work has been done in this direction in India. Therefore, the present study was under taken to assess the genetic relationship between different Jatropha accessions collected from different parts of Uttar Pradesh using RAPD markers.

MATERIALS AND METHODS

The experimental materials comprised of twenty accessions of Jatropha collection (CSJ-6, CSJ-38 and CSJ-42 from Allahabad; CSJ-10, CSJ-12, CSJ-13 and CSJ-47 from Sarsol; CSJ-15; CSJ-16, CSJ-17 and CSJ-53 from Lalitpur; CSJ-22 from Kanpur, CSJ-23, CSJ-24 and CSJ-ARIS-39 from Nawabganj, Kanpur; CSJ-29 from Kannauj, CSJ-36 from Kosambhi, CSJ-46 from Farukhabad; CSJ-51 from Sitapur and CSJ-LKO-48 from Lucknow) brought from Department of Forestry of C.S. Azad University of Agriculture & Technology, Kanpur. The leaves of same age of Jatropha accessions grown at medicinal garden near Department of Biochemistry were collected and washed thrice with sterilized water. The moisture was soaked with sterilized blotter paper and 100 mg leaves weighed for DNA isolation. The leaves were ground in pre-cooled mortar and pestle with liquid nitrogen and powder was transferred into 1.5 ml of CTAB extraction buffer [10 ml of 1 mol/L Tris-HCl (pH 7.6), 14 ml of 5 mol/L NaCl and 4 ml of 0.5 mol/L EDTA (pH 8.0) mixed together and dissolved 1g DDT and 1g CTAB in the solution and final volume was made upto 100 ml by adding ultra pure water] pre warmed at 65°C and incubated at 65°C for 15 minutes. The sample was centrifuged at 13,000 rpm at 20°C for 2 minutes. Supernatant was transferred in a fresh tube and rest of the protocol was followed according to the manufacturer of Simax Genomic DNA Extraction (SBS Genetech Co. Ltd., China). The isolated DNA was gel electrophoresed in 1% agarose gel, stained with ethidium bromide and visualized by gel documentation system (Uvite, UK) for knowing the quality and quantity.

The random primer kit (SBSA) used for amplification were procured from SBS Genetic Co. Ltd., China, PCR mix was set up by using Easy-Do PCR Pre Mix Kit (SBS Genetech, China) supplied in lyophilized form. The mix contained Taq DNA polymerase – 2 units, 10 x PCR reaction buffer - 5 µl. In this pre-mix about 50 ng DNA template and 50p mol of RAPD primer was made up by adding nuclease free water. The DNA amplification was performed in an Eppendorf Master Cycler Gradient (Eppendorf Hamburg) by using the programme; 1 cycle at 94°C for 5 minutes, 45 cycles at 94°C for 1 minute, 36°C for 1 minute and 72°C for 1 minute followed by final extension of 10 minutes at 72°C. The amplified DNA was analyzed on 1% agarose gel in 1 x TAE (40mM Tris-acetate and 1 mMEDTA, pH 8.0) by electrophoresis at 60 V for 2 hrs. The bands were visualized by adding ethidium bromide stain (0.5µg/ml) in the gel. The gels were photographed using Gel Doc System (Uvitec, U.K.). Reproducibility of the bands was confirmed by repeating the experiment twice (extraction and amplification).

RAPD bands were scored as present (1) or absent (0) for all the Jatropha accessions. Each band was assumed to represent a unique genetic locus. Its presence was interpreted as either a heterozygote or dominant homozygote and absence of the band as a recessive homozygote. The pattern and extent of RAPD variation were analyzed with respect to primer, polymorphic locus and cultivar. Data entry was done in to binary data matrix and statistical analysis was carried out using NTSYS-pc, 2.01 versions (Sneath and Sokal, 1973). Pair wise comparison of samples was used to estimate Jaccord's similarity coefficient (GS): a/n-d where a = number of +ve coincidence, n = total sample size and d = no of –ve coincidence. Genetic distance (GD) between pair of lines was estimated as 1-GS. Jaccord's similarity coefficient was used to penetrate dendrogram using unweighted pair of lines and was estimated with arithmetic mean-UPGMA (Rolf, 1993).

J. curcas accessions were analyzed using 8 random primers of which 6 produced reproducible polymorphic banding patterns. A total of 59 bands were scored out of which 58.89% were polymorphic. The number of bands generated per primer varied from 3 to 12 bands and a minimum of 3 bands with SBSA-06 primer while the maximum of 12 bands were observed with primer SBSA-04. The size of the amplified products varied from 1500 to 200 bp. The primer SBSA-01 generated 100% polymorphic bands. All the 8 primers used in the present investigation were found to display reliable amplification profile, which have been summarized .Some of the primers did not produce any scoreable bands in some Jatropha accessions. Two primers SBSA-06 and SBSA-08 could not generate any polymorphic bands. The average numbers of bands per primer were recorded as 7.37 and the average of number of polymorphic bands 4.87. Though, it can be explained as the capability of RAPD primers to amplify the less covered and highly repeated regions of DNA. Primer SBSA-01, 03, 05, 06, 07, 08 amplified similar bands of 700, 400, 900, 550, 1200, 800 bp respectively, in all the accessions of Jatropha used in the study. Primer SBSA-02, SBSA-03 and SBSA-04 have produced genotype specific bands that would be highly useful in identification of genotypes and designing future breeding strategies. Primer SBSA-02 and SBSA-03 amplified unique bands of 900 and 500 bp in CSJ-38. It has been observed that out of 400 RAPD primers tested, 339 primers produced amplification products of which 267 revealed polymorphic fingerprint patterns in 42 accession of *J. curcas* collected from different growing regions of the country along with a non toxic genotype from Mexico (Basha and Sujatha, 2007). Out of a total of 3,057 bands, 1,287 (42%) were polymorphic with an average of 4.82 polymorphic bands per primer. The total number of bands per amplification varied from 1 to 20 in the molecular size range of 100-3,500 bp.

Jaccord's genetic similarity co-efficient varied from 1.00 to 0.425. The highest genetic similarity co-efficient (1.00) was observed between *J. curcas* accessions CSJ-51, CSJ-13, CSJ-16 and CSJ-29. CSJ-15 has similarity with CSJ-36; CSJ-13 is similar with CSJ-16 and CSJ-29 and CSJ-16 resemble with CSJ-29. While the lowest value of (0.425) was measured in CSJ-LKO-48 accession. The maximum possibility of similarity coefficient 0.425 was observed in CSJ-38 with all 19 accessions of *J. curcas*. Similar wide ranges in similarity value have also been observed in many accession level studies (Nair *et al.*, 1999). The UPGMA dendrogram obtained from the cluster analysis using Jaccord's similarity coefficient ranged from 1.00 to 0.425. At 80% similarity level, all the accessions combined to form three major clusters. Cluster I comprised of 6 accessions CSJ-6, CSJ-46, CSJ-15, CSJ-36, CSJ-22 and CSJ-53. Cluster II comprised 10 accessions and cluster III comprised 3 accessions CSJ-24, CSJ-LKO-48 and CSJ-ARIS-39. It was interesting to observe that accessions CSJ-38 was distinct from all the other accession of Jatropha.

RESULTS AND DISCUSSION

The result of RAPD analysis of *J. curcas* accessions revealed that the genetic diversity is very low in the present accessions which are being cultivated in Uttar Pradesh. This suggests that there is urgent need to widen the genetic base of the *J. curcas* so that higher yield and oil may be obtained. In present study genetically different accession CSJ-38 has been found superior in seed length and breath. CSJ- 38 accession could be now identified by primers SBSA-02 (600 bp), SBSA-03 (900 bp) and SBSA-04 (500 bp) and can be utilized for widening the genetic base as well as for improving the other characteristics.

REFERENCES

Banerji, R., Choudhury, A.R., Mishra, G., Sudarshan, G., Verma, S.C. and Srivastava, G.S. 1985. *J. curcas* seed oil for energy. *Biomass* (8): 277 – 282.

Basha, S.D. and Sujatha, M. 2007. Inter and intra population variability of *Jatropha curcas* (L.) characterized by RAPD and ISSR markers and development of specific SCAR markers. *Euphytica* (156): 374 – 386.

Belaj, A., Trujilo, I., Rosa, R., Rallo, L. and Gimenez, M.J. 2001. Polymorphism and discrimination capacity of randomly amplified polymorphic markers in an olive germsplasm bank. *J. Ame. Soc. Hort. Sci.* 126: 64 – 71.

Deshwall, R.P.S., Singh, R., Malik, K. and Randawa, G.J. 2005 Assessment of genetic diversity and genetic relationships among 29 populations of *Azadaricta indica* A. Juss using RAPD markers. *Genet Resour. Corp Evol.* 52: 285 – 292.

Ginwal, H.S., Phartyal, S.S., Rawat, P.S. and Srivastava, R.L. 2005. Seed source variation in morphology, germination and seedling growth of *Jatropha curcas* Linn. in central India. *Silvae Genet.* 54: 76 – 80.

Kapteyn, J. and Simon, J.E. 2002. The use of RAPD for assessment of identity, diversity, and quality of Echinacea. In: Trends in new crops and new uses. (eds.) Janick, J. Wipkey, A , ASHS Press, Alexandria, V.A., pp 509 – 513.

Martin, G. and Mayeux, A. 1985. Curcas oil (*Jatropha curcas* L.) a possible fuel. *Agric. Trop.* (9): 73-75.

Nair, N.V., Nair, S., Sreenivasan, T.V. and Mohan, M. 1999. Analysis of genetic diversity and phylogeny in Saccharum and related genera using RAPD markers. *Genet. Resour. Crop Evol.* 46: 73 – 79.

Rolf, F.J. 1993 Numerical taxonomy and multivariate analysis system, version 2.01. *Exter Software. Setauket, NY.*

Sneath, P.H. and Sokal, R.R. 1973. *Numerical taxonomy*. V.H. Freeman & Company. San Francisco. pp. 573.

Takeda, Y. 1982 Development study on *Jatropha curcas* (Sabudam) oil as a substitute for diesel engine oil in Thailand. *J. Agric Assoc.* 120: 1-8.

Welsh, J. and McClelland, M. 1990. Fingerprinting genomes using PCR with arbitrary primers. *Nucleic Acids Res.* 18: 7213 – 7218.

Chapter - 4

ISOLATION OF ACAULOSPORA SPECIES FROM *JATROPHA CURCAS* L AND *PONGAMIA PINNATA* PIERRE

Madhavi Kamalvanshi, Anil Kumar, Anuradha Jha and Ashok Shukla

National Research Centre for Agroforestry, Gwalior Road, Jhansi - 284003
E-mail: nrcaf.madhavi@gmail.com

ABSTRACT

Jatropha curcas L. and *Pongamia pinnata* Pierre are gaining importance as biodiesel plants. India has 146 m ha of wasteland of which 33 m ha can be reclaimed for plantation of the bio-diesel crops, in addition to arable land that is being used for the plantation. The Indian government's US $ 300 million national bio-fuels programme has the potential to create the world's first large-scale national biodiesel industry. Looking to low yield of above mentioned plant species and ability of arbuscular mycorrhizal (AM) fungi to increase host biomass and yields under such conditions, spores were isolated from the rhizosphere soil of above mentioned plant species by using the method of Geredmann and Nicolson, the species were identified by the keys of Schenck and Perez. Rhizosphere soil collected from above mentioned plants was used to set trap cultures, which were maintained in greenhouse for 4 to 5 months. Sporocarps and spores of the AM fungi were isolated from these pots and were used to establish purified cultures.

Five species of *Acaulospora* (*A. dilatata, A. denticulata, A. longula, A. mellea* and *A. scrobiculata*) were isolated from rhizosphere soil samples. Spores are globose and subglobose either formed singly in the soil or borne in sporocarps. They are borne laterally on subtending hypha of large, terminal, relatively thin walled, sporogenous saccule. Terminal sporogenous saccules were upto twice the size of spore. The spores were composed of essentially two distinct and seperable wall groups. In the first group atleast part of outer is continuous with wall of subtending stalk and the wall may be pigmented, laminated and variously ornamented; while in other group the inner surface is composed of one or more walls that are usually membranous, hyaline, ornamental and may be laminated. The wall stains pink, red, or purple in *Melzer's* reagent. These isolated and purified cultures will be used for further studies.

Keywords: *Jatropha curcas, Pongamia pinnata,* arbuscular mycorrhizal fungi

INTRODUCTION

India has 146 m ha of wasteland of which 33 m ha can be reclaimed by plantation of the bio-diesel plants. *Jatropha curcas* L a member of the family Euphorbiaceac and *Pongamia pinnata* Pierre a member of the family Leguminosae are gaining importance as biodiesel plants. AM fungi form symbiotic association with most economically important plants. These fungi improve plant growth under low fertility conditions, confer tolerance against certain plant pathogens, improve water balance of the plants, contribute to the formation of soil structure and help plants to establish in new areas (Panwar and Vyas, 2002).

The role of arbuscular mycorrhizal (AM) fungi in enhancing plant growth and yield, resistance to drought and salinity, and tolerance to pathogens is well documented (Gianinazzi-Pearson and Diem, 1982). But very few references are available for *J. curcas* and *P. pinnata*. However, wide diversity exists within the group of fungi responsible for the formation of AM by most plants in the majority of terrestrial ecosystems. The selection of the most specific appropriate plant fungus association for each specific environmental and ecological situation is one of the main challenges in current research on AM.

MATERIALS AND METHODS

Selection of site: The study was conducted at National Research Centre for Agroforestry (NRCAF), Jhansi (24°11' N latitude and 78°17' E longitude), Uttar Pradesh (India). The region lies in agro-ecoregion 4 of India, Northern Plain and Central Highlands, Hot Semi Arid Ecoregion with Alluvium–derived soil (Sehgal *et al.,* 1990). Mean annual rainfall is 960 mm with an average of 52 rainy days per year. Mean maximum temperature ranges from 47.4°C (June) to 23.5°C (January) and mean minimum temperature from 27.2°C (June) to 4.1°C (December). The maximum-recorded temperature on a particular day often touches 47-48°C during summer. Soil samples were collected during 2007-08 from rhizosphere of 120 plants of *J. curcas* and *P. pinnata* (20 plant per species per location) from six locations viz., Hissar (Haryana; Western plain, hot arid with desert and saline soils N8E1), Hyderabad (Andhra Pradesh; Deccan plateau and Eastern ghats, hot semi arid with red and black soils K6D2), Jodhpur (Rajasthan; Western plain, hot arid with desert and saline soils M9E1), Jhansi, Lalitpur (Uttar Pradesh; Northern plain and Central highlands, hot semi arid with alluvium-derived soils N8D2) and Pantnagar (Uttrakhand; Western Himalayas warm sub-humid inclusion humid eco-region with brown forest and podzoice soils AL1C4).

Selection of crops: These crops (*J. curcas* and *P. pinnata,*) are important available bio-fuels crops grown in different wasteland location.

Therefore they were assessed for AM species and spore population in rhizophere soil.

Sampling: Rhizosphere soil was collected from selected plants to set trap cultures, which were maintained in greenhouse for 4 to 5 months. Sporocarps and spores of the AM fungi were isolated from these pots by using the method of Geredmann and Nicolson (1963) and were identified by the keys of Schenck and Perez (1990) and Vestberg (1998). Taxonomic identification of spores was provided by the Internationl collection of VAM fungi[1*]. Clearing and staining of root specimens was done following the method of Phillip and Hayman (1970). Mycorrhizal assay of Mycorrhizal colonization percentage was determined by grid line intersection method of Giovannetti and Mossae (1980). Root segments were selected at random from stained samples and mounted on microscopic slides in groups of 10. Twenty root segments from each sample were used for assessing length of cortical colonization in millimeters, at 100X. Average of the readings from a sample, were expressed as percentage root length colonization.

Spores were counted according to the method of Gaur and Adholeya (1994). Species richness and frequency of occurrence (%) of AM fungi associated with the studied plant species were calculated by the method of Khade and Rodrigues (2008).

Table 1. Characteristics of arbuscular mycorrhizal (AM) species isolated from rhizosphere of *Jatropha curcas* and *Pongamia pinnata* from different locations

AM species	Characteristics of isolated AM species												
	A[x]	B	C	D	E	F	G	H	I	J	K	L	M
A. denticulata	Yellow brown	(67.0-) 78.0 (-108.0)	(6.0-) 8.4 (-9.6)	2	-	-	+	+	-	-	-	-	Pitted
A. dilatata	Light yellow	(84.0-) 78.4 (-100.0)	(6.0 –) 6.6 (-7.2)	3	-	-	+	+	-	-	-	-	Pitted
A. longula	Pale yellow	(72.0-) 77.0 (-84.0)	(2.5 -) 4.8 (- 7.2)	2	-	-	+	+	-	-	-	-	Smooth
A. mallea	Honey	(64.8-) 74.4 (-91.2)	(4.8-) 6.4 (-8.4)	3	-	-	+	+	-	-	-	-	Smooth
A. scrobiculata	Yellow	(108 .0-) 115.0 (-130.0)	(4.8 -) 6.0 (-7.2)	2	-	-	+	+	-	-	-	-	Pitted

[x]A- Spore color, B- Spore size, C- Composite spore wall width (μ), D- Number of wall group, E- Hyphal attachment size, F- Pore size, G-Presence of hyphal terminus, H- Wall reaction to Melzer's regeant, I- Spore formation with in root, J- Presence of spores with extrametrical mycelium , K- Sporocarp, L- Sporocarp color, M- Surface ornamentation.

RESULT AND DISCUSSION

Occurrence of AM species

The results showed that different AM fungal species were present at different locations in arid and semi arid regions of India. Five AM fungal species of *Acaulospora* were isolated in trap cultures of *J. curcas and P. pinnata*. The characteristics of isolated AM species were given in Table 1.

In *J. curcas*, the highest frequency of occurrence was recorded for *A. scrobiculata* Trappe (67%) and is most dominant species followed by *A. denticulata*, recorded 16% frequency of occurrence. Maximum species richness (Table 2) was recorded at Hissar. In *P. pinnata* the highest frequency of occurrence was recorded for *A. scrobiculata* (50%), followed by *A. denticulata*, *A. dilatata*, *A. longula*, 16% frequency of occurrence. Maximum species richness (Table 2) was recorded at Hissar. *A. scrobiculata* is present in both plant species.

Table 2. AM species richness and frequency of occurrence in rhizosphere *J. curcas* and *P. pinnata*

AM species	*J. curcas*						*P. pinnata*						Frequency of occurrence (%)	
	[a]L_1	L_2	L_3	L_4	L_5	L_6	L_1	L_2	L_3	L_4	L_5	L_6	*J. curcas*	*P. pinnata*
A. denticulata Sieverding & Toro	+[b]	+	-	-	-	-	-	-	-	-	+	-	33	16
A. dilatata Mortron	-	-	-	-	-	-	-	-	+	-	-	-	0	16
A. longula Spain & Schenck	-	-	-	-	-	-	-	-	+	-	-	-	0	16
A. mellea Spain & Schenck	-	-	-	-	+	-	-	-	-	-	-	-	16	0
A. scrobiculata Trappe	+	+	+	-	+	-	-	-	+	-	+	+	67	50
Species richness	2	2	1	0	2	0	0	0	3	0	2	1		

[a]L_1- Hissar, L_2- Hyderabad, L_3- Jodhpur, L_4- Jhansi, L_5- Lalitpur and L_6- Pantnagar; [b]+ Present - Absent

CONCLUSIONS

The results revealed that different types of AM fungi are present in different location of India. Five species of *Acaulospora* (*A scrobiculata, A. diliatata, A. denticulata, A. mellea and A. longula*) were recorded in rhizosphere of selected TBO species. These will be isolated, purified and used for further studies. *A scrobiculata* was the most dominant genus, followed by *A. denticulata, A. dilatata, A. longula and A. mellea.* Among all the identified AM species, *A. scrobiculata,* was most abundant species at all studied locations.

ACKNOWLEDGEMENTS

The authors are thankful to Director, National Research Centre for Agroforestry, Jhansi for providing the facilities, and University Grant Commission, New Delhi, for financial support in the form of Post Doctoral Fellowship.

1*. Details available at <http:/ invam.caf.wvu.edu>

REFERENCES

Gaur and Adholeya, A. 1994. Estimation of VAMF spores in soil: a modified method. *Mycorrhiza News* 6 : 10-11.

Gerdemann, J.W. and Nicolson, T.H. 1963. Spores of mycorrhizal *Endogone* species extracted from soil by wet sieving and decanting. *Tran. Br. Mycol. Soc.* 46:235-244.

Gianinazzi-Pearson V. and Diem, H.G. 1982. Endomycorrhizae in the tropics. In: Dommergues Y.R. and Diem HG (eds) Microbiology of Tropical Soils and Plant Productivity. Junk, *The Hague*. 209-251.

Giovannetti, M. and Mosse, B. 1980. An evaluation of techniques for measuring vesicular arbuscular mycorrhizal infection in roots. *New Phytol.* 84:489-500.

Khade, W.S. and Rodrigues, B.F. 2008. Spatial variations of AM fungi in *Terminalia crenulata* Roth. from the Western Ghat region of Goa, India. *Mycorrhiza News* 20: 20-21.

Panwar and Vyas, A. 2002. Arbuscular mycorrhizal association in *Tarmarix aphyall* in Indian Thar Desert. *Mycorrhiza News* 14: 14-16.

Phillip, J.M. and Hayman, D.S. 1970. Improved procedure for clearing roots and staining parasitic and vesicular arbuscular mycorrhizal fungi for rapid assessment of infection. *Trans Br Mycol. Soc.* 55: 158-161.

Schenck, N.C. and Perez, Y. 1990. Manual for the identification of VA Mycorrhizal fungi. University of Florida, Ganinesville, Florida, USA.

Sehgal, J.L., Mandal, D.K. and Vadivelu, S. 1990. Agro-ecological regions of India. National Burea of Soil Survey & Land Use Planing, *Tech. Bull.*, Nagpur. 25.

Chapter - 5

CLONAL PROPAGATION OF *PONGAMIA PINNATA* (L.) PIERRE THROUGH STEM CUTTINGS

R. Nainamalai, Bhimi Ram and T.S. Rathore

Forest Research Centre, P.O. Hakimpet, Dullapally, Hyderabad – 500 014
E mail: rnainamalai@icfre.org

ABSTRACT

Pongamia pinnata is gaining lot of importance as a potential oil yielding tree. Since the oil extracted from its seed is used as biodiesel, raising of extensive plantations of this species in wastelands is essential. Identifying high yielding trees with good qualities, for large scale production are the base for any tree improvement programme. Vegetative propagation can ensure quick genetic gains by mass multiplication of superior genotypes and help in establishing clonal plantations in afforestation programmes. Amongst various methods of vegetative propagation, rooting of stem cuttings is the most effective method. Application of auxins increases the endogenous level and play significant role in the cell elongation, division and development of root primordia. Requirement of type of auxin and its concentration vary with the species. In the present study, an attempt was made to select the right growth regulators and its optimum concentrations. The cutting of pencil thickness with10-12 cm length was used. The base of the cutting is sterilized with 0.1 per cent bavistin solution to avoid fungal attack. Among the different auxins (IAA, IBA, NAA, NOA and 2,4-D) the IBA was found to be better even at lower dose (500 ppm). In this study wide range of concentrations (500-3000 ppm) of IBA were tested and it was revealed that 2500 ppm of IBA was found to be the most effective in terms of root length (16. 5 cm), root fresh weight (0.18 g), root dry weight (0.8g), number of roots formed (4.6) number of sprouts (2.67) and surviv-ability (66.7%). The cuttings treated with above and below this concen-tration, showed poor performance with respect to above parameters. The cutting of pencil thickness with 10-12cm length treated with IBA at 2500ppm shows better performance and can be used for mass multipli-cation of genetically superior clones in tree improvement programme.

Keywords: *Pongamia pinnata*, clonal propagation and biodiesel

INTRODUCTION

Pongamia pinnata, an arboreal legume that is a member of the subfamily *Papilionoideae, is* being considered native of Western Ghats. It is distributed throughout India and found primarily along the banks of rivers and streams. At a time when society is becoming increasingly aware of the declining reserves of fossil fuels for the production of petroleum oil, it has become apparent that bio-fuels are destined to make a substantial contribution to the future energy demands of the domestic and commercial economies. To this, *P. pinnata* will impact most significantly through the extraction of seed oil that can be used in the manufacture of biodiesel. The potential of *P. pinnata* oil as a source of fuel for the bio-diesel industry is well recognized (De and Bhattacharya, 1999; Azam *et al.,* 2005; Karmee and Chadha, 2005). Moreover, the use of vegetable oils from plants such as *P. pinnata* has the potential to provide an environmentally acceptable fuel, the production of which is greenhouse gas neutral, with reductions in current diesel engine emissions (Raheman and Phadatare, 2004). Importantly, the successful adoption of bio fuels is reliant on the supply of feedstock from non-food crops with the capacity to grow on marginal land and not destined to be used for the cultivation of food crops (Hill *et al.*, 2006). *P. pinnata* is a strong candidate to contribute significant amounts of fuel feedstock, meeting both of these criteria. The seeds of *P. pinnata* contain 30 to 40% oil (Natanam *et al.*, 1989; Nagaraj and Mukta, 2004), which can be converted to biodiesel (fatty acid methyl esters; FAMEs) by esterification with methanol in the presence of KOH.

According to Chaturvedi (2000) the seed production from seedling raised plants is variable. This variability in the seed production as well as quality/quantity of oil can be reduced or minimized to a great extent via clonal approach. From forest economic point of view, this is not a valuable species as the wood is of little commercial value. This may be the reason why systematic work on clonal planting of this species has not been done. Attention of foresters and environmentalists turned towards pongamia oil being used instead of diesel to run four wheelers. Hence, looking to the great demand for pongamia oil at present and in future the farmers would like to venture out for large scale pongamia plantation of high density with intensive management, than the traditional method of planting. As pongamia oil is in great demand and as most of the seed is used commercially, ways to propagate the plant by alternate methods are necessary. Therefore, an attempt was made to propagate this hitherto neglected plant through vegetative propagation.

MATERIAL AND METHODS

Pongamia is a hard wood species which can be propagated through hardwood cuttings. The cuttings of pencil thickness size with 10-12 cm length have been used. The cuttings were treated with the different growth regulators viz., IAA, IBA, NAA, NOA and 2, 4-D to see the effect on rooting and compared with control. The observations were recorded in terms of number of roots, number of sprouts, root length, root fresh weight and root dry weight. The data was analysed and subjected to statistical analysis to see the differences in the treatments. Among different plant growth regulators, IBA showed better performance and its required concentration was also studied. The cuttings were treated with IBA of different concentrations of 500 ppm, 1000 ppm, 1500 ppm, 2000 ppm, 2500 ppm and 3000 ppm.

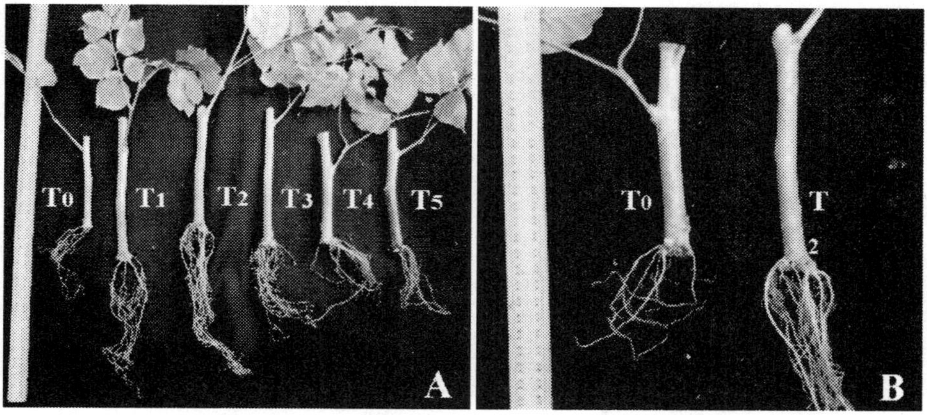

Fig. 1. Effect of growth regulators on rooting of cuttings. T_0 represents control and T_1 to T_5 represent different growth regulators (T_1-IAA; T_2-IBA; T_3-NAA; T_4-NOA; T_5-2-4-D). (A) Photograph comparing all the treatments with control; (B) Photograph comparing the best treatment (T_2) with control.

RESULTS AND DISCUSSION

Effect of auxins on root induction

Application of auxins increases the endogenous level and play significant role in the cell elongation, division and development of root primordia. Requirement of type of auxin and its concentration vary with the species. In the present study, among different auxins (IAA, IBA, NAA, NOA and 2,4-D) fried IBA was found to be better even at lower dose (500 ppm) (Fig. 1). It is evident that the cuttings treated with IBA showed better performance in terms of root length (22.33), root fresh weight (0.75), root dry weight (0.36), number of sprouts (2.00) number of roots (15.6), number of cuttings survived (93.33%). Out of the various growth regulators the cuttings treated with IBA showed maximum root length (22.33cm) followed by NAA (18.17cm). The cuttings treated with IAA, NOA, 2-4-D and control showed poor performance. In earlier studies on *Pongamia pinnata,* Karoshi and Hegde (2002) also found that IBA was the best auxin for rooting from stem cuttings of *Pongamia pinnata.* Most of the researchers have reported that IBA was the best auxin for induction of roots in the cuttings (Adil Al-Kinary, 1981; Khali *et al.*, 1996; Rana *et al.,* 1987; Bharadwaj and Mishra, 2002). Contrary to above observation, Kumar (2000) found that IAA was better than IBA in inducing roots from branch cuttings of *Acacia mangium,* whereas, Swamy *et al.* (2002), found that NAA was better in inducing roots in *Robinia pseudocacia* and IBA in *Grewia optiva*. Puri and Verma (1996) found that both NAA and IBA were effective in inducing roots in *Dalbergia sissoo.* The cuttings treated with IBA showed maximum root fresh weight, root dry weight, number of roots, number of sprouts as compared to the cuttings treated with other growth regulators. It was also noticed that the survivability of the cuttings was more in the IBA treated cuttings as compared to the control.

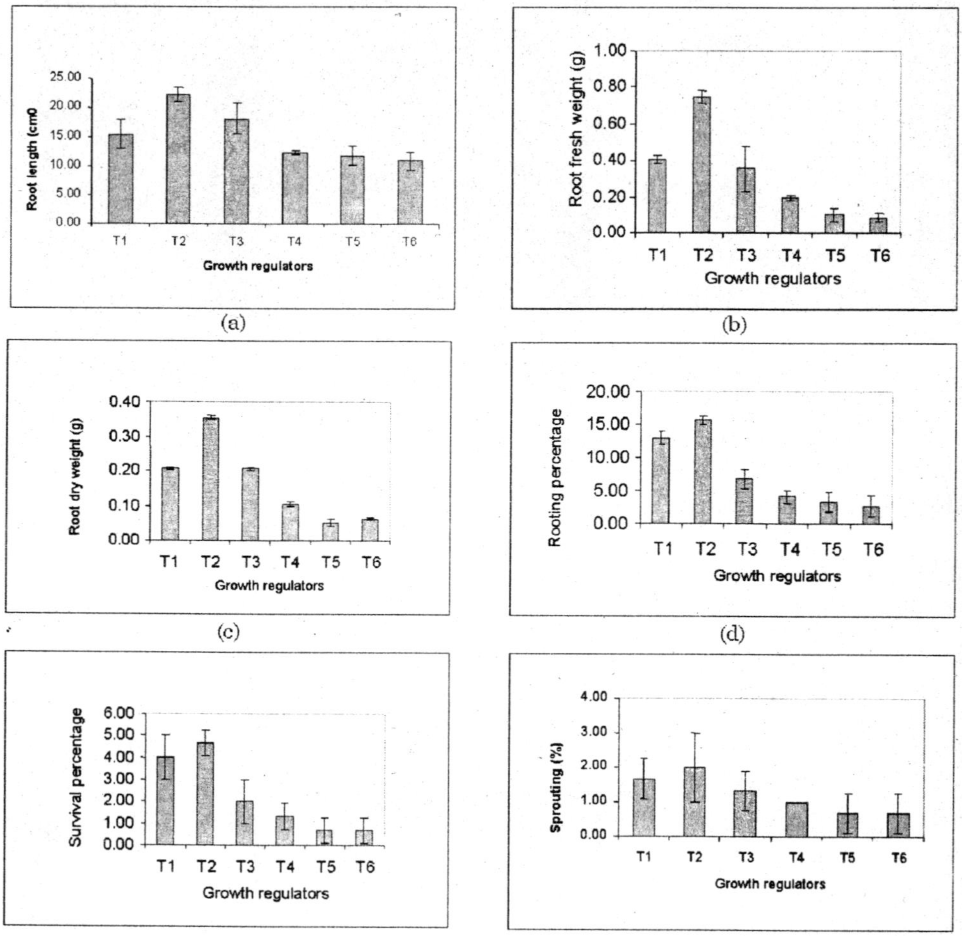

Fig. 2. Effect of growth regulators (a) rooting of cuttings (b) root fresh weight (c) root dry weight (d) rooting percentage (e) survival percentage (f) sprouting percentage

From the above cited observations by various researchers, it is evident that IBA is the most effective auxin in inducing roots in cuttings of most of tree species and woody plants, but in some species IBA may be less effective as compared to other auxins. This may be attributed to the endogenous auxin level and the physiological status of the cuttings. IAA is a natural auxin and photosensitive in nature, but considered weak as compared to synthetic auxins like IBA, NAA and NOA (Fig. 2 a, b, c, d, e and f).

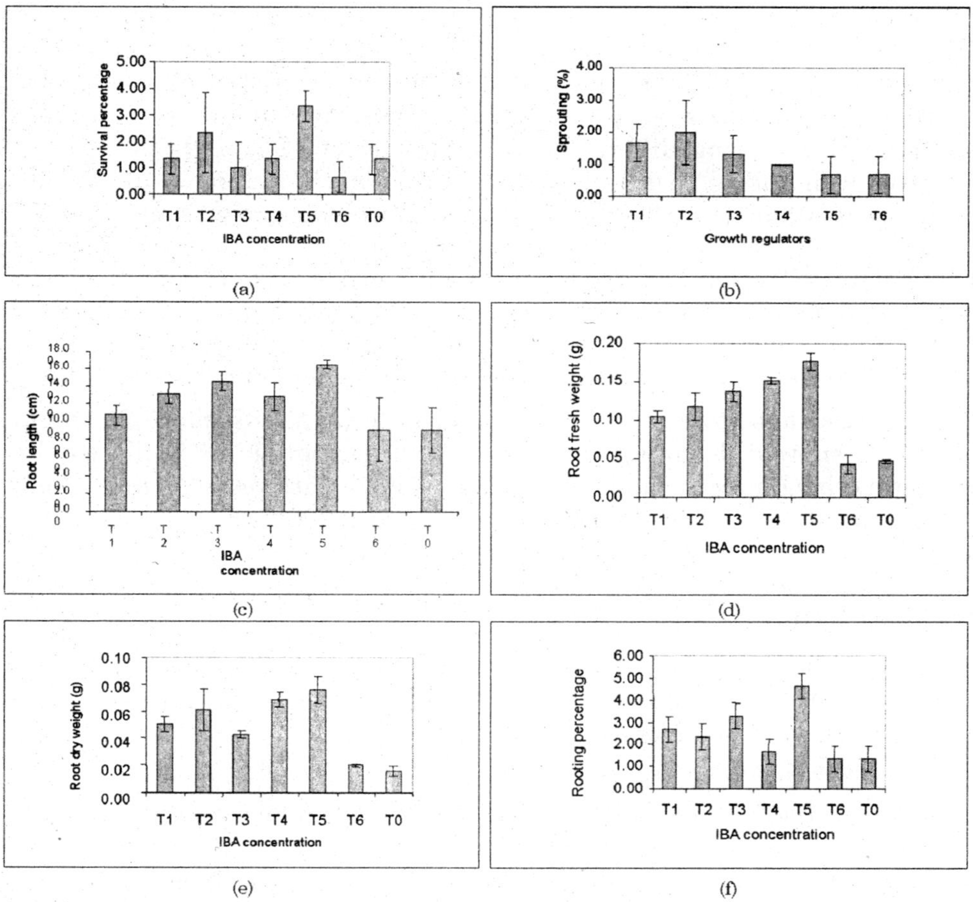

Fig. 3. Effect of IBA concentration (a) root length (b) root fresh weight (c) root dry weight (d) rooting percentage (e) survival percentage (f) sprouting percentage

Effect of different concentrations of IBA on root induction

In the present study wide range of concentrations (500-3000 ppm) of IBA were tested and it was revealed that 2500 ppm of IBA was found to be the most effective in terms of root length (16.5cm), root fresh weight (0.18 g), root dry weight (0.8 g), number of roots formed (4.6) number of sprouts (2.67) and survivability (66.7%) The cuttings treated with above and below this concentration, showed poor performance with respect to above parameters. The cuttings treated with none of the concentration were found to be very poorer than the other treatments. In the earlier studies also higher rate of rooting percentage was observed with 2500 ppm IBA in *Pongamia pinnata*. Karoshi and Hegde (2000) revealed that out of two concentrations (100 ppm and 200 ppm for 24 hr) of IBA, 100 ppm solution was the better for rooting from nodal hard wood branch cuttings of 20 cm in length and 1.5-2.0 cm in diameter in teak. They also observed that combined use of IAA + IBA was not effective as compared to IBA alone. Nanda *et al.* (1968) used 100 ppm NAA for rooting in branch cuttings from *T. grandis* and observed low rate of rooting. Optimum concentration of auxin varies with the species. Tchoundjeu *et al.* (2002) observed that among different concentrations of IBA, 100-200 μg was effective in inducing rooting in *Prunus africana*. Wendling *et al.* (2000), found that 1000 and 3000 ppm of IBA concentration was effective in inducing roots from different clonal cuttings of *Eucalyptus*. Banerjee and Prabhakar (1994) found differential requirement of IBA concentrations in inducing roots in cuttings of *Toona ciliata* (1500 mg/l) and *Acer oblongum* (2500 mg/l). Bharadwaj and Mishra (2002) found very high concentration (0.8%) of IBA effective in inducing root in *Acer oblongum* cuttings.

In accordance to our findings, higher concentrations above 2500ppm of auxins showed declining trend of rooting percentage or inhibited the root growth (Banerjee and Prabhakar, 1994; Puri and Verma, 1996; Panchasara *et al.*, 1998; Bharadwaj and Mishra, 2002) (Fig. 3 a, b, c, d, e and f).

REFERENCES

Azam, M.M., Waris, A. and Nahar, N.M. 2005. Prospects and potential of fatty acid methyl esters of some non-traditional seed oils for use as biodiesel in India. *Biomass and Bioenergy* 29: 293-302.

Chaturvedi, A.N. 2000. *Pers. Comm.*.

De, B.K. and Bhattacharyya, D.K. 1999. Biodiesel from minor vegetable oils like karanja oil and nahor oil. *Lipid-Fett* 101: 404-406.

Karmee, S.J. and Chadha, A. 2005. Preparation of biodiesel from crude oil of *Pongamia pinnata*. *Bioresource Technology* 96: 1425-1429.

Karoshi, V.R. and Hegde, G.V. 2002. Vegetative propagation of *Pongamia pinnata* (L.) Pierre: hitherto a neglected species. *Indian Forester* 128: 348-350.

Raheman, H. and Phadatare, A.G. 2004. Diesel engine emissions and performance from blends of karanja methyl ester and diesel. *Biomass and Bioenergy* 27: 393-397.

Hill, J., Nelson, E., Tilman, D., Polasky, S.M. and Tiffany D. 2006. Environmental, economic, and energetic costs and benefits of biodiesel and ethanol biofuels. *Proceedings National Academy of Sciences,* USA 103: 11206-11210.

Natanam, R., Kadirvel, R. and Balagopal, R. 1989. The effect of kernels of karanja (*Pongamia glabra* Vent) on growth and feed efficiency in broiler chicks to 4 weeks of age. *Animal Feed Science and Technology* 25: 201-206.

Nagaraj, G. and Mukta, N. 2004. Seed composition and fatty acid profile of some tree borne oilseeds. *Journal Oilseeds Research* 21: 117-120.

Chapter - 6

IN VITRO REGENERATION OF *JATROPHA CURCAS*

Pankaj Bhargava, S.K. Tiwari, Amit Pandey, M.P. Goswami and Shweta Mishra

Forest Genetics, Plant Propagation and Biotechnology Division,
State Forest Research Institute, Polipathar, Jabalpur
E-mail- drsktiwari@rediffmail.com

ABSTRACT

Biofuels are renewable and environmentally safe and refers to a non-petroleum based fuel obtained from the transesterification of vegetable oils or animal fats. Bioethanol and biodiesel are two important forms of biofuels. Non-edible oil obtained from the seeds of *Jatropha curcas* is the potential source of biodiesel. Therefore it is necessary to find out the varieties of Jatropha which contain maximum oil percentage. Tissue culture offers a reliable method to produce quality plants in large scale. Present paper highlights *in vitro* regeneration protocol for the species.

Keywords: Biofuels, in vitro regeneration, *Jatropha curcas*

INTRODUCTION

Due to rapidly diminishing reserves of the fossil fuels and increasing effects of pollution from these fuels, there is an urgent need of eco-friendly alternatives to fossil fuels. Many energy fuels are being investigated. The best alternative fuel of significance for the future is biofuel. Biofuels refers to a non-petroleum based fuel obtained from the transesterification of vegetable oils or animal fats. This special type of fuel is mostly produced from bio waste. The worldwide awareness regarding global warming has popularized the use of biofuel. They are highly effective in minimizing the extent of greenhouse gas emission. Owing to the limited availability of fossil fuels, the use of biofuels becomes all the more important. They are renewable and environmentally safe.

Non-edible oil seeds of forest origin are a promising option for producing biofuel. *Jatropha curcas,* the biodiesel plant is a potential source of such non-edible oil seeds. Among the many species, which yield oil as a source of energy in the form of biodiesel, *J. curcas* L. has been identified as most suitable oil seed bearing plant due to its various favourable attributes like hardy nature, short gestation period, adaptability in a wide range of agro-climatic conditions, high oil recovery and quality of oil etc. *J. curcas* (L.) (Ratanjot) or physic nut belongs to family Euphorbiaceae. It is a perennial, multipurpose and drought resistant large shrub or small tree. It is a native of tropical America and now acclimatized throughout Africa and Asia. It grows in a number of climatic zones in tropical regions of the world and can be grown in areas of low rainfall and in marginal/poor soil (Heller, 1996, Kureel, 2006, Kumar and Sharma, 2008). In India major states cultivating this crop are Andhra Pradesh, Gujarat, Madhya Pradesh, Maharashtra, Rajasthan, and Tamil Nadu.

The oil content is 35-40% in the seeds and 50-60% in the kernel. The oil can be used as a substitute for diesel and can be mixed with diesel at any ratio. Jatropha is usually propagated on mass scale both by seed as well as stem cuttings and for commercial cultivation normally it is propagated by seeds. It is necessary to find out elite plants of Jatropha with high seed yield and oil percentage and develop large number of such plants. These plants should be utilized for establishing plantations for biodiesel production. To meet the large scale demand and ensure easy supply of the elite material, there is a need to establish mass multiplication technique. Tissue culture offers a rapid, continuous and reliable method to produce quality plants on large scale.

Plant regeneration protocols using different explants have been described in *J. curcas* (Sujatha *et al.,* 1996 and 2006, Qin *et al.,* 2004, Rajore and Batra 2005). Nodal segments are important source tissue for micropropagation and plants raised from these are comparatively more resistant to genetic variation. Keeping the above facts in mind, experiments on refinement of *in vitro* regeneration protocol of *Jatropha curcas* were conducted and findings are presented in this paper.

MATERIALS AND METHOD

Apical and nodal explants were used from known germplasm having high oil percentage (40.8%). Apical buds of 1-3 cm length and nodal explants of 0.5 to 1.5 cm diameter were taken from nursery grown plants. The explants were washed under running tap water for 1 hour followed by treatment with 2-4% (v/v) Extran for 8-10 minutes with repeated washing with double distilled water. The explants were also treated with 0.2% (w/v) Bavistin solution for 8-10 minutes to minimize fungal load followed by repeated washing with sterilized double distilled water. The explants were then surface sterilized with 0.1% (w/v) mercuric chloride solution for 4-5 minutes followed by repeated washing with sterilized double distilled water. The explants were then trimmed and inoculated in MS medium supplemented with 3% sucrose, 0.8% agar-agar and different combinations and concentrations of plant growth regulators. pH of the medium was adjusted to 5.7 prior to sterilization under 121°C and 15 psi. for 30 minutes.

The cultures were initiated in 25 x 150 mm Borosil make rimless test tubes and kept in culture room at 25±2°C, with 85% RH, 16-hour-photoperiod and light intensity of 1000 lux. Observations were recorded at interval of 7 days. After optimum growth the regenerated shoots were sub cultured in 400 ml culture glass bottles on fresh MS medium with the same combination at every 25-30 days. The *in vitro* regenerated plantlets were hardened in mist chambers.

Table 1. Morphogenetic response of apical explants of *Jatropha curcas* on MS culture medium

MS Media composition	Percent response	No of days for morpho-genetic response	Average number of shoot induction per explant	Average length of shoots in cm	Morpho-genetic response of lateral shoot induction
MS+ No PGR (control)MS 1	0	No response	No response	No response	No response
MS+BAP 0.5 mg/lt and IAA 0.5mg/lt MS 2	36	21	2.3	1.1	poor
MS+BAP 1.0 mg/lt and IAA 0.5 mg/lt MS 3	46	21	2.8	1.3	poor
MS+BAP 1.5 mg/lt and IAA 0.5 mg/lt MS 4	58	23	3.6	2.4	moderate
MS+BAP 2.0 mg/lt and IAA 0.5 mg/lt MS 5	72	25	4.9	3.9	profuse
MS+BAP 2.5 mg/lt and IAA 0.5 mg/lt MS 6	54	25	4.0	2.6	moderate
MS+BAP 3.0 mg/lt and IAA 0.5mg/lt MS 7	57	22	3.3	1.5	poor

RESULTS AND DISCUSSION

Hormone free medium showed no morphogenetic response and the explants were turned yellow and finally died. Apical explants showed poor to moderate morphogenetic response (2 to 3 new lateral shoots) when medium was supplemented with BAP 0.5mg/lt to 3.0mg/lt along with IAA 0.5mg/lt. Very good morphogenetic response in terms of shoot induction was observed when medium was supplemented with BAP 2.0 mg/lt and IAA 0.5 mg/lt. On an average 5 to 6 new lateral shoots per explant were induced, within 21 to 25 days of culture period (Table 1 and

Fig. 1). For further multiplication, the *in vitro* regenerated shoots were sub cultured on the same culture media. After first sub culturing, the regenerated shoots showed profuse morphogenetic response in terms of shoot multiplication as well as shoot elongation. On an average 5 - 6 multiple shoots were induced within five to six weeks of culture period having the length of 3 – 4 cm (Table 3). In case of nodal explants good morphogenetic response was observed when medium was supplemented with BAP 5.0 mg/lt and IAA 3.0 mg/lt (Table 2 and Fig. 2). Callus formation was also observed when the medium was supplemented with BAP 4.0 mg/l and 2-4D 1.0 mg/l within 3-4 weeks. 3-4 green shoots are formed from callus initial within 4 – 6 weeks. Higher range of BAP and IAA showed poor morphogenetic response. The multiple shoots were separated under aseptic condition and were cultured for rooting. Profuse rooting was induced within 25 days when MS medium was supplemented with IBA 3.5mg/lt. The *in vitro* regenerated shoots were taken out from the culture vessels and were hardened in poly huts under high humid condition (90%) and temperature of 35±5°C on pure sand medium. The plant establishment rate was recorded more than 35% (Table 3).

Table 2. Morphogenetic response of nodal explants of *Jatropha curcas* on MS culture medium

MS Media composition	Percent response	No of days for morpho-genetic response	Average number of shoot ind-uction per explant	Average length of shoots in cm	Morpho-genetic response of lateral shoot induction
MS+ No PGR (control) MS 1	0	No response	No response	No response	No response
MS+BAP 5.0 mg/lt and IAA 0.5mg/lt MS 2	33	21	3.3	1.6	poor
MS+BAP 5.0 mg/lt and IAA 1.0 mg/lt MS 3	42.5	22	3.6	2.3	poor
MS+BAP 5.0 mg/lt and IAA 2.0 mg/lt MS 4	49	25	4.6	2.9	moderate
MS+BAP 5.0 mg/lt and IAA 3.0 mg/lt MS 5	65	25	5.3	4.1	profuse
MS+BAP 5.0 mg/lt and IAA 4.0 mg/lt MS 6	52.5	23	4.0	3.2	moderate

MS – Murashige and Skoog medium; PGR- Plant Growth Regulator;
BAP- 6-benzylaminopurine; IBA – indole 3- butyric acid;
IAA- indole-3-acetic acid

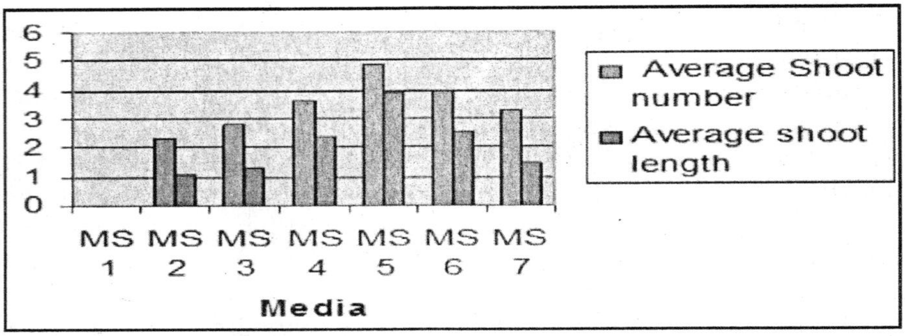

Fig. 1. Response of apical explants

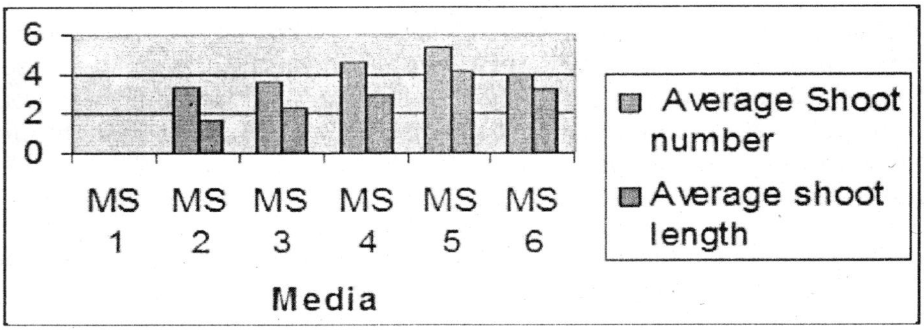

Fig. 2. Response of nodal explants

Table 3. *In vitro* rooting and hardening of plantlets

MS Media composition (treatments)	Number of culture bottles inoculated	Number of culture bottles responded	No of days for profuse rooting	Average length of roots in cm	Plant establishment rate
MS+IBA 3.5 mg/lt	100	64	35- 40	5 to 6	>35%

CONCLUSION

Regeneration is very important when mass production is taken into consideration. The present study described refinement of *in vitro* regeneration protocol of *Jatropha curcas* from apical and nodal segments with higher rate of multiplication. This protocol can be used to produce multiple copies of selected elite plants in large quantities, required for establishing commercial plantations for biodiesel production.

ACKNOWLEDGEMENT

Authors are thankful to NOVOD Board for financial assistance and Director, S.F.R.I., Jabalpur for providing laboratory facility and critical suggestions.

REFERENCES

Kumar, A. and Sharma, S. 2008. An evaluation of oil seed crop for industrial uses (*Jatropha curcas* L.): A review. *Industrial Crops and Products* pp 1-10.

Kureel, R.S. 2006. Prospect and potential of *Jatropha curcas* for biodiesel production. In *Biodiesel towards Energy Independence* (eds. Singh, B., Swaminathan, R. and Ponraj, V.), Rashtrapati Bhawan, New Delhi. p 374.

Murashige, T. and Skoog F. 1962. A revised medium for rapid growth and bioassay with tobacco tissue cultures. *Physiol. Plant.* 15:473-497.

Openshaw, K. 2000. A review of *Jatropha curcas*: An oil plant of unfulfilled promise. *Biomass Bioenergy* 19:1-15.

Sujatha, M., Makkar, H.P.S. and Becker K., 2006. Shoot bud proliferation from axillary nodes and leaf sections of non-toxic *Jatropha curcas* L. *Plant Growth Reg.* 47:83-90.

Sujatha, M. and Mukta, N. 1996. Morphogenesis and plant regeneration from tissue cultures of *Jatropha curcas*. *Plant Cell Tissue Organ Culture* 44:135-141.

Rajore, S. and Batra, A. 2005. Efficient plant regeneration via shoot tip explants in *Jatropha curcas*. *J. Plant Biochem. Biotechnol.* 14:135-141.

Qin, W., Wei-Da, L., Yi, L., Shu-Lin, P., Ying, X.U., Lin, T. and Fang, C. 2004. Plant regeneration from epicotyls explants of *Jatropha cucas*. *J. Plant Physiol. Mol. Biol.* 30:475-478.

Chapter - 7

SCOPE AND UTILIZATION OF BY-PRODUCTS OF BIODIESEL

Anubha Srivastav and A. K. Pandey

Centre for Social Forestry & Eco-Rehabilitation, Allahabad

ABSTRACT

Jatropha curcas seed oil holds immense opportunities for farmers and rural entrepreneurs for eco- friendly biodiesel. Biodiesel is a clean burning, bio-degradable, non toxic alternative fuel produced from renewable resources (animal fats and plant oils), that can be blended with petroleum diesel to create a biodiesel blend. The major hurdle for commercial biodiesel production is its higher cost of production compared to mineral based petroleum fuels. The actual cost of biodiesel production depends on the cost of procurement of seeds, scale of manufacturing, government policies related to taxation, utilization of by-products, utilization of seed cake etc. In India, it is estimated that cost of biodiesel produced by trans-esterification of oil obtained from *J. curcas* oil-seeds should be comparable with that of petroleum diesel. The byproducts of biodiesel from Jatropha seed oil are the seed oil cake and glycerol, which have good commercial value. Seed cake can be used as manure, as feedstock for biogas production, animal feed after detoxification, whereas glycerin can be utilized in pharmaceutical as well as cosmetic industries. This can also be used in defence, soap manufacture, and chemical industry.

The cost of biodiesel production can be reduced by proper utilization of by-products apart from improving transesterification process. The seed oil cake is very good compost being rich in plant nutrients for soils which are getting increasingly deficient in carbon and nutrients. It can also yield biogas, which can be used for cooking and the residue can be used as compost. Hence, oil cake will fetch good prices. Glycerol is also produced in the transesterification process of oil. Glycerol from biodiesel contains some peculiar impurities and may not be suitable to process according to the usual technologies to produce pharmaceutical product. The glycerol can be used for making soaps in the rural cottage industries. There is need, not only to develop purification technology for glycerol but also for its utilization as a raw material for the production of other chemicals. There is need to find the use of meal cake, which will be available in large quantities to reduce the cost of biodiesel. The cost recovery will be through sale of oil-cake and glycerol. In India research on biodiesel is in infant stage, there is an urgent need to make much sincere efforts for technological development for its production, utilization of by-products and evaluation in engine with respect to shortcomings, emissions, additive response etc.

Keywords: Bio-diesel, byproducts. oil cake, glycerol, utilization, cost recovery

INTRODUCTION

Biodiesel is methyl or ethyl ester of fatty acid made from virgin or used vegetable oils (both edible and non-edible) and animal fats. The main commodity sources for biodiesel in India can be non-edible oils obtained from plant species such as *Jatropha curcas* (Ratanjyot), *Pongamia pinnata* (Karanj), *Calophyllum inophyllum* (Nagchampa), *Hevea brasiliensis* (Rubber) etc. Biodiesel contains no petroleum, but it can be blended at any level with petroleum diesel to create a biodiesel blend or can be used in its pure form. Just like petroleum diesel, biodiesel operates in compression ignition (diesel) engine, which essentially require very little or no engine modifications because biodiesel has properties similar to petroleum diesel fuels. It can be stored just like the petroleum diesel fuel and hence does not require separate infrastructure. However, *Jatropha curcas* has been found most suitable for the purpose. It will use lands which are largely unproductive for the time being and are located in poverty stricken areas and in degraded forests. It will also be planted on farmers' field boundaries, fallow lands and in public lands such as along the railway tracks, roads and irrigation canals.

Process of transesterification

Transesterification also called alcoholysis, is the displacement of alcohol from an ester by another alcohol in a process similar to hydrolysis. This process has been widely used to reduce the viscosity of triglycerides. Triglycerides are readily trans-esterified in the presence of alkaline catalyst at atmospheric pressure and at a temperature of approximately 60 to 70°C with an excess of methanol. The mixture at the end of reaction is allowed to settle. The lower glycerol layer is drawn off while the upper methyl ester layer is washed to remove entrained glycerol and is then processed further. The excess methanol is recovered by distillation and sent to a rectifying column for purification and recycled. The trans-esterification works well when the starting oil is of high quality. However, quite often low quality oils are used as raw materials for biodiesel preparation. In cases where the free fatty acid (EFA) content of the oil is above 1%, difficulties arise due to the formation of soaps which promote emulsification during the water washing stage and at an free fatty acid content above 2% the process becomes unworkable.

Utilization of byproducts

The byproducts of biodiesel from Jatropha seed are the seed oil cake and glycerol, which have good commercial value. The seed oil cake is very good compost being rich in plant nutrients. It can also yield biogas, which can be used for cooking and the residue can be used as compost. Hence, oil cake will fetch good price. Glycerol is produced as a by product in the transesterification of oil. These by-products shall reduce the cost of biodiesel to make it at par with petroleum diesel. The cost components of biodiesel are the price of seed, seed collection and oil extraction, trans-esterification of oil, transport of seed and oil. As mentioned earlier, cost recovery will be through sale of oil-cake and of glycerol. The cost of biodiesel production can be reduced by proper utilization of by-products such as glycerol and meal cake apart from improving transesterification process. Glycerol from biodiesel contains some peculiar impurities and may not be suitable to process according to the usual technologies to produce product.

There is also need not only to develop purification technology for glycerol but also for its utilization as a raw material for the production of other chemicals as large quantity (200,000 tons/year) of the glycerol will be available even if 5% diesel is targeted to be replaced by biodiesel against the present glycerol demand in India, which is in the tune of 40,000 tons/year. There is a need to find the use of meal cake, which will be available in large quantities to reduce the cost of biodiesel. The meal cake may be used as fertilizer, after biogas production.

Economics of Jatropha biodiesel

It is estimated that cost of biodiesel produced by transesterification of oil obtained from *J. curcas* oil seeds shall be approximately same as that of petroleum diesel. The cost of biodiesel varies between Rs. 16.59 - 14.98 per liter. Assumptions are that the seed contains 35% oil, oil extraction will be 91-92%, 1.05 kg of oil will be required to produce 1 kg of biodiesel and recovery from sale of glycerol will be at the rate of Rs. 40 - 60 per kg. The price of glycerol is likely to be reduced with processing of such large quantities of oil and consequent production of glycerol. However, new applications are likely to be found creating additional demand and stabilizing its price.

Process flow chart

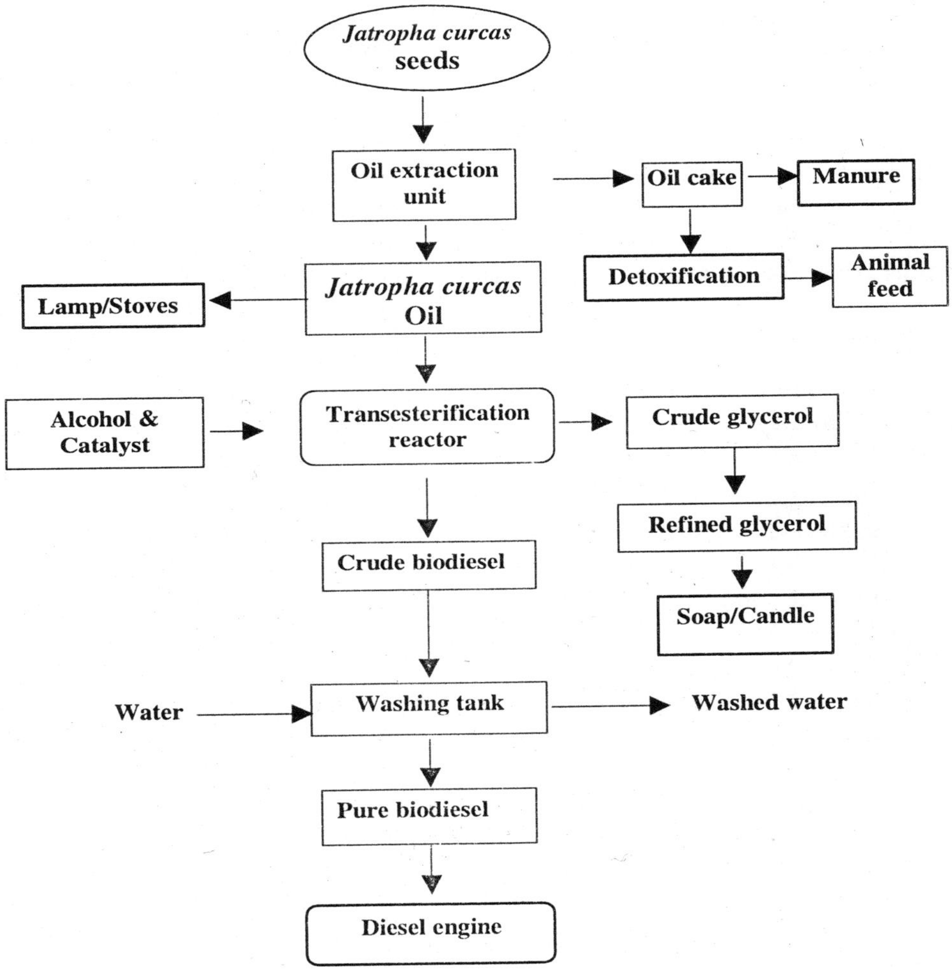

Research and development on technological development

In India research on biodiesel is in advanced stage, there is an urgent need to adopt vigorous programs on the technological development for its production, utilization of by products and evaluation in engine with respect to shortcomings, emissions, additive response etc. For efficient production of biodiesel, concerted R&D effort is needed to produce high quality feedstock material and to develop an improved, cost effective and efficient biodiesel production system. Biodiesel from different feed stocks may vary in composition, lubricity, oxidation stability, etc. It is desirable to carry out tests on biodiesel from all possible feed stocks available in India and generate comparative data on fuel composition emissions and materials compatibility, etc. Toxicological study is a prerequisite for introduction of any fuel. Procedure for detecting percentage of biodiesel in the blended fuel and to check adulteration of the fuel should also be developed. Emission norms for biodiesel vehicles may be similar to that of the conventional diesel vehicles. The other R&D issues which need attention are seed resource assessment, collection and their cryopreservation, increasing availability of seed, seed setting, inter-cropping, selection of high yielding crops, developing agro-technologies for different agro climatic regions, oil quality, biodiesel production technology using new catalyst systems like heterogeneous catalysts, lipase catalyst and supported catalysts on smart polymers, utilization of by-products apart from issues related to utilization of biodiesel as fuel.

REFERENCES

Anonymous 1995. Jatropha, Indian Council of Forestry Research and Education, Dehradun, pp-1-11.

Anonymous 2001. The Wealth of India, Vol. V, CSIR, New Delhi, pp. 293-295.

Chaudhari, D.C. and Joshi, D.N. 1999. *Jatropha curcas*- a multipurpose species for economic prosperity and wasteland development. *Advances in Plant Science Research in India* 9: 35-39.

Openshaw, K. 2000. Review of *Jatropha curcas*: an oil plant of unfulfilled promise. *Biomass and Bioenergy* 19(1): 1-15.

Parmathma, M., Parthiban, K.T. and Neelkanthan, K.S. 2004. *Jatropha curcas*, Forest College and Research Institute, Tamilnadu Agriculture University, Mettupalayam.

Singh, V.K., Dubey, K. and Srivastav, A. 2005. Jatropha – Ek Bahupayogi Paudha. CSFER, Allahabad.

Chapter - 8

ULTRADESICCATION- AN EFFECTIVE TOOL FOR THE CONSERVATION OF AN IMPORTANT BIODIESEL SEED, JATROPHA

J. Radhamani and Garima Srivastava

Division of Germplasm Conservation, National Bureau of Plant Genetic Resources, Pusa Campus, New Delhi-110012, India

E-mail: radhamani@nbpgr.ernet.in

ABSTRACT

Generally seed germplasm is dried to 3-7% moisture content (mc) before storage at subzero temperatures. Under the present investigations twelve accessions of jatropha collected from different regions of India were dried to 3.2 to 3.5% mc by equilibriating the seeds over saturated salt of lithium chloride ($LiCl_2$). The hermetically sealed ultradesiccated seeds were stored at ambient temperature to study the influence on the seed longevity. The results showed that eight of the twelve accessions under study showed a good survival of 78-94% even after storage of 20 months at ambient temperature. The biochemical aspects viz. electrical conductivity of seed leachates, lipid peroxidation and dehydrogenase enzyme activity correlated well with the seed viability.

Keywords: Ultradesiccation, Jatropha, seed viability, seed moisture, germplasm

INTRODUCTION

It is generally accepted that long term conservation conditions (low water content in the seeds and their maintenance at subzero temperature) prolongs the life span of seeds. However, the effect of extremely low water content on seed as a possible interaction of water content and temperature are disputed (Chenglion *et al.,* 1998). The potential benefits of ultra-desiccation and ambient temperature storage for extending the seed longevity has to be compared against the risk of seed deterioration. Drying seed to low water content may affect seed longevity or cause genetic alterations (Tao, 1992, Vartucci and Roos, 1990). Moisture content and storage temperature are the two key factors affecting the seed longevity. The internationally recommended standards are that seeds to be dried to 5-10% mc prior to storage (FAO/IPGRI, 1994). However, experiments using a number of different species have shown that longevity may be improved if seeds are dried to less than the recommended value (Ellis *et al.,* 1988, 1990a, 1995). On the other hand experiments have also shown that there are limits to the beneficial effects of drying, such that drying below an optimum or critical mc will not improve longevity and may have detrimental effects on seed longevity during storage (Nutile, 1964; Woodstock *et al.,* 1976; Vertucci *et al.,* 1994). The idea of ultradrying seeds (i.e. drying to moisture content less than 5%) was introduced as a means to reduce or eliminate the need for refrigeration (IBPGR, 1992; Ellis *et al.,* 1990b; FAO/IPGRI, 1994). However, before subjecting valuable germplasm to storage at such low seed moisture, the potential benefit and risk of ultradrying to seed longevity needs to be evaluated. Under the present investigation, the effect of ultradrying and its storage at ambient temperature of jatropha seeds was studied for twenty months.

MATERIALS AND METHODS

Freshly harvested seeds of jatropha were collected from different zones of India. The seeds of twelve accessions of jatropha were equilib-erated over saturated salt of $LiCl_2$ in desiccators for 15-20 days at 10°C (Rockland, 1969a). The seeds were equiliberated to moisture content of 3.2-3.5%. The moisture content (mc) was determined by low constant temperature oven method (ISTA, 1985) and is expressed on fresh weight basis. These seeds were hermetically sealed in three layered aluminium foil pouches and were stored at ambient laboratory condition. To determine the effect of ultra-desiccation on seed longevity, percentage germination and seed viability (triphenyl tetrazolium chloride, TTC) test at four monthly intervals was conducted for a period of twenty months. All the experiments were carried out in replicate of two with 50 seeds each. Germination percentage was determined after drying the seeds. Before conducting germination tests the ultradried seed samples were humidified at 90% RH in mist chambers for 8-10 h in order to avoid possible imbibitional injury. The percentage of hard, fresh and dead seeds was determined to calculate the germination percentage.

Biochemical parameters such as electrical conductivity of seed leach-ates by conductivity meter, lipid peroxidation (Heath and Packer, 1968) and dehydrogenase enzyme activity (Lakon, 1942) were undertaken in control and the stored samples. The seeds were removed at monthly interval till twenty months to study the corelation between the physio-logical and biochemical activities.

RESULTS AND DISCUSSION

Twelve accessions of Jatropha collected from diverse climatic conditions of India viz. Uttaranchal (western Himalayan regions with temperate conditions), Gujarat (hot and dry region), Uttar Pradesh (sub-humid and dry region), Rajasthan (western dry region), Tamilnadu (coastal plain) and Kerala (hot and humid tropical zone) were used under the present investigation. The mc of the fresh seeds were 12-18 percent on fresh weight basis, while the viability varied from 60-100 percent (Table 1 and 2). The highest 100 seed weight of 68 g and lowest of 47 gm was recorded in the seeds collected from Tamilnadu. The seeds of Gujrat and Akola were of medium size (50-59 gms/100 seeds) while the seeds from Uttaranchal, Uttar Pradesh, Rajasthan and Kerala regions showed bold and large seeds (above 60 gms/100 seeds). The seeds showed some difference in morphological and physiological characteristics (Table 1) indicating that the tree has a great diversity which exist almost in all climatic conditions. There is a wide variation in seed size and 100 seed weight, seed: kernel ratio and oil contents of seeds.

Table 1. Seed characteristics of Jatropha seeds collected from different states of India

National Identity	Source state	100 seed wt (gms)	Seed: Kernel ratio	Oil (%) Total seed basis	Moisture content (%)
IC 559360	Gujarat	59.12	1.75	31.56	12.8
IC 559358	Gujarat	50.06	1.67	30.12	13.2
IC 554538	Kerala	66.63	1.53	28.19	18.3
IC 564433	Kerala	62.65	1.98	31.12	18.8
IC 422125	Rajasthan	63.07	1.71	31.21	12.0
IC 555380	Tamilnadu	47.68	1.82	30.52	17.9
IC 555381	Tamilnadu	68.03	2.66	32.06	18.2
IC 558211	Maharashtra	57.32	1.77	29.7	16.4
IC 559370	Uttrakhand	60.6	1.78	23.67	15.8
IC 559368	Uttrakhand	59.44	1.94	26.35	14.4
IC 471303	Uttar Pradesh	60.38	1.71	31.5	13.6
IC 471326	Uttar Pradesh	66.62	1.39	30.5	14.8

The different seed accessions were equillibriated to 3.2-3.5% mc over standard salts of lithium chloride in different time durations (18-23 days). The accession from the southern zone viz. Tamilnadu and Kerala had taken a longer duration of 23 days as the fresh seed moisture content of the seed was 18% i.e. slightly higher than the other accessions (13%). The germination time period varied from 12-15 days in the fresh seeds which was extended to 18-25 days after ultradesiccation. The initial viability of seed was 60-100%. There was not much significant decrease in the seed viability till the 8 months of storage but it started declining thereafter (Table 2). The decline in viability was observed only in the accessions received from the southern India where the seeds might have been harvested during high humid conditions of September to October. In addition these seed samples also showed severe fungal infection responsible for lowering the seed viability.

Table 2. Effect of seed storage on the seed viability at different durations

National Identity	Viability (0 m)	Viability (8 m)	Viability (20 m)	TTC Staining patteren*
IC 559360	98	94	92	+++
IC 559358	100	96	94	+++
IC 554538	98	96	78	++
IC 564433	82	80	45	+
IC 422125	100	98	94	+++
IC 555380	92	90	50	+
IC 555381	60	54	32	+
IC 558211	82	80	55	+
IC 559370	98	98	88	+++
IC 559368	94	90	80	++
IC 471303	92	92	89	+++
IC 471326	94	90	86	+++

* + <70%, ++ >75%, +++ >85%

The percentage germination of Jatropha seeds gradually declined with storage time at ambient temperature but they all maintained well over a period of 20 months (Table 2). Most of the accessions stored showed 5 to 10% reductions over the control while the IC555381 showed a maximum reduction of 47% over control which had a very low initial viability of 60% in the fresh seeds (with fungal infection). So drying below 5% mc maintained viability well at ambient temperature over a period of time. The germination time period in the dried and stored seed samples was extended to 20 days which was about 14 days in the fresh seeds. There is a close correlation between the potential viability through TTC test and the actual germination in seeds. Therefore TZ test can be used as a viability test for Jatropha seeds, as has been earlier reported in *Pongamia pinnata* (Santosh *et al.*, 2008).

The results showed an increase in electrolyte leakage with the reduction in germination over storage time (Table 3). The maximum conductivity of 0.67µs/cm/g fwt was recorded in control seeds of IC555381 which had the least viability of 60% while the least value of electrical conductivity (0.03µs/cm/g f wt) was recorded in fresh seeds of three accessions IC422125, IC471303 and IC471326 where the viability was as high as 100%, 92% and 94% respectively (Table 2). The seeds with lower viability recorded the higher conductivity values showing a negative corelation of viability with conductivity. These studies are in confirmation with the earlier reports (Powell and Mathews, 1977). The increase in amount of electrolytes is found to be proportional to the seed deterioration, attaining maximum values when seeds lost viability (Table 2). Therefore, deterioration of seeds is primarily the loss of solutes resulting in increase in the electrical conductivity and therefore the maximum conductivity could be interpreted in terms of changes in membrane structure and loss of its chemical constituents of seed essential for its viability. The lipid peroxidation values were also found to be high in seeds with low viability which increased manyfolds (about 20 times) after storage. Lipid peroxidation also showed a negative corelation with the seed viability whereas, the dehydrogenase enzyme activity decreased with seed germination. The enzyme activity is found to be directly propotional to the seed viability (Table 3). Lipids are particulary susceptible to peroxidative attack and the byproducts of those reactions also affect proteins and nucleic acid functions. Present studies indicated a good correlation of increased electrolyte leakage with increased lipid peroxidation which indicates the destruction of membranes caused by peroxidation of lipid during ultra drying and storage. Inversely EC values, lipid peroxidation and loss of dehydrogenase enzyme activity in stored ultradry seeds might be the possible reason for loss in seed viability during storage.

In the present study, increase in the electrolyte leakage was observed with reduction in storage time which is in confirmation of the earlier report of Powell and Matthews (1977). The increase in the amount of electrolyte in seeds is found to be proportional to the seed detereoration, attaining maximum values when seeds lossed viability completly during storage. Changes in the membrane integrity (increase in EC values, lipid peroxidation in the stored seeds) depict that the storability of the Jatropha species in these 4 accessions might be the possible reason for the loss of seed viability during storage. Enhanced lipid peroxidation in biological sample is associated with loss of unsaturated bonds in lipids which can cause disturbance in the structure of membranes. A decrease in the dehydrogenase enzyme is probably associated with the lipid peroxidation in jatropha seeds. In conclusion it appears that the seed deterioration during storage is a sequential step of events occurring in the seeds leading to loss of seed viability, amongst which lipid peroxidation seems to be the most important event causing cellular degeneration through free radical damage. The free radicals in turn destroy nuclear membrane ultimately DNA and seed denaturation.

Table 3. Effect of ultradesiccation on biochemical activities of Jatropha seeds

National Identity	Electrical conductivity of seed leachates (µs/cm/g f wt)		Lipid peroxidation (TBA reactive product n mol / mg f wt)		Dehydrogenase enzyme activity (n moles/mg f wt)	
	Control	Treatment*	Control	Treatment*	Control	Treatment*
IC559360	0.19	0.20	0.01	0.02	1.68	1.30
IC422125	0.03	0.35	0.05	0.07	2.64	1.45
IC559370	0.30	0.38	0.04	0.05	1.85	1.38
IC559368	0.18	0.26	0.05	0.07	1.52	1.45
IC554538	0.22	0.29	0.05	0.08	2.6	2.54
IC555380	0.13	0.16	0.06	0.08	2.3	1.2
IC559358	0.15	0.16	0.08	0.09	2.43	2.13
IC555381	0.67	0.82	0.94	1.15	1.92	0.47
IC564433	0.15	0.22	0.07	0.09	2.0	1.48
IC558211	0.15	0.26	0.08	1.11	2.2	1.76
IC471303	0.03	0.05	0.05	0.08	1.61	1.03
IC471326	0.03	0.04	0.06	0.08	1.42	1.12

*Seeds stored for 20 months

Nevertheless, the results obtained in the study confirm earlier reports of a clear benefit of drying to low moisture content when seeds are stored at ambient temperature; the extent of improvement in longevity appears to be species specific, and the stage of physiological fruit maturity, time of harvesting and the region from where it is collected. Probably the four accessions might have not been harvested in the right time of physiological maturity which led them to loose viability during storage. Contrary to earlier reports (Ellis *et al.*, 1988, 1990; Vertucci and Roos, 1990; Dickie and Smith, 1995) no determintal effects of ultradry was observed in most of the accessions under study. At ambient storage most species showed maximum germination when seeds were sowed at moisture less than 4% (Singh *et al.*, 2003). So based on our data ultra drying of Jatropha seeds is beneficial for storing the seeds at least for one season which can be further experimented with more number of accession with longer storage periods to develop the true potential of ultradry technology and can be further utilized for the tree species where a farmer/ common people can easily store their seeds of one harvesting season to the next growing season. The moisture content of seeds can be precisely reduced to the extent possible or it can withstand desiccation with increased longevity in air dry storage. This has been tried by several workers. (Ellis *et al.*, 1988; Vertucci and Roos, 1990).

Despite large scale use, seed technologies for many tree species are inadequate and remained to be developed. Ultradrying of orthodox seeds offers an effective and economically viable alternative for long term *ex-situ* germplasm conservation.

ACKNOWLEDGEMENT

Authors acknowledge with thanks the valuable guidance and help provided by the Director, National Bureau of Plant Genetic Resources, and New Delhi. We are also thankful to Department of Biotechnology for the financial support.

REFERENCES

Chenglion, H.U., Yunlan Zhang, M. T., Xiarong, H.U. and Chaogu J. 1998. *Seed Science Research* (Suppl) 35:9.

Dickie, J.B. and Smith, R.S. 1995. Observations on the survival of seeds of *Agathis* spp. stored at low moisture contents and temperatures. *Seed Science Research* 5:5-14.

Ellis, R.H., Hong, T.D. and Roberts, E.H. 1988. A low moisture content limit to logarithmic relations between seed moisture content and longevity. *Annals of Botany* 6:405-408.

Ellis, R.H., Hong, T.D. and Roberts, E.H. 1990a. Moisture content and the longevity of the seeds of *Phaseolus vulgaris*. *Annals of Botany* 66: 341-348.

Ellis, R.H., Hong, T.D., Roberts, E.H. and Tao, K.L. 1990b. Low moisture content limits in relations between seed longevity and moisture. *Annals of Botany* 65:493-504.

Ellis, R.H., Hong, T.D. and Roberts, E.H. 1995. Survival and vigour of lettuce and sunflower seeds stored at low and very low moisture contents. *Annals of Botany* 76:521-534.

FAO/IPGRI. 1994. Genebank Standards. FAO/IPGRI, Rome, Italy 13.

Heath, R.L. and Packer, L. 1968. Photoperoxidation in isolated chloroplast. 1. Kinetics and stoichiometry of fatty acid peroxidation. *Archives of Biochem and Bio Physics* 125:189-198.

IBPGR. 1992. Annual Report 1991. International Board for Plant Genetic Resources. Rome, Italy.

ISTA. 1985. International Seed Testing Association. International rules for seed testing. Rules 1985. *Seed Science and Technology* 13:299-513.

Lakon, G. 1942. Topographiscer Nachweis der Keimfahigkeit der Getridfrichte durch tetrazolium sale. *Ber. Dt. Botanical Gazatte* 67: 299-305.

Nutile, G.E. 1964. Effect of desiccation on viability of seeds. *Crop Science* 4:325-328.

Powell, A.A. and Matthews, S. 1977. Deteriorative changes in pea (*Pisum sativum*) seed stored in humid and dry conditions. *Journal of Experimental Botany* 28:225-234.

Rockland, L.B. 1969. Water activity and storage stability. *Food Technology* 23:1241-1251.

Santosh, K., Radhamani, J., Singh, A.K. and Varaprasad, K.S. 2008. Investigating seed germination and storage problems in *Pongamia pinnata* L. *Current Science* 93 (7):910-911.

Singh, N., Singh, A.K. and Dhillon, B.S. 2003. Effect of ultradrying on *ex-situ* seed conservation. Seed conservation-Turning Science into practice. (Ed. Smith, R.D., Dickie, J.B., Linington, S.H., Pritchard, H.W., Probert, R.J.). Royal Botanical Garden. 797-805.

Tao, K. 1992. Genetic alteration in germplasm conservation. *Advances in Science and Technology of Seeds* 137-49.

Vertucci, C.W. and Roos, E.E. 1990. Theoretical basis of protocols for seed storage. *Plant Physiology* 94:1019-1023.

Vertucci, C.W., Roos, E.E. and Crane, J. 1994. Theoretical basis of protocols for seed storage III. Optimum moisture content for pea seeds stored at different temperatures. *Annals of Botany* 74:531-540.

Woodstock, L.W., Simkin, J. and Schroeder, E. 1976. Freeze drying to improve seed storability. *Seed Science and Technology* 4:301-311.

Biofuels and Environment

Chapter - 1

JATROPHA CURCAS: A SOLUTION FOR MOISTURE STRESS SHIFTING CULTIVATION AREAS OF NORTHEAST REGION

B.K. Pandey[1], H.O. Saxena[2], Y.C. Tripathi[3] and P. Hazarika[4]

[1]Centre for Social Forestry and Eco Rehabilitation, Allahabad
[2, 4]Rain Forest Research Institute, Jorhat (Assam), India.
[3]Advanced Research Centre for Bamboo and Rattan (ARCBR), Aizwal, Mizorum E-mail: pandeybk@icfre.org

ABSTRACT

The current practice of Jhum cultivation (shifting cultivation) in eastern and north eastern regions of India is an extravagant and ruthless form of land use system. The evil effects of shifting cultivation are devastating and far-reaching in degrading the environment and ecology of these regions. The earlier 15–20 year cycle of shifting cultivation on a particular land has reduced to 2–3 years now and fallow land in jhum cultivation area is increased significantly. This has resulted in large-scale deforestation, soil and nutrient loss, and invasion by weeds and other unwanted species. The indigenous biodiversity has been affected by shifting cultivation to a large extent. *Jatropha curcas,* a beneficial biofuel crop, can be introduced in shifting cultivation areas as a suitable agroforestry crop for generating better income to the jhumias in moisture stress areas of Northeast region because it is not labour intensive and easy to cultivate. It can be grown either in monoculture or as intercrop with other species like Aswagandha, Asparagus, pulses and vegetables etc., during initial growth period. Jatropha can also be intercropped with coconut or arecanut. It suits for all types of soil and requires minimal care after plantation. Therefore, potentiality of growing Jatropha in agroforestry system as viable alternative to sustain economy and environmental need of the region is discussed.

Keywords: Jhum cultivation, fallow land, jhumias

INTRODUCTION

Shifting cultivation is the most common primitive way of food production system, which is still in vogue among tribal communities in various underdeveloved countries of the world. In India, around two million tribal people are engaged in shifting cultivation in 15 states and approximately 11 million hectares of the forest land is affected by this practice in states of North-East, Orissa, Andhra Pradesh, Madhya Pradesh, Bihar, Kerala, Sikkim, Tamil Nadu, Maharastra and Karnataka. In the hilly region of North-East India, Shifting cultivation (locally known as Jhum cultivation or Jhuming) is very common for food production affecting an area of 25,6083 sq. km which accounts for 7.76% of the country's total geographical area. During the last few decades shifting cultivation had been under conflicting views on the issue of its being hazardous or otherwise.

The commercial plantation of Jatropha could be a viable alternative to the practice of jhum and could make a difference in the lives of jhumias because it is not labour intensive and is easy to cultivate. Jatropha is a very promising and economical biofuel plant for plantaion, agroforestry, mixed plantation and suitable for fallow land in jhum cultivation areas and also waste land developments. During recent years, Jatropha has gained more importance mainly due to its potential in biodiesel production. Seeds of the plant particularly grown in tropical and semi-arid region are reported to contain 50-55% of fatty oil which can be used as automobile fuel after chemical modification i.e transesterification.

Oil extraction and its chemical conversion into biodiesel can be commercially viable only when the oil content is not less than 20-25% and it requires moisture stress conditions. Since, a major part of northeast region, have high soil moisture level, possibility of high oil yield in Jatropha grown in the region is very limited. In fact, farmers, especially in Ri-Bhoi district in Meghalaya, have been growing Jatropha as a fencing crop. It was only recently that its value was realised after the oil content of its seed was found to be as high as 30-35%. This much oil percentage is viable for the commercial exploitation of Jatropha in the Northeast region.

In Meghalaya, *J. curcas* is locally known as diengsong dkhar and phok seems to be a wonder crop. It can grow on any type of soil, its climatic requirements suits most areas of the state, it's quick growing, not browsed by animals and has an average lifespan of 50 years.

About the plant

The genus Jatropha belonging to the family Euphorbiaceae, is one of the prospective biodiesel crop yielding genera. It is a morphologically diverse genus comprising 160-175 species of shrubs, rhizomatous shrubs, herbs and small trees. About nine species of Jatropha have been recorded in India. Out of these, important ones are *Jatropha curcas*, *Jatropha gossypifolia*, *Jatropha glandulifera*, *Jatropha multifida* and *Jatropha podagrica* and among these also, *Jatropha curcas* is the most important species yielding biodiesel.

Jatropha curcas Linn. is a large woody shrub, 3-4 m high, commonly known as Jangali arand or Ratanjyot, believed to be native of South America and Africa. Its other vernacular names are *Physic nut* in English, *Bongalibhotora* in Assamese, *Jamalgota* in Punjabi and *Jahazigaba* in Oriya etc.

Uses

The plant is mainly valued for its fatty oil obtained from its seeds by solvent extraction or mechanically by hydraulic pressure by screw press and known in the trade as curcas oil. Seeds of the plant resemble castorseed in shape, but are smaller in size and dark brown in colour. The oil content has been reported up to 60% of the seed weight and it has a low viscosity as compared to the castor oil. The oil is semi-drying which can be put into a variety of commercial uses.

The plant has medicinal properties and is also used for the manufacture of lubricants, soaps candles and cosmetics. According to the latest report, curcas oil is an efficient substitute fuel for diesel engines, thus can be used as a bio-diesel to supplement the diesel requirements of the country. Jatropha oil cakes serve as an effective organic fertilizer and a viable substitute for chemical fertilizers that adversely affect the chemical balance and soil fertility.

Soil and climate

The crop is undemanding in soil type, can grow on stony, gravely or shallow, even on calcareous soils can also tolerate high temperature and salinity. It can be grown over a wide range of arid and semi-arid climatic conditions. Whilst Jatropha grows well in low rainfall conditions (200 mm), it can also tolerate high rainfall (up to 1200 mm) conditions.

Cultivating Jatropha

The plant grows readily from cuttings or seeds. Cuttings strike root so easily that the plant can be used as an energy producing living fence post. In seed plants, full growth requires some what more time compared to plants raised through cuttings. Best time for planting branch cuttings is July and August. The shorter cuttings of 10 and 20 cm length produce poor sprouting whereas 60 cm long cuttings produce good sprouting. Cuttings planted in July-August under a moderate temperature (25-30°C) and relative humidity (60-80%) produce height up to 1.5-1.8 meters and exhibit multiple branches and branchlets, densely covered with broad leaves. Flowering and fruiting of the plants takes place by winter in the next year. Direct seed sowing can also be done where adequate irrigation facilities are available in early stage. Plants raised through the seeds during the same period (July and August), do not flower well and attain maximum height up to 0.8-1 meter only.

Intercropping

It can be grown in monoculture or as intercrops with other species like Aswagandha, Asparagus, pulses vegetables etc., during initial growth period. Jatropha can be intercropped with coconut or arecanut channel along the bunds of water channel.

Fertilizer

Although Jatropha is adapted to low fertility sites and alkaline soils, better yield is obtained on poor quality soils if fertilizers with small amounts of calcium, magnesium and sulphur are used. Micorrhizal associations have been observed with Jatropha and are known to aid the plants growth under conditions where phosphate is limiting. In general application of super phosphate @ 150 kg/ha and alternated with one dose of 40:100:40 kg/ha NPK at 6 monthly intervals improves the yield. The application may coincide with rainy season or ensure proper irrigation immediately after application of fertilizers.

Weeding

The field should be kept free from weeds at all times. Around 3-4 weedings in the initial period are enough to keep the field free from weeds until the crop reaches the grand growth period stage.

Pests and diseases

Jatropha is highly disease-resistant species and no serious pest and diseases are reported to affect this plant. However, if bark eater and pod borer are found, they can be controlled with herbal pesticides such as mixtures of Vitex, Neem, Aloe, Calotropis, *Clerodendron interme* etc. or Rogor @ 2 ml/lit of water. Collar rot may be a problem in the beginning and can be controlled with 0.2% COC.

Harvesting

Jatropha propagated from stem cuttings starts flowering in second year and the plants propagated from seeds in the third/fourth year. Generally, harvesting takes place during August – December (depending on climatic condition and locality). The flowering in Jatropha depends on the location and agro-climatic conditions. Generally the plant flowers in hot and rainy seasons and sets fruit in winter when it is leafless. Fruits mature in 2-4 months after flowering. Each fruit contains 2-3 seeds. The mature yellow capsules may be collected by hand plucking and the seeds are separated from the capsules manually or mechanically. Seeds may be cleaned by blowing or winnowing in direct sunlight for 4 to 5 days (6-10% moisture) before packing.

Yield and economics

Grown up Jatropha plants yields 4-6 kg dried seeds per plant with an estimated yield of 10-12 tonnes of seeds per hectare in well irrigated plantations from 6th year onwards whereas 4-6 tonnes of seeds from the 6th year onwards under rainfed condition. With the minimum market price of Rs.10/-kg of seeds an essential gross return of Rs. 1, 00000-1, 20000/- per hectare per annum under well irrigated conditions and 40,000-60,000/- per hectare per annum under rainfed conditions, could be realized.

Biodiesel - Fuel of future

Use of biodiesel in automobile sector has already been started. The Indian Oil Corporation has conducted trials of running buses on biodiesel. Successful trail run of Delhi-Amritsar Shatabdi Express on biodiesel has also established biodiesel as a fuel of future. Government of Haryana state has identified three depots in Rewari for playing Haryana Roadways buses on blended fuel. Mixing of non-edible oil, extracted from Jatropha would cut India's import dependence. Diesel constitutes 80% of fuel sales and mixing non-edible oil in it would certainly help to cut the import dependence. No modifications in vehicle engines are required for use of biodiesel. Use of biodiesel has already been started in US, Austria, Belgium, France, Germany, Italy and Sweden. Several automobile manufacturers like Mahindra and Mahindra and Ashok Leyland have endorsed biodiesel as fuel for their vehicles.

However, availability of non-edible oilseeds is a major constraint in beginning commercial supplies. Planning Commission has asked states to grow more Jatropha on wasteland and semi-rain fed areas. Production of biodiesel has the potential to replace significant quantities of fossil fuels in many transport applications. The widespread use of ethanol in Brazil has shown that biofuels are technically feasible on a large scale. In USA and Europe biofuel production (ethanol and biodiesel) is increasing, with most of the products being marketed in fuel blends, e.g. E20 is 20% ethanol and 80% petrol and has been found to be suitable for most spark ignition engines without any modifications.

REFERENCES

Farooqi, A.A., Srinivasappa, K.N. and Kennedy, R.R. 2004. Jatropha - The Biofuel Plant. *I Map Today* pp. 4-9.

Ranjan, R. and Upadhyay, V.P. 1999. Ecological problems due to shifting cultivation. *Curr. Sci.* 77(10): 1246-1250.

Shyam Sunder 2006. *Jatropha curcas* for biodiesel, organic farming and health. J.V. Publishing House, M.G.H. Road, Jodhpur.

Singh, J. and Chandra, A. 2001. Afforestation: A pragmatic approach to shifting cultivation. Proceedings of the workshop on Application of Improved Technology for Afforestation in Northeast India held at RFRI, Jorhat, Assam, India on 24-26th April, pp.80-86.

Times News Network 2007. Meghalaya farmers stumble upon wonder fuel plant. Times of India, April, 2007, pp. 2.

Chapter - 2

JATROPHA CURCAS - A POTENTIAL SPECIES TO RECLAIM SILICA MINING AREA

Kumud Dubey*, A.K. Pandey* and K.P. Dubey**

*Centre for Social Forestry and Eco-Rehabilitation, 3/1 Lajpat Rai Road, Allahabad

**Chief Conservator of Forest, Southern Zone, Allahabad

e-mail: dkumud@yahoo.com,dkesheo@yahoo.co.in, csferald@yahoo.com

ABSTRACT

Reclamation of wasteland is a huge challenge. About 75 million ha of wasteland need recovery through revegetation. Several initiatives have been taken up in recent years in different parts of the country to reclaim these wastelands. *Jatropha curcas* is an important bio-fuel crop. Its hardy nature and wide adaptability to varied climate and soil have made this species, a potentially useful species to reclaim wasteland. The Shankargarh area, a part of Allahabad District, is famous for silica mines and for the good quality of silica deposits. These mines are situated in the Vindhyan hills of the Allahabad District. Extensive open cast silica mining is being performed since more than three decades, causing great damage to the forest as well as productivity of the region. Continuous open cast mining has scarred the local landscape, disrupted ecosystem and destroyed micro flora of the area. Therefore, there is an urgent need to reclaim these abandoned areas by using such plants, which can easily establish themselves in such adverse conditions and also support the economy of the local people as well. *Jatropha curcas* has been found most suitable for the purpose. Under NOVOD Board sponsored National Networking Programme, a provenance trial was established to study the performance of provenances *of J. curcas*. For this, twenty three provenances were procured from different parts of country. Field trial of these provenances was established in the silica mining area of Shankargarh. The survival, growth performance and seed production data were recorded. Provenance STP-21 from Shankargarh, Allahabad performed best followed by STP-23 from Jabalpur with respect to seed production. The best provenance will be recommended for the reclamation of such areas.

Keywords: Reclamation, silica mining area, *Jatropha curcas*, bio-fuel

INTRODUCTION

Land degradation has become a major stumbling block to world food security with about 2000 million ha equivalent to fifteen percent of the Earth's land area have been degraded through human activities. Mining of mineral wealth is a major cause of land degradation. Silica sand is a major mineral being extensively used in glass industries. Its resources are distributed in the states of Andhra Pradesh, Bihar, Gujarat Haryana, Karnataka, Madhya Pradesh, Maharashtra, Rajasthan and Uttar Pradesh. The Shankargarh area, a part of Allahabad District is famous for its silica sand mines owing to good quality of the silica sand deposits. These mines are situated in the Vindhyan hills of the Allahabad District and extensive open cast silica mining activities are being performed since over three decades, causing great damage to the forest as well as productivity of the region. Therefore, reclamation of this mining area becomes a priority to counter environmental hazards and to restore the ecological balance. Reclamation of such mining area, proper selection of species that will adapt to climatic and soil conditions is a critical step. Moreover, the selected species should also alleviate the poor socio-economic status of the local people. In the present study, performance of *J. curcas* provenances has been studied in the Shankargarh silica mine area with regard to the perceived advantage of Jatropha for its capability to grow on marginal lands and its ability to reclaim problematic lands and restore eroded areas.

Jatropha curcas, physic nut, is adapted to higher temperatures and grows well in well drained soils receiving an average rainfall of 300 to 1000 mm. It can grow almost on any type of soil whether gravelly, sandy or saline and thrives even on the poorest stony soils and rock crevices. It is a drought resistant perennial. It can however tolerate slight frosts and years without rainfall and is well adapted to marginal soils with low nutrient content. Yields vary from 0.5 to 12 t/ha/yr depending on soil and rainfall conditions. Jatropha often starts yielding from the first year onwards, but yields are maximum from plants that are about 3-5 years old. The productive age of Jatropha is over 30 years. It is not eaten by animals and when planted as a fence repels rodents and has phyto-protective action against pests and pathogens and thus provides additional protection to intercropped plants. In India *J. curcas* is found in almost all states. Being drought tolerant, it can be used to reclaim degraded areas (Francis *et al.*, 2005, Dubey *et al.*, 2005, Jones and Miller, 1992). Characteristics of soil of degraded area where *J. curcas* fruited after nine months were reported by Becker and Makkar (2008). The species has already found favour for biodiesel production. *J. curcas* is still a wild plant; therefore, initiation of systematic selection of seed sources/ provenances suited to different climatic conditions is a prerequisite for sustainable and commercially profitable utilization of this plant for biodiesel production as well as for reclamation of degraded lands. For the selection of suitable provenance, a field trial was laid out to study the growth performance and suitability of different provenances of *Jatropha curcas* in the silica mined area.

MATERIALS AND METHODS

Site selected for the study was situated at Shankargarh in Allahabad District. The area was degraded due to open cast silica mining and other biotic interferences. Area involved in mining in Shankargarh includes 40 villages covering about 150 sq km. Overburden in the area is not uniform and varies from place to place. The soil has low moisture retaining capacity. The area suffers from high temperature and moisture stress conditions during most of the period of the year. For Jatropha field trial, seeds of twenty three provenances were procured from the different research organizations from all over India. Seedlings were raised at Padilla Research Nursery, Allahabad of the Centre. The seeds were cleaned manually. Before sowing, the seeds were soaked in water for 12 hours at ambient temperature and sown in polythene bags filled with soil, Farm Yard Manure (FYM) and sand in 1:1:1 ratio in the month of February. The germination started within a week. The weeding, hoeing and irrigation of the seedlings of Jatropha provenances were done regularly. The plantation trial was established in Randomized Block Design with three replications and 2m x 2m spacing in the month of August. Growth data (height and girth) were recorded regularly.

RESULTS AND DISCUSSION

The growth data viz. plant height, girth and seed yield after four year of plantation are prsented in Table 1. The data revealed Jhansi (NRCJ-42) STP18 provenance, followed by Pantnagar (PJ Sel-1) STP4 and Jhansi (NRCJ-47) STP19 were best with respect to height growth. The highest growth in girth was observed in provenances Pantnagar (PJ Sel-1) STP4, Jhansi (NRCJ-42) STP18, Jhansi (NRCJ-47) STP19 in decreasing order. The lowest increment in collar diameter was observed in Pantnagar (PJ Sel-1) provenance. Overall growth performance with respect to height, collar diameter and no of branches was found best in Pantnagar (PJ Sel-1) STP4, Jhansi (NRCJ-42) STP18, Jhansi (NRCJ-47) STP19 and Pantnagar (PJ Sel-2) STP5 provenances. The carbon sequestration by these species is more as far as productive biomass is concerned. Therefore, these provenances may be considered for reclaiming such sites. As far as seed yield was concerned, provenance CSFER1STP21 is best, followed by STP22 and STP23.

Table 1. Performance of different provenances in field trial at Shankargarh Silica mining area after four year of plantation.

Provenance	Height (cm)	Girth (cm)	Branch number	Seed yield in g per plant
Akola PKVJ-MKU-1(STP1)	171.54	23.54	17	531.33
Akola PKVJ-DHW-1 (STP2)	179.08	22.08	18	517.5
Rahuri (RJ-117) STP3	160.69	22.14	12	494.83
Pantnagar (PJ Sel-1) STP4	201.88	29.46	21	408.66
Pantnagar (PJ Sel-2) STP5	191.25	23.04	22	500.08
Jabalpur (Silviculture nursery) STP6	178.46	24.01	17	421.83
Jabalpur (Barah Experimental area) STP7	163.42	21.21	12	448.75
SFRI, Jabalpur (Indore) STP8	173.69	21.60	13	470.16
SFRI, Jabalpur (Sagar) STP9	152.46	16.92	8	373.91
Coimbatore (TNMC-2) STP10	139.00	19.96	11	427.33
Coimbatore (TNMC-3) STP11	153.10	21.26	14	453
Coimbatore (TNMC-4) STP12	150.21	21.21	15	457.16
Coimbatore (TNMC-5) STP13	176.13	23.79	15	455.58
Coimbatore (TNMC-7) STP14	151.25	22.67	20	452.33
Coimbatore (TNMC-22) STP15	150.29	22.79	20	497.41
Kanpur STP16	149.42	22.36	25	448
Jhansi (NRCJ-41) STP17	170.67	23.33	27	461.91
Jhansi (NRCJ-42) STP18	218.42	28.75	26	470.5
Jhansi (NRCJ-47) STP19	199.83	27.25	24	492.66
Jhansi (NRCJ-53) STP20	176.08	25.00	21	479.58
CSFER1 STP21	162.58	24.57	19	977.25
STP22	92.21	22.69	21	679.5
STP23	157.33	21.17	17	586

Mining is essentially a destructive developmental activity where ecology suffers. Unfortunately, in most regions of the world, underground geological resources are superimposed by biological resources i.e. forests. This is particularly evident in India. Hence mining operations necessarily involve deforestation at large. It also affects the economic and social life of local communities, who depend on the natural resources of the locality for large portions of their income. Shankargarh, where these silica mines are situated, is a backward area despite rich mineral resources it has and despite the revenues that the state gets from mining. Jatropha is a perennial, hardy, drought resistant and valuable multi-purpose crop used to alleviate soil degradation, desertification and deforestation (Henning, 1996). It is an economically important species and helps to increase rural incomes, self-sustainability and alleviate poverty. Thus jatropha offers the option both to cultivate wastelands and to produce vegetable oil suitable for conversion to biodiesel and other economically important byproducts as an alternative resource of income (Francis *et al.,* 2005). Globally there are large chunks of degraded land available for planting *J. curcas*. In India and China alone about 150 million ha of degraded lands are available where plantation of *J. curcas* has the potential to not only reduce green house gas emissions but also create opportunities of employment and socio economic development for impoverished local people. Moreover, only degraded land, where profitable food production would not be possible, is anticipated to be used as it does not affect the food production and seems to be an ecologically and socio-economically viable option for biodiesel production. Carbon sequestration by Jatropha plantations is an added attraction as sources of "carbon dioxide emission credits". In previous reclamation studies of silica mining site suitability of other species had also been studied (Dubey *et al.,* 2006), but through Jatropha plantations, three major objectives viz. wasteland reclamation, biofuel production, and socio-economic development of people inhabiting near degraded areas can be achieved simultaneously.

REFERENCES

Becker, K. and Makkar, H.P.S. 2008. *Jatropha curcas:* A potential source for tomorrow's oil and biodiesel. *Lipid Technology* 20 (5):104-107.

Dubey, K., Singh,V.K. and Dubey, K.P. 2006. A step towards reclaiming silica mining site at Allahabad district, Uttar Pradesh, India. Proceedings of the *10th Billings Land Reclamation Symposium* June 4 – 8, 2006 Billings, Montana "Reclamation: Supporting Future Generations". Edited by Richard I. Barnhisel, published by BLRS and ASMR, R.I. Barnhisel (ed.) 3134 Montavesta Rd., Lexington.pp.182-190.

Dubey, K., Khan, M.R., Srivastava, A. and Singh, V.K. 2005. Oil from wasteland-the Jatropha project in India. National Conference on "*Management of Land Resource & Land Use Towards sustainable Development*" October 18-19, held at Institute of Environment and Development Studies, Bundel Khand University, Jhansi, U.P. India.

Francis, G., Edinger, R. and Becker, K. 2005. A concept for simultaneous waste-land reclamation, fuel production, and socio-economic development in degraded areas in India: Need, potential and perspectives of *Jatropha* plantations. *Natural Resources Forum* 29 (1):12.

Henning, R. 1996. Combating Desertification: the Jatropha project of Mali, West Africa. *Arid lands news letter* No.40, Fall/Winter, The CCD, Part I: Africa and the Mediterranean.

Jones, N. and Miller, J.H. 1992. *Jatropha curcas*: A multipurpose species for problematic sites, The World Bank, Washington DC USA.

Chapter - 3

EFFECT OF FROST ON *JATROPHA CURCAS* PLANTATIONS

V.R.R. Singh, Dinesh Kumar, Sanjay Singh, Ritesh Sharma, Harish Chandra and Kshitij Malhotra

Silviculture Division, Forest Research Institute, Dehradun

ABSTRACT

Jatropha curcas is being perceived as one of the most important biofuel species in India and massive planting programmes of this species are underway in different states of the country. This hardy and drought tolerant species can be raised in marginal lands with very little input. This species is known to be tolerant of most of the biotic and abiotic stresses encountered in the wastelands. The species has, therefore, been planted on a large scale in Uttarakhand as well as other states in north India. However, the authors have observed severe damage due to frost to *J. curcas* plants in young plantations in Uttarakhand during 2007 and 2008. The leaves of the affected plants dried up and turned brownish. The branches showed pulpy appearance which later dried and turned black. The entire plant developed signs of frost injury; these symptoms progressed from the top to the base of the plant. Plants with well developed root system recovered from frost injury and sprouted from ground level in the ensuing spring season.

The effect of frost was more severe on plants planted in the recent monsoon: plants with 40-55 cm height and less than 25 mm collar diameter at the planting time were completely wiped out, while plants having 50-90 cm height and 25-35 mm collar diameter showed 35 per cent mortality. Plants in the second year of field plantation (150-175 cm height and 40-58 mm collar diameter) showed signs of frost injury above 15 cm height from ground level. However, all of them resprouted but assumed a bushy appearance. Plants in the third year of field planting (230-295 cm height and 63-80 mm collar diameter) also met with the same fate and assumed bushy appearance. Plants that were growing in shade or under canopy of tree remained protected against frost. Such observations were made in various parts of Uttarakhand viz. Dehradun, Rudraprayag, Rishikesh, Haridwar, Augustyamuni and Betalghat (Nainital). Similar phenomenon is likely to occur in other parts of north India as well where frost occurs during the winter season. Frost has therefore posed a major challenge in establishing plantations in these areas. Measures to protect plantations from this damage are required.

Keywords: Frost, resprout, challenge, bush, protection, abiotic stress

INTRODUCTION

Jatropha curcas is multipurpose non-edible oil-yielding perennial shrub originated in tropical America and West Asia. It is commonly known as physic nut or purging nut. *J. curcas* belongs to the family Euphorbiaceae and has the tendency to produce latex. It is not browsed by animals. This hardy and drought tolerant crop can be raised in marginal lands with little input. *J. curcas* is a succulent plant that sheds its leaves during the dry season. It is, therefore, best adapted to arid and semi-arid conditions (Srivastava, 1999). *Jatropha* can withstand only a very light frost that causes complete loss of its leaves, and sharp decline in seed yield (Benge, 2006). Very cold areas do not suit to *Jatropha* (Singh *et al.*, 2007). Quantitative data about resistance or susceptibility of this species to frost are not reported in the literature.

Frost is quite common in Dehradun; severe frost occurs in the month of January and February and causes severe damage to regeneration of sal (Kumar, 2004). Desiccation injury in winter and frost injury occurring in various seasons constitute the greatest limiting factors for growing forest trees individually or in combination. From an ecological point of view, these injuries are also known to be the most important factors affecting distribution of plants (Toda, 1970). Frost injury has been reported in many other species also. Seedlings of White ash (*Fraxinus americana*) in the United States and Canada were injured by both late spring and early fall frost (Clausen, 1982). Early winter frost caused vascular necrosis and wilt of the young olive trees. This unusual and severe damage was related with the lack of frost hardiness due to warm temperatures during the previous autumn (Hern´andez *et al.*, 1998).

The present study was carried out to study the effect of frost in plantations of different ages of *J. curcas*. An experiment was also carried out to study the effectiveness of the protection measures and post-frost treatment on *J. curcas* plantation. Results of both studies are presented in this paper.

MATERIALS AND METHOD

Plantations of *J. curcas* on which this study is based, were established at Premnagar (March 2004), Raipur (July 2005) and FRI campus (July 2006) by Forest Research Institute, Dehradun and at Rishikesh (July 2004) and Bhopalpani (July 2004) in Dehradun district by Forest Department Uttarakhand. In the plantation established by Forest

Research Institute at Premnagar, irrigation was given four times in the first year in the month of March, April, May and January. Weeding was carried out at the plantation site twice in first year of establishment in the month of August and February. Second year onwards irrigation was provided to the plants in month of May and April and weeding was carried out once in year in the month of August. In the plantation of Raipur and FRI campus, irrigation was given in month of November, January, March, April, and May. Weeding was given in the plantation in month of August and April. The plantations established by Forest Department were maintained by local forest staff, irrigation was given twice in a year in month of April and June and weeding was carried out in the month of August. Frost is common in Dehradun but it was observed in *J. curcas* plantation sites for the first time in year 2007. To study the impact of frost injury, collar diameter and height of the plants affected by frost were recorded. The mortality in plantation was determined by recording the mortality in the entire plantations. The collar diameter and height were measured through random sampling of 25 percent intensity.

To study the effect of post frost treatment and frost protection on *J. curcas* an experiment was carried out in plantation established at Prem Nagar, Dehradun (30 20' 15"N latitude, 77 57' 40" longitude and 600 m altitude). This experiment was conducted in three year old plantation. The plantation was established by seedlings in April 2004 and was affected by frost in year 2007. Four treatments were given to the plants that survived the frost injury of year 2007. In treatment T1 the branch damaged from frost of year 2007 were removed in April 2007 and frost protection was provided to the plant in month of December 2007. Frost protection was provided by covering the above ground part of plant with dry grass. In treatment T2 the branch damaged from frost was removed in April 2007 and no frost protection was provided to the plant. In treatment T3 the branch damaged from the frost were not removed and no frost protection was provided to the plant. In treatment T4 those plants were taken that have managed to escape from frost in previous year. In second year i.e. year 2008 frost was observed in month of January and February. Growth data was taken in the month of July 2008 to study the effect of frost on the growth and yield of the plant.

OBSERVATIONS

The authors have noticed frost injury in *J. curcas* in Dehradun district in Uttarakhand. It was first observed in March 2007. Absolute maximum temperature had risen to 25.7°C in the last week of February 2007 but absolute minimum temperature suddenly dropped to 4°C on 1st March 2007 thereafter. Though this condition of low temperature prevailed for few days only and within a few days the stem of *J. curcas* which had sprouted in favourable temperature by that time turned black, all the leaves dried and turned brownish. The branches had become pulpy in appearance which ultimately dried and turned black. The entire plantation showed signs of frost injury which progressed from top to base of the plant (Fig. 1, 2 and 3). This effect of frost, characterised by blackening of leaves, branches and stem, was observed for two successive years i.e. 2007 and 2008 (Graph 1 and 2). Even *Lantana camara* growing near *J. curcas* research plots of Forest Research Institute, Dehradun was affected (Fig. 5). It is note worthy that this type of frost injury was not observed in the area in 2005 and 2006 (Graph 3 and 4).

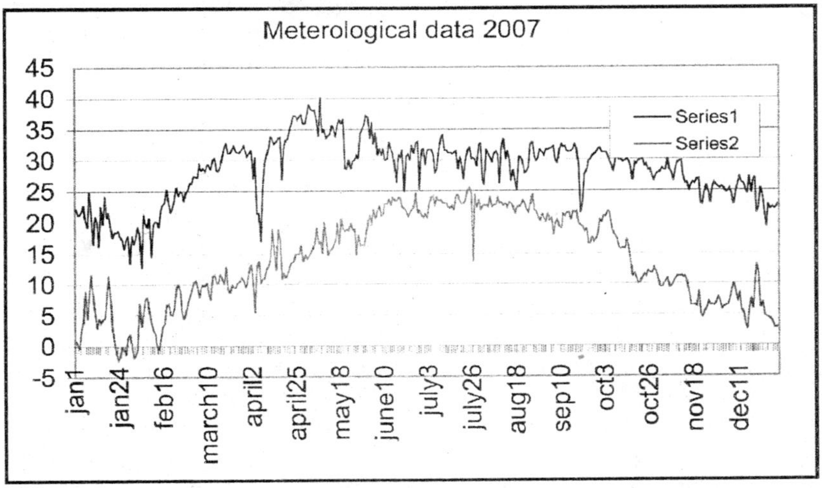

Graph 1. Meteorological data 2007

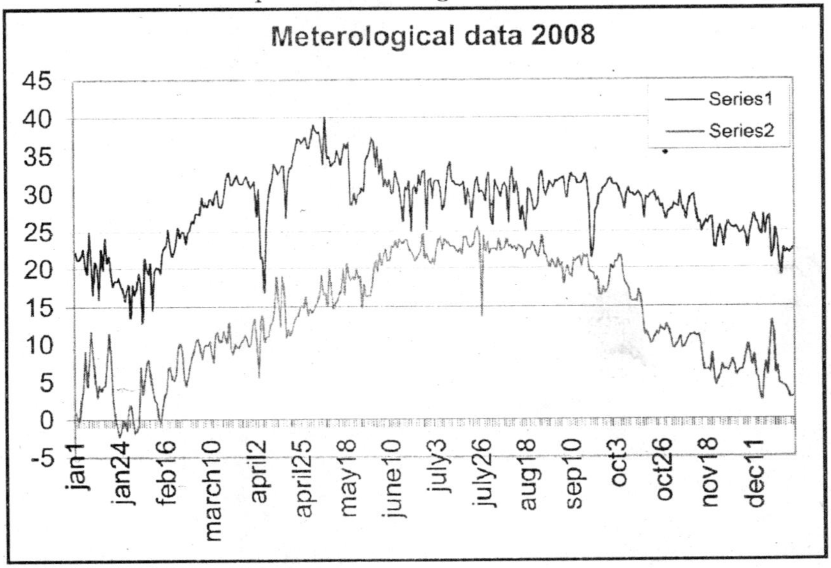

Graph 2. Meteorological data for the year 2008

Fig. 1. Young one year old plantation of *J. curcas* completely wiped out by frost

Fig. 2. *J. curcas* plantation in third year affected by frost

Fig. 3. Close-up of frost affected stem of *J. curcas*

Fig. 4. Frost-affected plant of *J. curcas* resprouted during spring

Fig. 5. Frost-affected Lantana near *J. curcas* plantation

Fig. 6. Bushy appearance of *J. curcas* plant after recovery from frost

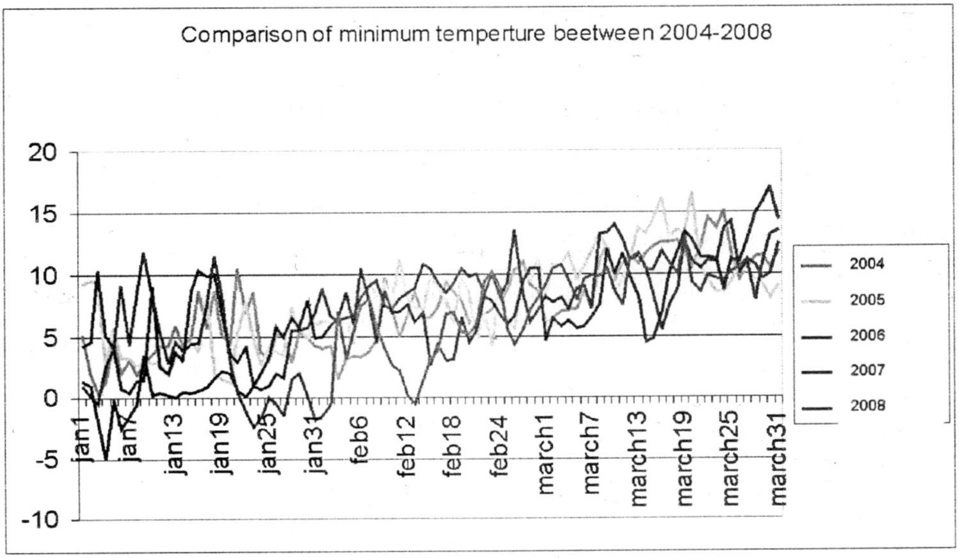

Graph 3. Comparison of Minimum temperatures of years 2004-2008

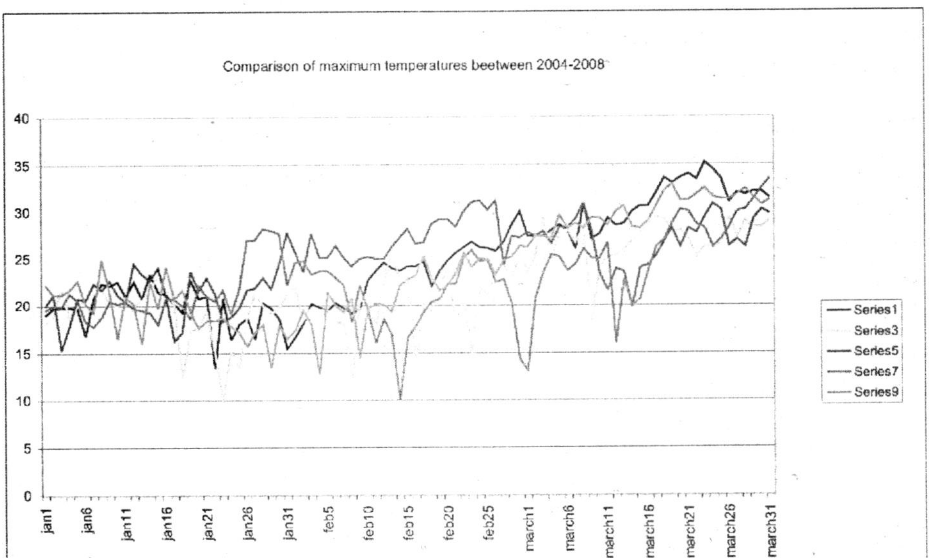

Graph 4. Comparison of maximum temperature of year 2004-2008

The effect of frost was more severe on plants planted in the recent monsoon. There was differential response of plants in the first year of growth in the field trial. Plants that had 40-55 cm height and less than 25 mm collar diameter at the planting time were completely wiped out (Fig. 1). While plants of the same age having 50-90 cm height and 25-35 mm collar diameter showed mortality of 35 per cent. Plants in the second year of field plantation having 150-175 cm height and 40-58 mm collar diameter showed signs of frost injury up to the height of 15 cm from ground level. However, all of them recovered from frost after resprouting in the ensuing spring season, but they assumed a bushy appearance. Plants in the third year of field planting with 230-295 cm height and 63-80mm collar diameter also met with the same fate and assumed bushy appearance (Fig. 6). The observation of study is summarised in Table 1.

Table 1. Effect of frost on plants of *J. curcas* in field plantation

Diameter class	Height class of plants*	Age of plantations at time of frost	Response of plantation to frost
Less than 25mm	40-55 cm	First year	Entire plantation wiped out
25-35mm	50-90 cm	First year	35% mortality
40-58 mm	150-175 cm	Second year	No mortality although shoot system was affected. Plants resprouted in spring season
63-80mm	230-295 cm	Third year	No mortality although shoot system was affected. Plants resprouted in spring season

*measured in August 2006

Table 2. Comparison of three year growth of the frost affected plantation.

Parameter		Year		
		2006	2007	2008
Diameter		50.00	64.54	79.26
Height		163.69	210.40	165.12
Branch number		2.85	7.52	10.90
Height of first branch from ground level		49.28	24.78	0
Crown	E-W		152.60	142.08
	N-S		149.08	139.53

If we compare the growth data of three years recorded for the plantation established in Premnagar, Dehradun in year 2005, it is observed that the average height of the plantation in year 2006 in month of January was recorded to be 163.69cm and it has increased to 210.40 cm in year 2007 (Table 2). However, in year 2007 the plantations faced frost injury and in year 2008 after 9 month of frost injury the average height of plantation was recorded to be 165.12 cm which is very near to the average of first year. Average diameter of the plantation for the year 2006 was recorded to be 50 mm in second year and the diameter increased to 64.54 mm in third year. However, the average diameter of the branches increased to 79.26mm. There was a significant increase in the average number of branches in year 2008, it was recorded to be 10.90, while in year 2006 it was only 2.85 while in year 2007 it was 7.52. Also after the frost injury the plant showed branching at the ground level, while in year 2007 it was recorded to be 24.78 cm from ground level. There was a decrease in the crown spread after the frost injury. The average crown spread for the year 2008 was recorded to be 142.08 x 139.53 cm (E-W X N-S) while in the year 2007 it was recorded to be 152.60 x 149.08 cm.

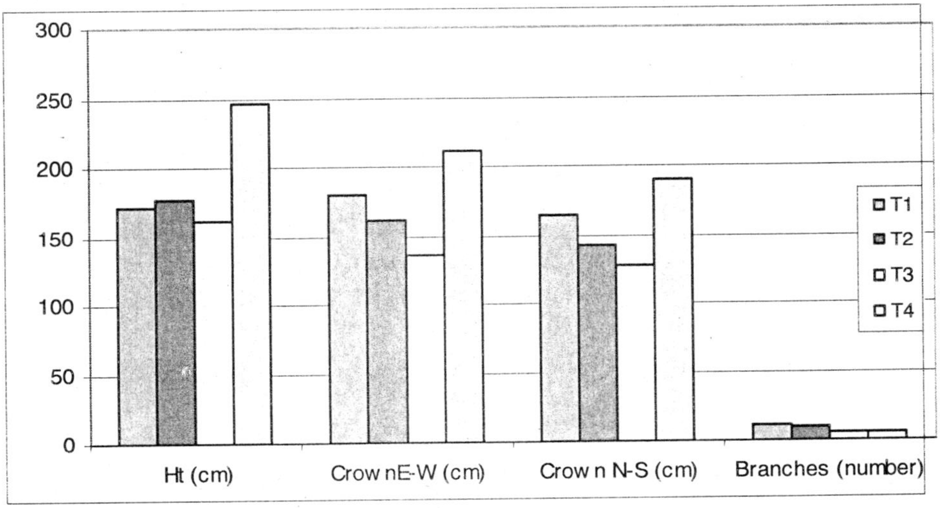

Graph 5: Vegetative growth data for the four treatments given to frost affected plants

Four treatments were given in the frost affected plants to study the effects of frost and post-frost treatments on the frost injured plant. Of the four treatments the highest height was recorded for the plants that escaped from the frost, the average height of T4 was 246.56 cm while the second best was for treatment T2 where the height was 172.48 cm followed by T1 with average of 172.48 cm, the lowest average height was found to be of treatment T3 with the value of 162.23 cm (Graph 5). However, T1 recorded highest average branch number of 11.4 branches per plant and second best was T2 having 9.43 branches per plant followed by T3 with 6.20 branches per plant. The lowest number of branches was recorded for T4 with 5.02 branches per plant. The authors would like to clarify here that only primary branch i.e. branches arising from the main axis of plant were counted in the study. The secondary and tertiary branches (branches arising from primary and secondary branches respectively) were not accounted. However, the number of secondary and tertiary branches has been included in the parameter of crown spread. The highest crown was recorded for T4 with average crown spread of 210.74 x 190.56 cm (E-W X N-S), the second best crown was recorded for treatment T1 with average crown spread of 180.20 X 164.50 cm followed by T2 161.92 X 142.63 cm and the lowest value for the crown spread was recorded for treatment T3 with the average value of 135.37 X 126.17 cm. The highest average seed yield was recorded in treatment T4 with the value of 589.11g, the second highest average yield was recorded for treatment T1 with value of 126.55 g, followed by T3 with the average yield of 100.24 g (Graph 6). The lowest yield was recorded for treatment T2 with average yield of 90.39 g. The study shows that the control plants that managed to escape from frost were best performer. Among the different treatments given to the frost affected plants the best overall performance was of treatment T1.

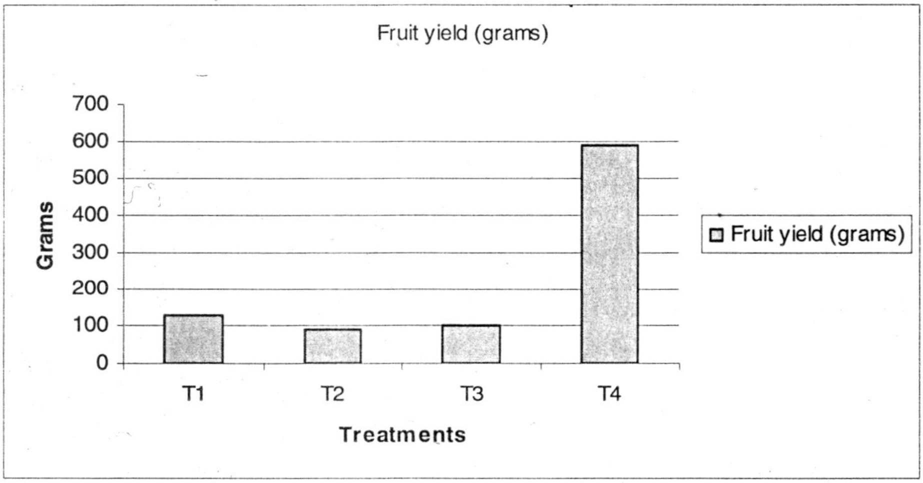

T1-Branch damaged from frost removed + frost protection.

T2-Branch damaged from frost removed + no frost protection.

T3-Branch damaged from frost not removed + no frost protection.

T4-Plants escaped from frost damage.

Graph 6: Seed yield data of the four treatments given to frost affected plants

RESULTS AND DISCUSSION

It is found that the plants with well developed root system recovered from frost injury and sprouted from ground level in spring season (Fig. 4). The plants in first year of the plantation having collar diameter of 35-45mm and height ranging between 50-90 cm showed only 35% of mortality. It can be concluded that the plants used for the purpose of plantations should have height more than 50 cm and collar diameter of 35mm; it is also recommended that the plants should be planted in rainy season so that they develop a good root system. The plants with well developed root system recovered from the frost injury as the frost affected only the above ground parts of the plant. However the plantation of 1 year age due to under developed root system could not sustain frost injury. It was found that the plantation established from seedlings showed more frost resistance than plantations established by cuttings. It is worth mentioning that the plants growing in shade or under canopy of tree remained unaffected by frost and showed no signs of frost injury.

The comparison of three year growth data of the frost affected plantation shows that there was a decrease in average height of the plantation though there was increase in number of branches and also the height of first branch from ground level decreased to zero. The plant assumed a bushy appearance. This may be one of the reasons for bushy nature of the tree in hill regions (frost pockets). It can be concluded that there is significant effect of frost on vegetative growth of *J. curcas*.

It is apparent from the experiment that there is significant effect of frost protection and post frost treatment on growth and yield of plants. The plants which escaped the frost injury in year 2007 showed best vegetative growth and highest seed yield. If we compare the treatments given to the frost affected plants, T1 was best in overall vegetative growth. Thus it can be concluded that in the plantations of region susceptible for frost injury, frost protection should be provided to the plant if the temperature goes below 5°C and if the plantation is affected from frost injury, the frost affected portion should be removed. However, the plants with well developed root system recover in favourable climatic conditions but there is a significant effect of frost on growth and yield of plantations.

ACKNOWLEDGEMENT

The authors wish to acknowledge Department of Biotechnology, Govt. of India for providing financial support for establishing the plantations mentioned in the paper.

REFERENCES

Benge, M. 2006. Assessment of the potential of *Jatropha curcas* (biodiesel tree) for energy production and other uses in developing countries. July 2006 and updated August 2006, ECHO's website.

Clausen, K.E. 1982. Variation in frost injury to white ash families in an Ontario plantation. *Can. J. For. Res.* 12(2): 440–443.

Hern´andez, M.E.S., D´avila, A. R., Algaba, A.P. de, L´opez, M.A.B. and Casas, A.T. 1998.Occurrence and etiology of death of young olive trees in southern Spain. *European Journal of Plant Pathology* 104: 347–357.

Kumar, P. 2004. Economics of soil erosion: issues and imperatives from India. Concept Publishing Company, New Delhi, India. p. 80.

Singh, R.A., Kumar, M. and Haider, E. 2007. Synergistic cropping of summer groundnut with *Jatropha curcas* –A new two-tier cropping system for Uttar Pradesh. *SAT e-journal,e journal.icrisat.org*.5(1).

Srivastava, R.L. 1999. Study in variation in morpho-physiological parameters with reference to oil yield and quality in *Jatropha curcas* Linn. Ph. D. Thesis, Forest Research Institute (Deemed University), Dehradun, India

Toda, R. (Ed.) 1970. Abstracts of Japanese literature in forest genetics and related fields. Vol.1, Part A. Noorin Syuppan Publishers, Tokyo, Japan. 362 p.

153-154

Biofuels: Challenges and sustainability issues

Chapter - 1

WHY DIGESTERS EXCLUSIVELY FED WITH CATTLE DUNG DID NOT SUSTAIN IN BIOGAS PRODUCTION?

P.C. Suryawanshi[1*], M.R. Patil[1], A.B. Chaudhari[2] and R.M. Kothari[3]

[1]Jain R & D Laboratory, Agri Park, Jain Hills, Jain Irrigation Systems Ltd., Jalgaon 425001
[2]Department of Microbiology, School of Life Sciences, North Maharashtra University, Jalgaon 425001
[3]Rajiv Gandhi Institute of IT and BT, Bharathi Vidyapeeth University, Dhanakawadi, Pune 411043 E-mail: biogaslab@jains.com

ABSTRACT

Higher organic loading rate (OLR) of substrate cattle dung (CD) prior to stabilization of digester in respect of pH and alkalinity: acidity ratio led to reduction in respect of pH due to excessive production of fatty acids thus, creating problems for survival of methanogens, less biogas production and poor methane content. Secondly, seasonal fluctuations in the composition of CD especially, C:N ratio leading to rapid overloading and its continuation for longer period rendered their failure. Taking these factors into consideration, smooth biogas production at 25 L scale has been demonstrated by gradually increasing OLR from 0.5-6.0 over 28 weeks and controlling C:N ratio, pH, alkalinity:acidity ratio, with consistent 24 ± 4 L biogas/ day with 66 ± 3% methane content.

Keywords: Biogas production, methane, organic loading rate, cattle dung

INTRODUCTION

Khadi and Village Industries Commission (KVIC), Mumbai did set-up millions of biogas units in rural areas of India, using locally available substrate cattle dung, with an objective to render villages self-reliant in energy for cooking and water heating and providing effluent slurry for improving soil fertility. Assessment of this mission spread over three decades indicated that few biogas units remained functional, while majority failed (Marchaim, 1992). Probing the reasons of failure from literature, reports and visits to biogas plants lead to certain clues for trouble in operating them and eventual failure. However, they required to be experimentally verified leading to identification of parameters for their smooth operations. Efforts gone into verification constitute the theme of the present article.

MATERIALS AND METHODS

Cattle dung substrate

It was procured locally on as is basis @ Rs. 500/- per ton. Its biochemical analysis is summarized in Table 1.

Table 1. Biochemical analysis of cattle dung (CD)

Analytical parameter	Ingredient %
Total solids (TS)	60±3.2
Volatiles solids (VS)	95.2±3.7
Total organic carbon (TOC)	59.8±1.4
Total carbohydrates	34.5±3.1
Total lignin	10.3±1.3
Total Kjeldahl nitrogen (TKN)	1.3±0.3
Total proteins	7.6±0.6
Total fat	NA
C: N ratio	45.4
Total phosphorus	0.6±0.04

NA = Not analyzed; analysis is an average of 10 estimates on dry weight basis.

Digester

A circular, fixed-dome, digester (height 32.0 cm, internal diameter 29.6 cm and 25 L capacity) mimicking the design of KVIC digesters, fabricated from PVC, with an inlet and outlet provision was fitted with an agitator (automated at 100 rpm for 3 min at 30 min frequency) to provide uniformity of pH and temperature in the substrate slurry and its close contact with a consortium of microbes from the inoculum involved in anaerobic digestion (AD).

Inoculum

It comprised of an effluent from the ongoing biogas plant fed with fruit waste, where cattle dung was used as an inoculum. It was 10% (v/v) of the substrate slurry volume in the digester.

Experimental set-up

Anaerobic digestion was carried out at 35±0.5°C and pH 6.8-8.0. The digester was fed initially with cattle dung at the organic loading rate (OLR=kg of VS/m^3/day) of 0.5. This rate was continued until 30 days, after which it was increased by 0.5 at an interval of 30 days and thereafter by 1.0 at 30 days interval.

Analytical methods

Analysis for cattle dung and its digester slurry was carried out for TS, VS, TOC, TKN, lignin and phosphorus as per Clesceri *et al.* (2000). Volatile fatty acids (VFAs) and methane analysis was done by gas chromatography (Perkin Elmer, *Clarus 500*), equipped with flame ionization detector (FID) and thermal conductivity detector (TCD). VFAs were analyzed using FID, capillary column (Elite WAX ETR, 30 m x 0.32 mm x 1.0 µm) and helium as a carrier gas. The detector, injector and oven temperature were programmed at 260°C, 240°C and 80-230°C, respectively. Methane was analyzed using stainless-steel packed column (HAYSEP-Q, 80/100 mesh, 9' x ⅛" OD). The temperature of detector and injector were 100°C and the oven 40°C.

For collection of biogas, water displacement method was used (Kirtane *et al.*, 2009). The measurement of biogas was carried out after every 24 h using a calibrated transparent fiber glass container of 20 L capacity.

RESULTS AND DISCUSSION

Functioning of KVIC digesters

In KVIC digesters, acid forming (acidogens) and methane forming (methanogens) archaea co-exist in the same environment, being a single phase anaerobic digester (AD). The substrate is first acted upon by hydrolytic and acidogenic bacteria present in the top portion of the digester, resulting in the production of VFAs, which in turn acted upon by methane formers at the bottom, resulting in the production of biogas.

Merits and demerits of cattle dung as a substrate

As summarized in Table 1, cattle dung (CD) has (i) 95.2% VS rendering it an attractive substrate for biogas production, (ii) 34.5% TC, 7.6% proteins and 0.6% phosphorus for promoting microbial growth, (iii) 45.4 C:N ratio for expediting AD to speedy biogas production and (iv) 10.3% lignin, which is inhibitor to microbial growth and hence a rate limiting factor in biogas production.

Functioning profile of the digester

In the light of literature survey, reports and feedback from the farmers, who did not succeed in operating biogas plants, functioning of the digester was consciously organized with a gradual increase in OLR, allowing it to stabilize for pH in the optimal range, monitor every 24 hr for volume of biogas and its methane content so that either it gave an indication of smooth working or provided data where cause and effect relationship for biogas production could be ascertained.

On the above premise, during the operation of the digester, while initial OLR was 0.5, it was subsequently increased to 1.0 and subsequently @ 1.0 per month, until it reached to 4.0 in 4 months. At 24 hr interval, % TS, % VS, OLR, pH, alkalinity: acidity ratio, C: N ratio, biogas production and its methane content were monitored, data were summarized on weekly basis and presented in Table 2.

Table 2. Profiles of OLR, pH, alkalinity:acidity ratio, C:N ratio, biogas and methane as a function of digestion

Week	OLR (kg of VS/m^3/day)	pH	Alk:acid ratio	C: N	Biogas (L)	CH_4 %
1	0.5	7.5	3.8	14.4	10±1.3	55±2.4
		7.3	4.0	14.4	11±2.1	58±2.8
		7.5	4.3	20.0	17±4.0	61±3.9
		7.0	3.6	18.3	16±3.3	62±2.0
5	1.0	7.1	5.3	18.8	17±3.2	65±3.3
		7.4	5.3	13.9	18±1.2	62±3.7
		7.2	5.1	14.3	19±2.5	62±1.9
		7.5	5.6	15.0	19±2.9	64±2.0
9	2.0	7.4	8.7	16.1	21±2.9	64±2.1
		7.7	8.6	14.2	21±4.4	66±2.0
		7.6	5.0	13.9	19±1.6	63±2.0
		7.6	6.2	17.3	19±1.5	66±1.4
13	3.0	7.4	5.0	17.9	21±3.6	68±1.6
		7.5	4.9	16.6	25±2.2	65±1.4
		7.7	5.9	14.6	23±3.8	73±2.9
		7.4	6.0	16.7	20±1.0	68±2.5
17	4.0	7.6	5.7	16.9	23±2.4	67±3.8
		7.2	6.8	15.2	24±3.5	69±4.0
		7.2	5.7	15.4	28±4.4	70±3.5

		7.1	6.4	16.3	29±4.1	70±2.0
22	5.0	7.3	8.2	15.4	27±2.6	69±2.0
		7.3	6.0	13.6	25±3.7	70±1.4
		7.4	6.9	15.2	28±3.8	70±2.4
		7.7	6.0	14.8	23±3.5	70±2.0
26	6.0	7.5	6.4	16.1	28±5.2	67±1.7
		7.6	6.3	17.2	28±2.7	68±2.2
		7.4	6.6	17.4	27±3.2	70±3.1
		7.7	6.0	16.8	28±4.2	70±2.1

Table 3. Physico-chemical characteristics of digestate of CD

Parameter	CD
OLR (KgVS/m³/day)	0.5-6.0
pH	7.4
Acidity:alkalinity ratio	5.8
C:N ratio	15.7
Total solids (%)	5.4
Total in put volatile solids (%)	75.3
Total out put volatile solids (%)	56.2
Total VS reduction %	19.2
Biogas production (L)	24±4
Methane content (%)	66±3

Analysis of parameters for normal functioning of the digester

Effect of OLR and C: N ratio: It is well known that higher C: N ratio slows down the rate of digestion (Ramamurthy *et al.*, 1998). Therefore, it has a direct bearing on increase in OLR. In case of CD, increase in OLR also depends upon the rate at which 23.5 % lignin is degraded. Higher content and complexity of lignin renders it difficult to degrade. Moreover, content and complexity of lignin in CD also depend upon the quality of feed fed to cattle. The green grass available during monsoon has less lignin and relatively simple structure, in turn providing less and simple lignin metabolites in CD for anaerobic digestion. Similarly, de-oiled cake of cotton seed and byproducts from pulse processing industry, which constitute feed in winter are equally poor in lignin content and hence their digestion is also easy. On the contrary, dry fodder available during summer, being rich in lignin and complexity in structure is difficult to digest. Accordingly, the composition of CD for lignin content and complexity varies from season to season, depending upon the feed given to cattle (Ramamurthy *et al.*, 1998). This change in substrate composition has been totally overlooked in determining the OLR, which upon overloading caused fluctuation in pH and souring of the digester. Its recurrence over prolonged period caused failure of digesters. Lack of operating knowledge of the farmers, little training and casual follow up by KVIC made the mission failure.

Effect of pH: The pH of the digester medium has a direct influence on the functioning of acid formers and methane formers. Primarily, while acid formers are relatively robust to reduction in pH, methane formers in contrast are extremely sensitive. Secondly, the rate of multiplication of acid formers is 30-120 min, while the rate of multiplication of methane formers is 12-48 hr (Gerardi, 2003). Thirdly, while acid formers are optimally active in pH 6.0-7.0 range, methane formers are optimally active in 7.0-8.0 range (Suryawanshi *et al.,* 2009). Therefore, any fluctuation in pH is either inhibitory to acid formers or detrimental to methane formers. Unfortunately, separation of phases in a single phase digester of KVIC is not possible. As a result pH has to be maintained in the compromised range so as to harmonize activities of both types of microbes for sustainable production of biogas.

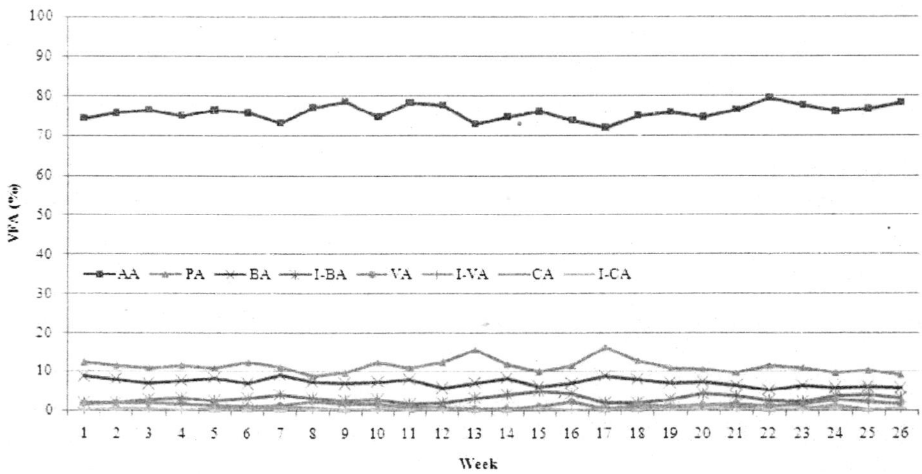

AA: acetic acid; PA: propionic acid; BA: butyric acid; I-BA: iso-butyric acid; VA: valeric acid; I-VA: iso-valeric acid; CA: caproic acid and I-CA: iso-caproic acid.

Fig. 1. Profile of VFAs in CD3

Effect of alkalinity: acidity ratio: It determines the buffering capacity of the digester medium, which in turn determines the stability of pH. The change in OLR depends upon the rate of conversion of VS in to VFAs and the level of VFAs in turn depends upon the rapidity at which they are converted in to biogas. Thus, in this chain of reactions, OLR, VFA level and biogas production are inter-connected and any unilateral increase in OLR beyond the capacity of methane formers increases acidity, predominantly due to excessive accumulation of propionic acid at the cost of acetic acid (Boone and Xun, 1987; Fukuzaki *et al.*, 1990) which is detrimental to the functioning of biogas digester. Since pH in failed digesters was not maintained, possibility exists for failure in digester due to imbalanced pH profile. These requirements have been met in running the present digesters (Fig. 1) and therefore, their operation was smooth as judged from the continuously increasing generation of acetic acid vis-à-vis other fatty acids, in turn production of biogas and enrichment in its methane content.

Trained personnel: Majority of the digesters in India failed, while majority in China succeeded (Marchaim, 1992). Probing the causes of our failure and success of Chinese revealed that their program was manned by trained personnel, who had operating knowledge of biogas production, while our program was manned by a casual training of 1-2 days by KVIC, with the effect our operators had no operating knowledge, leave alone trouble shooting.

Increase in OLR: Our farmers did not have enough knowledge on stabilization of digesters by controlling OLR. Initially, the biogas started flowing with certain OLR of CD as the substrate. Non-familiarity with the process coupled with the greed to have more biogas, more CD was added, thus increasing the OLR. The digesters could digest additional OLR up to a limit, by virtue of the substrate to act as an inoculum. Further increase in the OLR added extra burden on methane formers, which affected their transformation into methane, in turn creating accumulation of VFAs, reducing the pH further, inhibiting methane formers still more, with the result its continuation first led to souring of the digester and keeping it unattended ultimately led to its failure. These are the events faced by the farmers in the given chronological order. Therefore, the OLR increase has to be gradual, cautious and increased only after stabilization of pH, biogas output and methane content.

Effect of seasonal changes: Some consideration will required to be taken into account for content and complexity of lignin in CD from season to season. Broadly, OLR could be rapid in monsoon, a bit slower in winter and yet slow in summer.

Improving buffering capacity of the digester: Another way of enhancing the buffering capacity of the digester is to add de-oiled cake of cotton seeds in guarded lots, which adjusts C:N ratio in favor of digestion and imparts stability to pH fluctuations, pre-empts effect of rapid increase in OLR, thereby reduces the chances of upset digesters, taking it forward for smooth production of biogas.

Activation of enzymatic processes: Addition of calculated amount of ferric ore provides activators to several enzymes operating during digestion.

CONCLUSION

Implementation of the factors described in preceding section will render smooth functioning of digester for higher biogas production and its methane content.

ACKNOWLEDGEMENT

The authors are grateful to Mr. B.H. Jain, Founder Chairman, JISL for providing state-of-the-art R & D facilities to undertake the present work.

REFERENCES

Boone, D.R. and Xun, L. 1987. Effects of pH, temperature and nutrients on propionate degradation by a methanogenic enrichment culture. *Appl. Environ. Microbiol*. 53:1589-1592.

Clesceri, L.S., Greeberg, A.E. and Eaton A.D. 2000. Standard methods for the examination of water and wastewater. American Public Health Association (APHA), 20th Edn., Washington, DC, USA.

Fukuzaki, S., Nishio, N., Shobayashi, M. and Nagai, S. 1990. Inhibition of the fermentation of propionate to methane by hydrogen, acetate and propionate. *Appl. Environ. Microbiol*. 56:719-723.

Gerardi, M.H. 2003. The microbiology of anaerobic digesters-Wastewater microbiology series. Wiley-Interscience, New York, USA.

Kirtane, R.D., Suryawanshi, P.C., Patil, M.R., Chaudhari, A.B. and Kothari, R.M. 2009. Optimization of organic loading rate for different fruit wastes during biomethanization. *J. Sci. Ind. Res*. 68:252-255.

Marchaim, U. 1992. Biogas processes for sustainable development. FAO report, pp. 1-267.

Ramamurthy, V., Sharma, R.K. and Kothari, R.M. 1998. Microbial conversion of ligno-cellulosic wastes into soil conditioners. In: Advances in Biotechnol (Ed. A. Pandey), Educational Publ., New Delhi, pp. 433-438.

Suryawanshi, P.C., Kirtane, R.D., Chaudhari, A.B. and Kothari, R.M. 2009. Conservation and recycling of pomegranate seeds and shells for value addition. *J. Renew. Sustainable Energy* (In Press).

Chapter - 2

HANDLING AND PROCESSING OF JATROPHA SEEDS FOR LONG-TERM CONSERVATION

J. Radhamani and Garima Srivastava

Division of Germplasm Conservation, National Bureau of Plant Genetic Resources, Pusa Campus, New Delhi-110 012
E-mail - radhamani@nbpgr.ernet.in

ABSTRACT

Jatropha is an important oilseed crop with greater promise for production of industrial and biofuel. The potential can be further increased with intensification of cultivation, genetic improvement and domestication of natural genetic resources. Variability that would help in selection and distribution of germplasm/selected species/clones to all stake holders (co-operators) to promote cultivation and sustainable use. The National Gene Bank holds 1813 number of accessions in Gene Bank which have been collected and assembled from different climatic zones of India. There is lot of seed variability in morphological characteristics of seeds assembled from different zones. The oil content varied from 26% to 40% on fresh weight basis. The seed viability also varied significantly in different accessions. The physiologically matured seeds were harvested at about 65-70 days. The fresh seed showing moisture content of 16-22% on fresh weight basis can be desiccated to 5-6% moisture content. The seeds are orthodox, so can be dried and packed in three layered aluminium foil pouches for long- term conservation.

Keywords: *Jatropha curcas,* long term storage, National Gene Bank, germplasm, accession

INTRODUCTION

The genus Jatropha belongs to tribe Joannesieae of Crotonoideae in the Euphorbiaceae family and contains approximately 170 known species. Dehgan and Webster (1979) revised the subdivision made by Pax (1910) and now distinguish two subgenera (*Curcas* and *Jatropha*) of the genus Jatropha, with 10 sections and 10 subsections to accommodate the Old and New World species. Although most of the Jatropha species are native to the New World, approximately 66 species are native to the Old World. Dehgan and Webster (1979) offered a key to the infrageneric taxa but this should not be considered as final since information is still lacking on many species. No complete revision of the Old World Jatropha exists. Hemming and Radcliffe-Smith (1987) revised 25 Somalian species, all of the subgenus Jatropha, and placed them in six sections and five subsections. *Jatropha multifida* L. and *J. podagrica* Hook. of section *Peltatae*, *J. integerrima* of section *Polymorphae*, and *J. gossypiifolia* of section *Jatropha* are well known and cultivated throughout the tropics as ornamental plants.

MATERIALS AND METHOD

Jatropha seeds are orthodox in nature. To minimize the alteration in genetic structure and loss of viability in germplasm accession during storage we follow genebank standards as recommended by FAO/IPGRI (1994) in relation to various factors. Seed viability is assessed by germination by 'between paper method' in paper towel. The first count was taken on the 7th day of plating while the final count is recorded on 15th day of plating. Seed moisture content (mc) was estimated by low constant temperature oven method at 103 $\pm$ 2°C for 17 $\pm$ 1 hour. The seed samples are accepted only if initial viability is $\geq$ 85 percent. The moisture content of the seed was determined by gravimetric method i.e., low constant temperature oven method where seed is ground coarsely and used for moisture estimation (ISTA, 1985). Jatropha is a cross pollinated crop with large size seeds and hence 500-1000 seeds were cleaned and dried to 5% moisture content at 15°C and 15% RH for approximately 2-3 weeks, monitored for their viability and stored in trilayered aluminium foil pouches with label on it including its national identity and minimum passport data. These samples were arranged according to the accession number and are stored at -18°C in long term storage. The passport data along with the national identity is documented. All the packed seed packets are labelled arranged in the baskets and were stored in modules at -18°C with its location entered in the data base for its easy retrieval.

RESULTS AND DISCUSSION

Seed storage problems are more common in India, because a large part of the country has a predominantly hot and humid, tropical and subtropical climate with great variation in temperature and relative humidity across the year. Therefore, a complimentary approach would be appropriate to ensure the dynamism and safety, and economical and effective conservation. The base collections are being stored in modules maintained at -18°C. Such a low temperature minimises metabolic activities and is expected to enable the seed to retain the viability for 50-100 years without any change in genetic structure. Seed is processed after validating physical and genetic purity, assessment of viability and moisture content. In most crops seed samples with more than 85% seed viability are conserved. However, recognizing inherent problems which limit the harvest of physiologically mature seeds of the same age in certain crops the initial viability standards have been lowered down to 65-70. For long term storage the seed moisture content is brought down to 5-7% (ISTA, 1985).

Germplasm conservation

The National Gene Bank (NGB) is responsible for the conservation of seeds on a long term basis, as base collection for posterity. In addition it provides technical support to the network in planning, development and operation of medium term gene bank facilities, in human resource development and in provision of accessions for the regeneration of active collections. It is supported by a network of 11 regional stations in different agro-ecological zones of the country and 57 crop based, National Active Germplasm Sites.

Identification of physiologically mature seed and collection of material

In *J. curcas* the flowering takes place in the month of July-August and the fruit matures in the month of October-November. The fruits are generally collected in the month of October-November. The excised seeds without any mechanical damage were used for the experiment. Seed collectors should aim to ensure that seeds are collected/ harvested at peak quality so that their longevity in the seed bank is optimal. Ideally the seeds should be collected when they are on the point of natural dispersal. Since the flowering in Jatropha is asynchronous all the fruits do not mature at the same time. So the harvesting should be done very carefully.

Under the experiment the fruits were immediately tagged after anthesis and were collected at different stages of development. The colour of the capsule remained green up to 40 days of development and turned yellow within 50-60 days and ultimately black at about 67-70 days of development. Whereas the, seed colour changed from creamish to brown and ultimately deep brown/black colour at maturity. The size in terms of length and diameter increased with maturity. Significant increase occurred upto 50th day. Thereafter the diameter of the capsule remained same.

Extraction of seeds

Germplasm collection in Jatropha species is undertaken by selecting the fully ripened, physiologically mature fruits which can be assessed by some of the field indicators such as picking up the fruits when they just begin to open, brownish black in colour, fully ripened (about 70 days after anthesis), seed moisture content 16-18 percent on fresh weight basis, as shown in Fig.1. The fruits should be picked up by shaking the branches or with basket attached to a stick directly from the tree. Collection of fallen fruits from the tree should be avoided to avoid fungal infection.

Fig. 1. Freshly harvested fruits of *Jatropha curcas* at different maturity levels

Processing of seed for long-term conservation

Ex-situ conservation of orthodox seeds has been successful in the genebanks where collections of germplasm/seed accessions will retain very high viability for many decades when dried to low moisture contents (3-7%) and stored hermetically sealed moisture impervious containers at sub-zero temperatures (-20°C). Following steps were involved in processing the seed material for genebank which have been depicted in the flow chart (Fig. 2).

Germplasm augmentation

The Jatropha constitutes about four percent of total oilseed germplasm collections of NGB. Germplasm diversity was assembled of native as well as exotic through specific collection missions. Under augmentation of germplasm the seed samples must be well cleaned to reduce the plant bulk, without incurring any damage to the collection. This also helps in reducing the risk of disease carry-over and contaminants of the collection, especially seeds of other species or inert matter.

Receipt of the seed material

On receipt of the seed material the list is compared with the data base based on its collector number/ National Identity to identify duplicates. If found duplicates rejected otherwise are processed further for storage.

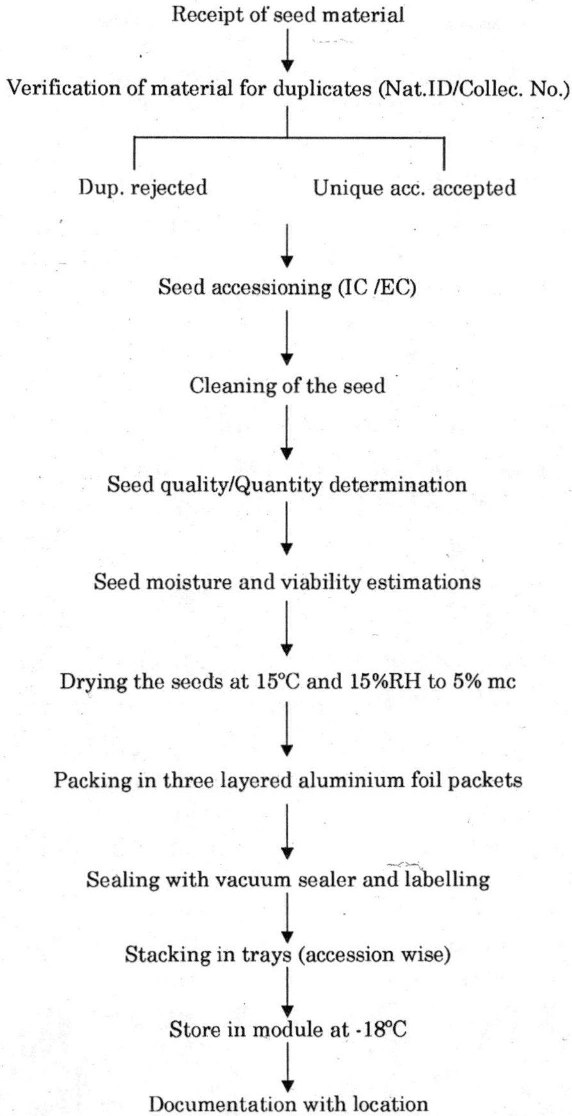

Fig. 2. Flow chart for the processing of seed material in NGB

Accessioning

Germplasm is augmented through two main processes, introduction of exotic species and exploration and collection of indigenous species in various crops. Seed materials on receipt are assigned the accession numbers by NBPGR. These numbers are unique and are maintained all through. Introduced Germplasm are assigned Exotic Code (EC) numbers while Indigenous Germplasm collected from within the country are assigned Indigenous Code (IC) numbers.

Physical inspection

Each accession is inspected, and if needed cleaned. Only insect, weed, disease free, clean and pure seeds are processed for conservation.

Seed quantity determination

For most collections, a 4 or a 7 decimal place balance (dependent upon seed size) is used to weigh five samples each of 50 seeds or fruits. The remainder of the collection is then weighed.

Seed viability assessment

(i) Seed viability (Germination method)

Fresh seeds of *Jatropha* showed 19.8% moisture content with a range of 0-96% viability. The seed viability also varied in different accessions. Three different types of substrata were used to assess the seed viability by germination method.

- Over sand (Fig. 3)
- In between the folds of germination towel (Fig. 4)
- In petri plates over the moistened filter paper discs

The standard germination test was conducted by placing the seeds in replicates of 50 seeds each in all the tests and the seed samples were kept at 27 $\pm$ 2°C with 80-90 percent relative humidity in the seed germinator. In the fully mature seeds, the radical started emerging on the third day of plating. The seeds were considered germinated only when both radical and plumule emerged completely. The percentage germination was calculated based on the count of normal seedlings. Seedling with stunted growth or very slow growth was considered abnormal.

Fig. 3. Seed germination of *Jatropha curcas* over sand in poly bags

Fig. 4. Seed germination of *Jatropha curcas* in between the folds of Germination towels

Germination percentages were recorded as per ISTA guidelines. The normal and the abnormal seedlings were also counted separately for calculating the viability percentage. The seedlings were evaluated on the 15th day of plating and the speed of germination was calculated by counting the number of fresh emergence on each day till the final count. The seed germination of the fresh seed was 0-96%.

Amongst the treatments the seeds put for germination over the sand as well as the once plated in between the folds of germination paper responded well, while the seeds sown in the petri plates did not show good growth. Plating the seeds in between the folds of germination paper is much simpler and economical method and has been used for all the experiments.

Viability and moisture testing

Viability is tested by triphenyl tetrazolium chloride (TTC) method which showed a good staining pattern in the viable seeds.

Drying, packaging and cold storage

After cleaning and. quantity determination is completed, the accessions are moved to a further drying room which is maintained at 15% relative humidity (RH) and 15°C, an internationally accepted regime for drying collections. Seeds in muslin cloth bags are placed loosely in slatted plastic crates to facilitate drying. Seed-lots remain within the room for one to 3 weeks by which time they should have reached equilibrium moisture with the air. Before being packaged ready for storage, their moisture status is checked. Jatropha seeds are packed once they attain a moisture content of 5 percent. The passport data along with the National identity is documented. The minimum data is printed on the labels which are pasted on the respective packed seed packets. The processed accessions are arranged in plastic baskets and placed in cold storage vault at –18°C at a defined location in the genebank.

Seed conservation

The seeds are best stored at subzero temperature with 3-7 percent moisture content (IBPGR, 1985) depending on species as the acceptable standards. Conservation of genetic resources is a high priority in the national context, as a global plan of action activity or a convention on biological diversity requirement. The National Plant Genetic Resources programme in India has already developed elaborate guide lines for sending germplasm for long term conservation in its Gene Bank, which maintains international standards and also ensures long term viability of the material conserved after due processing. The following points are check listed before sending the material for *ex-situ* conservation

i. Well developed and physiologically matured seeds.

ii. Free from insects, weeds and disease

iii. Clean and free from under sized, shrivelled, immature and discolour seeds.

iv. Properly labelled and packed to avoid damage during transit

v. Untreated with chemicals.

vi. Send to gene bank at the earliest possible (soon after harvest).

vii. Accompanied with the minimum passport data.

Documentation

Data documentation is the most important and dynamic activity. Correct and reliable recording of data, its documentation and information transfer is as important as proper handling of germplasm itself. Proper documentation of plant genetic resources allows the best use of data, it's easy retrieval and usefulness. This way the documentation of plant genetic resources is very useful in disseminating information on germplasm holdings more effectively and for a specific purpose enabling one to a search information having desired traits so that the selected accessions could be used for breeding programmes for developing new traits. It also acts as a source of information to assist in planning and operation of any gene bank activity. For documentation of gene bank holdings in NGB, two types of information files are used: passport data descriptors and gene bank management descriptors.

The passport data descriptors include Name of crop, Taxonomic code, Cultivar, National_Id, collector_no, other_Id, location in the Gene bank and Gene Bank management descriptors include seed quantity, seed viability, seed moisture content etc. Being the most important component of National Gene Bank, it is accomplished efficiently through a well managed computer network

Management of Jatropha seeds in NGB

Presently LTS at National Gene Bank holds 1813 accessions of Jatropha mainly comprising of 3 exotic and 1800 of indigenous collections. Maximum collections were assembled/collected from Southern India (40.4%) followed by Northern India (34.9%) (Fig. 5).

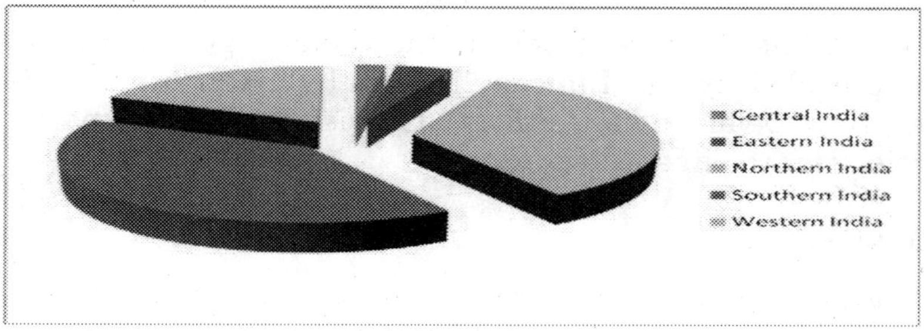

Fig. 5. Regionwise collection status

Germplasm registration and utilization

India has a very strong crop improvement programme undertaken by the Indian Council of Agricultural Research. During the process of germplasm manipulations to meet the breeding requirements, research and experimentation to develop superior genotype for specific or multiple traits, many materials are developed, which give good superiority over the existing varieties. In addition, there are raw germplasm, identified or developed by scientists (landraces/traditional varieties), with one or more outstanding traits, such as resistance to disease(s), high nutritional quality, bold grain etc. or with traits of academic or scientific importance (mutants), but do not qualify to get notified because of poor agronomic performance. Four germplasm accessions of Jaropha have been registered till date (Table 1).

Table 1. Status of registered germplasm in NGB

S. No.	Botanical name	National identity	Donor/other identity	Trait
1	*Jatropha curcas*	IC541650	BAAS-51	High oil content (40.6%)
2	*Jatropha curcas*	IC537939	SNES-45	High oil content (42%)
3	*Jatropha curcas*	IC427819	Soumya	Pink flowers with continuous flowering
4	*Jatropha curcas*	IC427820	Swetha	White flowers with continuous flowering

FUTURE PERSPECTIVE

There is an urgent need to assess the actual genetic diversity in the germplasm to be used in breeding programmes. It is imperative that a core collection is developed to facilitate the plant breeders in making an appropriate choice to ensure rapid genetic improvement abroad genetic base.

ACKNOWLEDGMENTS

Authors acknowledge with thanks the valuable guidance and encouragement provided by the Director, National Bureau of Plant Genetic Resources, New Delhi, India. Authors also wish to thank Department of Biotechnology for the financial support.

REFERENCES

Dehgan, B. and Webster, G.L. 1979. Morphology and infrageneric relationships of the genus *Jatropha* (Euphorbiaceae). University of California Publications in Botany, Vol. 74.

FAO/IPGRI. 1994. Genebank Standards. FAO/IPGRI, Rome, Italy. pp 13.

Hemming, C.F. and Radcliffe-Smith, A. 1987. A revision of Somali species of Euphorbiaceae. *Kew Bull.* 42:103-122.

International Board for Plant Genetic Resources. 1985. Cost effective long-term seed stores, IBPGR, Rome. 38 pp.

ISTA. 1985. International Seed Testing Association. International rules for seed testing. Rules 1985. *Seed Science and Technology* 13:299-513.

Pax, F. 1910. Euphorbiaceae-Jatropheae. In: Das Pflanzenreich IV. (A. Engler, ed.). Verlag von Wilhelm Engelmann, *Leipzig* 147(42):1-148.

Chapter - 3

JATROPHA CURCAS (L.) - SEED PRODUCTION AND SEED HANDLING

R. Anandalakshmi, B. Gurudev Singh, Rekha R. Warrier and V. Sivakumar

Institute of Forest Genetics and Tree Breeding, P.B. No. 1061, R.S. Puram, Coimbatore- 641 002

ABSTRACT

Oil crisis one of the prime issues to be addressed by our nation, *Jatropha curcas* highlights as a promising species of concern. Seeds of *J. curcas* yield oil ranging from 30 to 35% that is utilized for production of biofuel or directly employed as straight vegetable oil for energy. Unfortunately, the large scale planting programmes could not give substantial outputs due to the problem of low seed production in this species. At the same time, when targeting for seeds as end product, appropriate post harvest technology is felt 'the need of the hour'. Hence research on flowering and fruiting phenology was conducted to study the reproductive efficiency, appropriate maturity stage for seed collection along with suitable post harvest techniques for improving oil yield and quality. Reproductive efficiency studies were conducted over a period of two peak flowering seasons in a plantation at Anaikatti (Tamilnadu). The study revealed each inflorescence having an average of 80 flowers which constituted female flowers of only 3%. Of this 3% female flower, 68 per cent of the flowers succeeded in producing fruits suggesting high reproductive efficiency. Since the ratio of female flowers to male flowers in an inflorescence is low, there is a need to device methods to increase female flower ratio. The post harvest study revealed that Jatropha fruits should be harvested at its yellow or black pulpy stage instead of dry stage to obtain oil with superior physicochemical properties. The Lowest Safe Moisture Content (LSMC) to which seeds can be dried to enable maximum oil recovery with good quality was found to be 6% on fresh weight basis. Oil from seeds dried in an oven (40°C) and also seeds dried in shade for 10 days under ambient conditions were found to be superior. Efficacy of seed coat removal showed that oil extracted from kernel was significantly high both in terms of percentage and volume than from whole seed suggesting need to separate the shell or seed coat from seed so as to restrain loss of oil through adsorption by seed coat and thereby maximize oil recovery.

Keywords: Jatropha, seed, biofuel, post harvest, oil, processing

INTRODUCTION

For developing countries like India, rising world prices of crude oil and petroleum gives maximum cause for concern (Banerjee and Srivastava, 2006). At present, India imports about 73% of total consumption (124 million tonnes (MT)) of crude and petroleum products. As per Planning Commission estimates, the requirement of petrol is expected to grow to 12.5 million tonnes and the demand for diesel is likely to touch the level of about 66 million tonnes in 2011-2012. With the growing rate of energy consumption and that India has only about 700 million tonnes of economic reserves of crude oil left, need to reduce dependence on imports and search for better alternatives has become indispensable (Kureel, 2006).

Launching of a national biofuel policy by the Indian Government and adoption of specifications for biodiesel for use in India (IS 15607) by the Bureau of Indian Standards (BIS) (Tuli, 2006) highlights the importance of quality seeds, seedlings and oil of Jatropha. Under the Micro-mission on production and demonstration of quality planting material of Jatropha supported by the Department of Biotechnology (DBT), Government of India (GoI), work on agronomic practices, selection of elite trees, clonal propagation, tissue culture, etc. is being undertaken. National Oilseeds and Vegetable Oils Development (NOVOD) Board is exploring and augmenting the potential of Tree Borne Oilseeds (TBO's) in the country by undertaking activities like nursery raising, plantation, installation of preprocessing and post-processing facilities, capacity building, need based research etc. Indian Railways, the single largest consumer of high speed diesel (HSD) and a few state road transport undertakings in the country are also making efforts to substitute mix biodiesel with HSD (Singh, 2006).

Jatropha curcas (L.), known as physic nut, a member of the Euphorbiaceae family is believed to have its origin at Mexico and Central America. Among many tree borne oils as a source of biodiesel, *J. curcas* has been found fairly suitable owing to its short gestation period and hardy nature. Jatropha grows under a wide range of rainfall regimes from 200-1500mm. Ideal would be 600mm in moderate climates and 1200 mm in hot climates (Biswas *et al.*, 2006). The seed yield of Jatropha has been estimated to range from 0.75-2 kg/year per plant after three to four years of plantations (Biswas *et al.*, 2006). The viscous oil from *curcas* seeds can be transesterified to produce biodiesel (Kadlaskar and Dewan, 2004) while the filtered oil can also be directly used in many types of engines (Pak and Allexi, 1994). Plantations of *J. curcas* is now being established by both government and private sectors on large scale which in turn demands efficient handling of voluminous seeds for improved oil yield and quality.

Seeds being the potential source for the success of biofuel programme, invites special care and investigation. Fixation of standards for seed is an important pre-requisite in production of quality seeds (Paramathma and Srimathi, 2006). Quality seeds of Jatropha assure not only production of quality seedlings but also provide good quantity and quality oil. Hence a comprehensive study was conducted to standardize appropriate post harvest techniques suited to *J. curcas* for obtaining maximum oil with desirable physico-chemical characteristics.

MATERIALS AND METHOD

Studies on reproductive efficiency

Studies on flowering and fruiting phenology were conducted to identify the reproductive efficiency and appropriate maturity stage for seed collection. In a three year old Jatropha plantation at Anaikatti, 17 plants were selected and tagged for the study during two flowering season. Flower production per inflorescence, number of female flowers and fruits produced per inflorescence were recorded tree-wise. Observations were made every week for 14 weeks. The physical details of the site are given below

Specifics and agroclimatic details of the study site

Location	Anaikatti, Tamil Nadu
Specifics	10° 30' N; 77° 15' E
Elevation	450 msl
Annual Mean Temperature range	25-35 °C
Annual Mean Rainfall	500 mm
Soil Type	Laterite soil

The observations over two peak flowering seasons indicate that each inflorescence has an average of 80 flowers of which female flowers constitute only 3% (Table 1). Of this 3%, 68 per cent of the flowers produce fruits. This indicates the high reproductive efficiency of the species with reference to the selected locality and there is a need to device methods to increase female flower production in *J. curcas*. Moreover, it was observed that plant no. 8 and 12 did not produce any female flowers over the two peak seasons. In depth studies on this line could help us in establishing dioecy in *J. curcas*. This could further open up avenues for breeding programmes.

Fruits appeared 2 weeks after bud initiation. Fruit development prolonged for 8 weeks. During the 6th week the green fruits started turning yellow. The ripening process prolonged for 2 more weeks. By the end of 8 weeks all fruits turned black pulpy, beyond which drying process set in. This trend was found similar to the previous year study.

It was also observed that during the second flush in 2006 (October), the flowers produced were only male and female flowers were totally absent in the 17 tagged plants. This suggests poor seed yield during the second season.

Table 1. Flowering and fruiting phenology of *Jatropha curcas* over two eak seasons at Anaikatti

	Flowers/ inflore-scence	Female flowers/ inflorescence	Fruits/ inflores-cence	Female flowers (%)	Fruits (%)	Fruiting success (%)
Mean values for 2006	90.59	4.76	3.41	5.102	3.645	59.678
Mean values for 2007	65.294	1.294	1.294	1.957	1.957	76.471
Grand Mean	77.942	3.027	2.352	3.530	2.801	68.074

Studies on seed handling

Seed collection

Fruits of *Jatropha curcas* collected during the peak fruiting season (July) from 20 trees each in different locations viz. Attapady (11°14' N; 76°48' E), Coimbatore (11°N; 77° E), Palani (10°26' N; 77°30' E), and Sathyamangalam (11°30'N; 77°17'E), Trichy (10°16' N; 78°15' E) Mettupatti (10°44'N; 77° 31'E) and Bhavanisagar (11°25' N, Longitude: 77°4'E) were used for different experiments.

Oil content estimation

In the present study, whole seeds were used for oil extraction. The seeds were ground using mortar and pestle and 20 g of coarse seed powder was taken for oil extraction. Commonly used semi-continuous method in Soxhlet apparatus was applied, using petroleum ether (boiling point: 40-60°C) as solvent for extraction of oil. Petroleum ether in the flask was evaporated, the extracted oil was allowed to cool, weighed and expressed as percentage. The percentage of oil in the sample was calculated as follows:

$$\textbf{Oil content (\%)} = [(B-A)/C] \times 100$$

A = weight of clean dry flask (g)

B = weight of flask with fat (g)

C = weight of sample (g)

Characterization of oil

Specific gravity of oil was determined at 30°C using specific gravity bottle. The viscosity (at 30°C) of oil samples was measured using Engler Viscometer and expressed in Degree Engler. The oil samples were subjected to chemical characterization for Acid value, an important indicator of vegetable oil quality (Kardash and Tur'yan, 2005). Acid value is expressed as the amount of Potassium hydroxide (in mg) necessary to neutralize free fatty acids contained in 1 g of oil. Saponification number (Horowitz, 1975) expressed in mg of Potassium hydroxide absorbed per gram of oil, is a measure of fatty acid chain length in oils. Iodine value (Horowitz, 1975) expressed in grams of Iodine absorbed per 100 g of oil, indicates the degree of unsaturation and peroxide expressed in milli eq / kg of oil is an index of fatty acid oxidation.

Statistical analysis

The experiments were carried out in Completely Randomized Design with four replications each. The data were subjected to Analysis of Variance and means that exhibited significant differences were compared by Duncan's Multiple Range test ($\alpha = 0.05$) (Panse and Sukhatme, 1995).

Fig. 1. Flowering and fruit development stages in *Jatropha curcas*

Experiment 1: Harvesting maturity

Fruits of *J. curcas* collected from four different populations were used for this study. The bulk of each source was separated into three different sub lots based on the fruit pericarp colour namely yellow (8 weeks following anthesis), black pulpy (10 weeks following anthesis) and dry (12 weeks following anthesis) which is considered in the present study as an indicator of maturity stage (Fig. 1). The yellow and black pulpy fruits were manually processed to extract seeds, washed in running water and surface dried for one day at 30 ± 1°C. The dry fruits were processed in a hand operated custom made decorticator which breaks open the dry fruits to release the seeds. The seeds thus obtained from different fruit maturity stages were tested for oil content and corresponding physico-chemical properties.

Experiment 2: Moisture reduction studies

Black pulpy fruits of *Jatropha curcas* were manually processed to extract the seeds. The processed fresh seeds were tested for moisture content on fresh weight basis by oven dry method (ISTA, 1999). The seed moisture content was estimated inclusive of seed coat. Four samples from the same seed lot were drawn and dried to different target moisture contents (9, 8, 7, 6 and 5%). The target weight for drying the seeds was determined using the formula given by DFSC (1999). The seeds were subjected to shade drying at ambient conditions (30±1°C; RH 65%) to attain the targeted weight. The actual moisture contents of the seeds were also determined as per standard procedures recommended by International Seed Testing Association and found to be 9.209, 8.116, 7.283, 6.245 and 4.979%. The desiccated seeds were then tested for oil parameters.

Experiment 3: Drying method and duration

Extracted seeds from black pulpy fruits of *J. curcas* were subjected to three different drying methods namely sun drying (32 ± 1°C), shade drying (30±1°C; RH 65%) and oven drying at 40°C. Each drying method was conducted for 3 different durations and studied the variation in oil content and characteristics.

Experiment 4: Seed coat removal

Seeds collected from three populations in Tamil Nadu were used for this study. The oil percentage in different parts of the seed was estimated. 10 g of seed was weighed and its shell and kernel were separated. Based on the weights of shell and kernel, calculated oil percentage was compared with whole seed oil percentage.

RESULTS AND DISCUSSION

Effect of fruit maturity

The values for oil content and physical properties namely viscosity, specific gravity and refractive index are given in Table 2. The quantity of Jatropha oil extracted from seeds of yellow, black pulpy and dry fruits viz. 30.82%, 30.0% and 29.06% respectively did not show significant difference. There was no variation in specific gravity of oil obtained from the three treatments. Specific gravity is used to describe the weight or density of a liquid compared to an equal volume of fresh water at 4°C (39°F) whereas viscosity is a measurement of a fluids resistance to pouring and there is no correlation between viscosity and specific gravity (De Clerck, 2007). Viscosity of oil was found to vary noticeably among the different fruit maturity treatments. The lowest viscosity was recorded by oil from yellow fruits followed serially by black pulpy and dry fruits. Relative viscosity depends on many factors such as the method of extraction, age, storage, degree of saturation and extent of oxidation of the oil being analyzed. In the present study, an increasing viscosity was observed with increasing maturity despite consistency in the specific gravity of the oil suggesting that the fatty acid composition of the oil could have been altered which could have resulted during the process of maturation showing a tendency towards development of undesirable factors in oil. The Refractive index was highest (1.465875) for yellow stage while the other two treatments showed no significant difference in their refractive indices. Greater the Refractive index, better the clarity of oil.

Table 2. Effect of fruit maturity on Jatropha oil content and its physical properties

Maturity stage of fruit	Oil %	Specific gravity	Vicosity at 30°C (Deg. Engler)	Refractive index
Yellow	30.82[a]	0.9140[a]	9.925[b]	1.465875[a]
Black pulpy	30.00[a]	0.8974[a]	11.459[a]	1.465750[b]
Dry	29.06[a]	0.8995[a]	12.822[a]	1.465625[b]
	S.e.d.=1.274	S.e.d.=0.0098	S.e.d=0.0782	S.e.d=0.0000589
	C.D.=2.584	C.D.=0.0198	C.D.=0.1587	C.D.=0.0001195

Means with the same letter do not differ significantly as per DMRT at 5% level of significance.

* S.e.d. = Standard error of deviation and C.D. = Critical difference

The values for chemical properties of oil are presented in Table 3. The means for acid value of Jatropha oil from yellow, black pulpy and dry fruits were found to be 5.85, 5.06 and 4.88 mg of KOH/g oil respectively showing no significant variations among maturity stages. But oil chemical properties like saponification number, iodine value and peroxide value were appreciably influenced by fruit maturity. The highest saponification number (135.3 mg of KOH/g oil) was recorded by black pulpy fruits. Iodine number was low for yellow and black pulpy fruits corresponding to 114.6 and 111.6 g. of iodine/ 100g oil. On the other hand, lowest peroxide value (3.59 milli Eq/kg oil) was recorded by yellow fruits followed by black pulpy (4.94 milli Eq/kg oil) and dry fruits (5.04 milli Eq/kg oil).

Table 3. Effect of fruit maturity on chemical properties of Jatropha oil

Maturity stage of fruit	Acid value (mg of KOH/g oil)	Saponification value (mg of KOH/g oil)	Iodine value (g of iodine/ 100g oil)	Peroxide value (milli Eq/Kg oil)
Yellow	5.85[a]	126.1 [a]	114.6 [b]	3.59 [b]
Black pulpy	5.06[a]	135.3 [a]	111.6 [b]	4.94 [a]
Dry	4.88[a]	119.3 [a]	120.3 [a]	5.04 [a]
	S.e.d.=0.577	S.e.d.=6.58	S.e.d=2.72	S.e.d.=0.381
	C.D.=1.171	C.D.=13.35	C.D.=5.52	C.D.=0.772

Means with the same letter do not differ significantly as per DMRT at 5 % level of significance.

Saponification value is inversely proportional to the mean molecular weight of the glycerides in the oil. Highest saponification value was noticed in black pulpy fruits (135.3) indicating that unsaponificable matters like sterols, phospholipids, waxes, terpenes etc. were higher in yellow and dry fruits. The higher the saponification number of a fat free from moisture and unsaponifiable matter, the more desirable is its quality. Oil from both yellow and black pulpy fruits showed significantly lesser values for iodine value while peroxide value was lowest for yellow fruits. Studies in Longan (*Dimocarpus longan* Lour.) fruits suggests that desiccation-induced browning of longan pericarp associated with maturation may be related to decrease in activities of reactive-oxygen-scavenging enzymes and amounts of endogenous antioxidant substances, an accumulation of active oxygen, an increase in membrane lipid peroxidation, an injury of the integrity of cellular membrane structure (Lin *et al.*, 2005). Similar observations have also been recorded in our study wherein low peroxide values have been recorded in yellow fruits.

This suggests that with most of the chemical changes in the oil properties tending to occur during maturity drying, free fatty acids are subject to auto-oxidation which triggers rapid rancidification of the oil. This is also corroborated with the peroxide values showing an increasing trend from yellow to drying stages. With increasing auto-oxidation, fatty acids break down into hydrocarbons, ketones, aldehydes, and smaller amounts of epoxides and alcohols which results in increased viscosity of the oil as was observed in the present study. Substantial quantum of desirable properties such as higher saponification number, lower iodine and peroxide values in oil from yellow and black pulpy fruits contribute to their superiority over oil from dry fruits.

Effect of moisture content

The values obtained for oil percentage and its physical properties have been presented in Table 4. Ranking of the treatments was based on oil percentage. The oil content was found to be significant with values as high as 23.98% and 26.18% respectively, in 6% and 5% MC when compared to other treatments. Between the two, the quality characteristics of oil were relatively low in 5% MC than that of seeds with 6% MC. The oil physical properties, Specific gravity (0.8580) and Refractive Index (1.463) which are the important desirable characteristics of a fuel oil were found to be low and promising in seeds with 6% MC. The viscosity of the fuel affects the pumping characteristics and the atomisation of the fuel during injection (droplet spectrum, geometry of the injection stream). As a result of the more difficult flowing, pumping and atomising behaviour, high viscosity may lead to cold-start problems. An increase in viscosity may be an indicator of excessive oxidation or polymerisation of the oil (Thuneke and Kern, 2002).

Table 4. Effect of seed moisture on oil content and physical characteristics of Jatropha oil

Seed Moisture content (%)	Viscosity at 30°C	Specific gravity	Refractive Index (nD)	Oil %
9 (9.209)	9.322 [c]	0.8775 [a]	1.466 [c]	20.59 [b]
8 (8.116)	10.285 [b]	0.9095 [b]	1.466 [c]	21.28 [b]
7 (7.283)	8.893 [a]	0.8965 [b]	1.465 [b]	21.26 [b]
6 (6.245)	9.714 [c]	0.8580 [a]	1.463 [a]	23.98 [a]
5 (4.979)	9.607 [c]	0.8565 [a]	1.465 [b]	26.18 [a]
	S.e.d.=0.148	S.e.d.=0.0157	S.e.d.=0.0002	S.e.d.=1.217
	C.D.=0.316	C.D.=0.0334	C.D.=0.0006	C.D.=2.593

Means with the same letter do not differ significantly as per DMRT at 5 % level of significance.

The chemical characterization of oil under different treatments has been furnished in Table 5. Favourable status of chemical properties namely, low Acid value (2.81) and low Peroxide value (2.0) as evident from the results assign 6% MC seeds superior to all other treatments. However, better ranking with respect to high saponification number (171.11) and low iodine value (118.30) were recorded for 7% and 9% MC treatments respectively. The acid value is an indicator of the content of free fatty acids in plant oil. High acid value in the fuel leads to corrosion, abrasion and deposits in the engine. Furthermore the free fatty acids may react with the alkaline components of the lubricating oil and affect its lubricity. Peroxide value can be an useful early indicator of oxidative deterioration and a decrease in the effectiveness of the oil's own antioxidants. During storage, peroxides are generated by oxygen access. A rise of viscosity and gummy deposits can be the consequence. Furthermore there are indications that plant oils damaged by oxidation may lead to thickening of the oil and therefore to engine damage. The iodine value is an indicator of the amount of fatty acids with double bonds in the oil. A low iodine value indicates a high degree of saturation of the oil. This value gives information about the possible creation of deposits at the injector nozzles and in the combustion chamber, about the storage stability of the oil and its tendency to polymerise, and about the possibility of adverse effects on the oil (Thuneke and Kern, 2002).

Table 5. Effect of seed moisture content on chemical characteristics of Jatropha oil

Seed Moisture Content (%)	Acid value (mg of KOH/g oil)	Sap. value (mg of KOH/g oil)	Iodine no. (g. of iodine /100 g oil)	Peroxide value (milli Eq/ kg oil)
9 (9.209)	4.49 [a]	164.09 [b]	118.30 [b]	1.8 [a]
8 (8.116)	4.49 [a]	168.30 [b]	145.30 [a]	1.4 [a]
7 (7.283)	3.37 [a]	171.11 [a]	149.36 [a]	2.6 [b]
6 (6.245)	2.81 [a]	158.48 [c]	140.86 [a]	2.0 [a]
5 (4.979)	5.61 [b]	168.30 [b]	147.20 [a]	2.2 [b]
	S.e.d.=0.893	S.e.d.=3.9	S.e.d.=3.85	S.e.d.=0.292
	C.D.=1.903	C.D.=8.31	C.D.=8.21	C.D.=0.623

Means with the same letter do not differ significantly as per DMRT at 5 % level of significance.

The lowest safe moisture content (LSMC) for *Azadirachta indica* A. Juss (Neem) seeds was found to be 10.9% (Varghese and Naithani, 2001). In *Helianthus annuus* L. (Sunflower) seeds it has been found that desiccation to seed moisture between 26 and 33% may increase oil content by 1% while treatment between 55 and 65% may reduce oil content by 3%. Radic (2006) has opined that reduction in seed moisture increases the contents of oil and similar results were found in the present study.

Effect of drying method and duration

The oil content and their corresponding physico-chemical properties obtained for the three drying regimes have been furnished in Table 6 and 7. All the values for oil content and oil characteristics showed significant variation. Both shade drying for 10 days and oven drying at 40°C for 1 day were found to be statistically on par recording 26.18 and 25.48% oil respectively. Following ranking based on oil content these two treatments were compared with each other for superior oil quality. Except for viscosity (oven drying-8.89; shade drying 10 days-9.61), iodine number (oven drying-111.78; shade drying 10 days-147.20) and peroxide value (oven drying-2.2; shade drying 10 days-1.4) the two treatments did not vary significantly.

Table 6. Effect of drying method and duration on oil content and oil characteristics

Drying method & duration	Viscosity at 30°C	Specific gravity	Refractive Index (nD)	Oil %
Control (fresh)	9.32 [c]	0.878 [b]	1.466 [b]	20.59[b]
Sun drying 10 days	9.57 [b]	0.899 [a]	1.467 [a]	21.78 [b]
Sun drying 20 days	10.43 [a]	0.889 [a]	1.466 [b]	20.42 [b]
Sun drying 30 days	8.89[d]	0.897 [a]	1.465[c]	21.26 [b]
Shade drying 10 days	9.61 [b]	0.857 [b]	1.465 [c]	26.18[a]
Shade drying 20 days	9.71 [b]	0.858 [b]	1.463[e]	22.45 [b]
Shade drying 30 days	10.29 [a]	0.910 [a]	1.466 [b]	21.28 [b]
Oven drying at 40°C for 1 day	8.89 [d]	0.854 [b]	1.465[c]	25.48 [a]
Oven drying at 40°C for 3 days	9.18 [c]	0.852 [b]	1.464[d]	20.08 [b]
Oven drying at 40°C for 7 days	9.61 [b]	0.856 [b]	1.465[c]	21.93 [b]
	S.e.d.=0.123	S.e.d.=0.0139	S.e.d.=0.0002	S.e.d.=1.110
	C.D.=0.251	C.D.=0.0283	C.D.=0.0004	C.D.=2.267

Means with the same letter do not differ significantly as per DMRT at 5 % level of significance.

Table 7. Effect of drying method and duration on chemical characteristics of Jatropha oil

Drying method & duration	Acid value (mg of KOH)	Sap. value (mg of KOH)	Iodine no. (g. of iodine)	Peroxide value (milli Eq/kg)
Control (fresh)	4.49[a]	164.09 [a]	118.30[d]	1.8 [b]
Sun drying 10 days	2.81 [b]	154.28 [c]	125.63[c]	2.2[a]
Sun drying 20 days	5.61 [a]	171.11 [a]	131.47[b]	1.6 [b]
Sun drying 30 days	3.37 [b]	171.11 [a]	149.36[a]	2.6 [a]
Shade drying 10 days	2.81 [b]	168.30 [a]	147.20 [a]	2.2 [a]
Shade drying 20 days	5.61 [a]	158.48 [c]	140.86 [b]	2.0 [b]

Shade drying 30 days	4.49[a]	168.30[a]	145.30[a]	1.4[c]
Oven drying at 40°C 1 day	2.24[b]	164.09[a]	111.78[d]	1.4[c]
Oven drying at 40°C 3 days	2.81[b]	162.69[b]	128.17[c]	2.2[a]
Oven drying at 40°C 7 days	4.49[a]	166.90[a]	140.22[b]	2.0[b]
	S.e.d.=0.780	S.e.d.=3.967	S.e.d.=3.460	S.e.d.=0.2781
	C.D.=1.593	C.D.=8.101	C.D.=7.066	C.D.=0.5679

Means with the same letter do not differ significantly as per DMRT at 5 % level of significance.

The suitability of fats and oils as biofuel results from their molecular structure and high energy content. The lesser viscous oil from oven drying method helps better blending with diesel and also easy flow of the fuel. Long-chain, saturated, unbranched hydrocarbons are especially suitable for conventional fuel purposes. The higher the iodine value, the more unsaturation is present in the fat or oil. Hence saturated fatty compounds are preferred as they are more stable and also known to have higher cetane number (Knothe *et al.*, 1996). Lesser the peroxide value lower is the oxidation, and hence minimal rancidity was noticed in oven drying method. Hence on comparison of the two treatments, low values for viscosity, iodine number and peroxide value recorded by oven drying at 40°C for 1 day marks its superiority over shade drying.

Efficacy of seed coat removal

The experimental results are represented in Table 8. Fig. 2 depicts the variation in oil content recorded by treatments namely, kernel, shell and whole seed. Studies revealed that shell has very negligible quantity of oil. The oil extracted from kernel was found to be significantly high both in terms of percentage and volume when compared to that from whole seed. This suggests that it is essential to separate the shell or seed coat from seed in order to restrain loss of oil through adsorption by seed coat and thereby maximize oil recovery.

Table 8. Recovery of Jatropha oil from different parts of seed

			Oil %		
Source of variation	Degrees of freedom	Sum of squares	Mean square	Variance ratio	F probability
Seed part	2	5615.68	2807.84	127.94	< 0.001
Residual	33	724.26	21.95		
Total	35	6339.94			

S.e.d.= 1.913 ; C.D.= 3.891

Oil Volume					
Source of variation	Degrees of freedom	Sum of squares	Mean square	Variance ratio	F probability
Seed part	2	59.9286	29.9643	202.25	< 0.001
Residual	33	4.8892	0.1482		
Total	35	64.8178			
S.e.d.= 0.157 ; C.D.= 0.320					

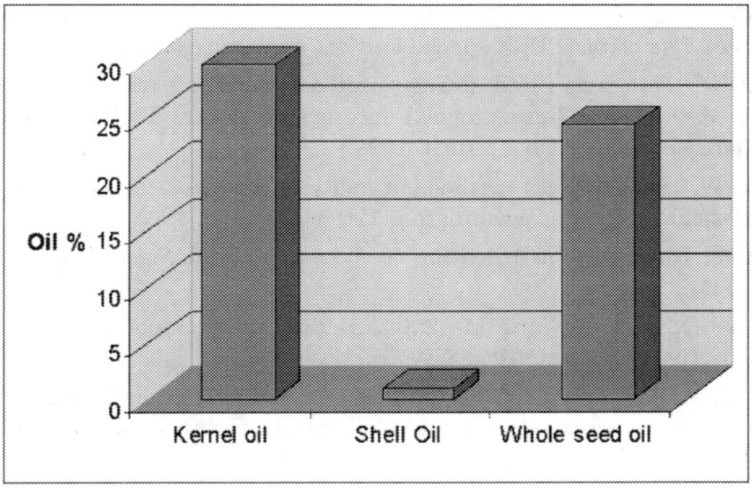

Fig. 2. Variation in oil content in different parts of *Jatropha* seed

CONCLUSION

From the present study, it is evident that to obtain good quality oil, *J. curcas* fruits need to be harvested at yellow or black pulpy stage of fruit maturity and is safer to avoid collecting completely dry fruits. At the same time it has to be understood that the time required for attaining the different maturity stages depends on environmental conditions. Drying Jatropha seeds to the lowest safe moisture content of 6% helps better oil recovery with desirable characteristics. For Jatropha either oven drying at 40°C for 1 day or shade drying for 10 days can be practiced as suitable drying techniques. Removal of seed coat and use of kernel alone instead of whole seed was found to enhance oil recovery in Jatropha seeds. Practice of standardized technologies would not only help obtaining quality oil but also provide better returns to the farming communities and industries.

ACKNOWLEDGEMENT

The authors thank the Department of Biotechnology for extending financial assistance to conduct the present study.

REFERENCES

Banerjee, M. and Srivastava, S. 2006. *In vitro* regeneration of *Jatropha curcas* (Ratanjyot): prospects for biofuel production and medicines. *Indian Journal of Botanical Research* 2(2):195-200.

Biswas, S., Kaushik, N. and Srikanth, G. 2006. Biodiesel Technology and Business Opportunities - An Insight. In: Brahma Singh, Swaminathan, R. and Ponraj, V. (eds). *Biodiesel Conference Towards Energy Independence – Focus on Jatropha*, 9-10 June, 2006, Rashtrapati Nilayam, Bolaram, Hyderabad.

De Clerck, D. 2007. Understanding viscosity and how it affects pump performance. *The primer*, FRX special edition, summer issue, pp.3.

DFSC. 1999. Desiccation and storage protocol: handling and storage of recalcirant and intermediate tropical forest tree seeds. DANIDA Forest Seed Centre. Newsletter No. 5, April, 1999.

Horowitz, W. 1975. Official methods of analysis of AOAC- Association of Official Analytical Chemists, Washington, 12th ed., p. 488, 490.

ISTA. 1999. International rules for seed testing. *Seed Science and Technology* 27:30-35.

Kadlaskar, D. and Dewan, M. 2004. Biofuel from eroded soils. The India Perspective. *German News*, XLV:15.

Kardash, E. and Tur'yan, I. 2005. Acid Value determination in vegetable oils by indirect titration in aqueous- alcohol media. *Croatica Chemica Acta* 78 (1): 99-103.

Knothe, G., Dunn, R.O. and Bagby, M.O. 1996. Biodiesel: The use of vegetable oils and their derivatives as alternative diesel fuels. In: Fuels and Chemicals from Biomass. Washington, D.C.: American Chemical Society. http://www.biodiesel.org/reports/GEN-162.doc.

Kureel, R.S. 2006. Prospects and potential of *Jatropha curcas* for biodiesel production. In: Brahma Singh, Swaminathan, R. and Ponraj, V. (eds). *Biodiesel Conference Towards Energy Independence – Focus on Jatropha*, 9 – 10 June, 2006, Rashtrapati Nilayam, Bolaram, Hyderabad pp. 43-93.

Lin, H.T., Xi, Y.F. and Chen, S.J. 2005. The relationship between the desiccation-induced browning and the metabolism of active oxygen and phenolics in pericarp of postharvest longan fruit. *Journal of Plant Physiology and Molecular Biology* 31(3):287-97.

Pak, M. and Allexi, A. 1994. Testing engines fuelled with Physic oil. *Landtechnik*, 49 (1): 29-30.

Panse, V.G. and Sukhatme, P.V. 1995. Statistical methods for agricultural workers. Indian Council of Agricultural Research, New Delhi, 359pp.

Paramathma, M. and Srimathi, P. 2006. Seed standards for quality seedling production. In: Brahma Singh, Swaminathan, R. and Ponraj, V. (eds). *Biodiesel Conference Towards Energy Independence – Focus on Jatropha*, 9 – 10 June, 2006, Rashtrapati Nilayam, Bolaram, Hyderabad, pp. 97-128.

Radic, V. 2006. Effect of maturation period on seed quality; optimum time for desiccation in Sunflower (*Helianthus annuus* L.) genotypes. *Helia*, 29(44): 145-152.

Singh, M.J. 2006. Biodiesel an environmentally friendly alternate fuel experience of Indian Railways. In: Brahma Singh, R. Swaminathan and V. Ponraj. (Eds.) *Biodiesel Conference Towards Energy Independence – Focus on Jatropha*. Paper presented at the conference held at Rashtrapati Nilayam, Bolaram, Hyderabad on 9 – 10 June, 2006, pp. 289-290.

Thuneke, K. and Kern, C. 2002. Pflanzenölbetriebene Blockheizkraft-werke – Teil 1. "Materialien 171", Hrsg.: Bayerisches Staatsministerium für Landesentwicklung und Umweltfragen (StMLU).

Tuli, D.K. 2006. Biodiesel production through transesterification. In: Brahma Singh, Swaminathan, R. and Ponraj, V. (eds). *Biodiesel Conference Towards Energy Independence – Focus on Jatropha*, 9 – 10 June, 2006, Rashtrapati Nilayam, Bolaram, Hyderabad pp. 285-288.

Varghese, B. and Naithani, S.C. 2001. Desiccation stress in neem seeds: Physiological and biochemical considerations. The project on handling and storage of recalcitrant and intermediate tropical forest tree seeds. DANIDA Forest Seed Centre Newsletter, March 2001, pp. 16-19.

Chapter - 4

BIOFUELS: PERFORMANCE CHARACTERISTICS OF GASOLINE AND DIESEL PASSENGER CARS

Sunil Pathak, Yograj Singh, Vineet Sood and S.K. Singal

Indian Institute of Petroleum, Mohkampur, Dehradun-248008, Uttranchal
E-mail: spathak@iip.res.in

ABSTRACT

Recent concerns over the environment and scarcity of fuel supply have promoted interest in the development of the alternative sources for petroleum fuels. Many researchers have reported the use of biodiesel as a fuel in diesel engine, thereby achieving reduction in harmful exhaust emissions as well as a comparable engine performance as that with diesel. However, study on chassis dynamometer and in the field related to fuel economy, acceleration was undertaken in order to understand the impact of blending vegetable oil esters with the diesel in vehicles. For the gasoline vehicle the ethanol is being blended with gasoline in proportion of 5% at the national level. Parallelly, the efforts are also being made to use higher level of ethanol blending in gasoline vehicles. The present study was carried out on a diesel passenger car on chassis dynamometer using diesel, 10%, 20% and 30% blend of biodiesel (JME) and (JEE) fuels and on the gasoline passenger car using up to 10% ethanol. The vehicle performance study have been carried out with the aim of obtaining comparative measures of the constant speed fuel consumption, driving cycle fuel economy, WOT power and fuel consumption. The emission of the CO_2 was also computed using the fuel economy data on the above-mentioned fuels. The vehicle acceleration tests were also carried out on actual road condition using global positioning systems (GPS) and the results for maximum speed, time elapsed for a distance of 1 km and also the acceleration time for reaching speeds of 0~60 km/hr and 0~90 km/hr were recorded including the performance and mass emission studies on a gasoline passenger car on chassis dynamometer. The acceleration results were also given up to 30% blending in the gasoline.

Keywords: Performance characteristics, biofuels, passenger cars

INTRODUCTION

The vehicle population in India is growing at the rate of 5% and more per annum. In the recent few years, the diesel driven commercial vehicle population is increasing rapidly and resulting in higher consumption of the diesel fuel. So, the extensive research has been carried out in India for the use of biodiesel in diesel-fuelled vehicles. The blends of the methyl ester and ethyl ester were tried in a passenger car and the various performance parameters of the vehicle were analyzed to quantify the difference as compared to diesel in order to establish the feasibility of using these blends in vehicles without any engine modifications. The study was focused on vehicle performance parameters like acceleration, fuel economy etc.

The large population of the two-wheelers and the gasoline passenger car consumes almost 70% of the gasoline fuel in India. So, the ethanol blending has been done for partial replacement of gasoline. The large number of the sugarcane based industries is a good source for ethanol at reasonable prices. The 5% blending has already been done and the efforts are being made to use higher level of ethanol blending in the gasoline. A study was done on a gasoline passenger car which consists of mass emission test in the laboratory and the acceleration test in the actual field condition

MATERIALS AND METHODs

Fuel matrix

The Fuel matrix is given in Table 1. The blending was done on volume basis.

Table 1. Fuel Matrix

S.No.	Fuel
	Diesel car
1.	Diesel
2.	JME (10%)+Diesel
3.	JME (20%)+Diesel
4.	JME (30%)+Diesel
5.	JEE (30%)+Diesel
	Gasoline car
1.	Gasoline
2.	E-10

Test vehicle

The brief specification of the diesel passenger car and the gasoline car are given in Table 2.

Table 2. Technical specification of the vehicle

S.No.	Item	Details/Specifications
Diesel Passenger Car		
1.	Engine Type	Water cooled, Indirect injection, EGR and oxidation catalyst
2.	Capacity	1405 cc
3.	Compression ratio	21:1
4.	No. of Cylinder	4 inline
5.	Max. Power	45 kW at 4500 rpm
6.	No. of Gears	5 forward, 1 Reverse
Gasoline Passenger Car		
1.	Kerb weight, kg	854
2.	Valve train	SOHC
3.	Capacity	1086 cc
4.	No. of Cylinder	4
5.	Max. Power	63 PS at 5500 rpm
6.	Max. Torque, Kgf.m	9.8@3000

Vehicle performance study

Diesel passenger car

The vehicle performance parameters like constant speed, fuel consumption on top gear and one lower gear, driving cycle fuel consumption and WOT power were measured using the chassis dynamometer in laboratory at controlled conditions of the ambient and fuel temperatures. The vehicle acceleration test was also carried out on actual road condition. The route selected was plain road, reasonably leveled and the data of speed vs time was observed for a stretch of 1km on different fuels.

Gasoline passenger car

The mass emission test was carried out using modified Indian driving cycle on gasoline and ethanol blend and the acceleration was carried in actual field condition.

RESULTS AND DISCUSSION

Fuel characteristics

The measured properties of some of the test fuels are given in the Table 3.

Table 3. Fuel properties of diesel and JME and JEE blends

S.No.	Fuel	Cetane number	Calorific Value (MJ/kg)	Density (kg/m³)	Kinematic Viscosity (cst)	Flash Point (deg c)
1.	Diesel	49	47.40	0.82	2.7	50
2.	JME-5%	49.3	-	0.842	2.93	-
3.	JME-10%	49.3	-	0.844	3.05	-
4.	JME-20%	53.6	-	0.847	3.09	-
5.	JME-30%	-	-	0.849	3.28	55
6.	JME-100%	58.4	40.98	0.87	4.35	>250
7.	JEE-5%	51.2	-	-	-	-
8.	JEE-10%	52.5	-	0.83	2.91	52
9.	JEE-20%	53.3	-	-	-	-
10.	JEE-30%	-	-	0.8441	3.19	56
11.	JEE-100%	58	-	0.87	4.74	185

Performance studies

Diesel passenger car

Constant speed fuel consumption

The result of the constant speed fuel consumption on Vth gear and IVth gear engaged conditions are given in Fig. 1 & 2 respectively. It is observed that at the IVth gear position, the difference in the fuel consumption for the entire fuel matrix at all the speed points is negligible. However, at the Vth gear position marginal difference in the fuel consumption was observed with increased percentage of the biodiesel in the diesel. The reason could be that the energy content of the biodiesel is less as compared to diesel but the density is higher and the fuel consumption has been measured on the volume basis. So the effect is neutralized. It is also observed that the fuel consumption increases with the increase of speed.

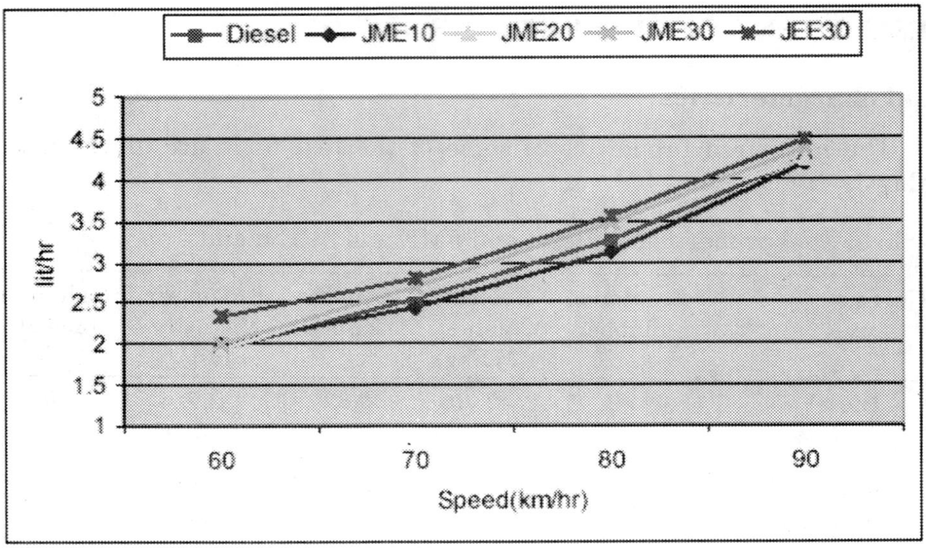

Fig. 1. Constant speed fuel consumption (Vth Gear)

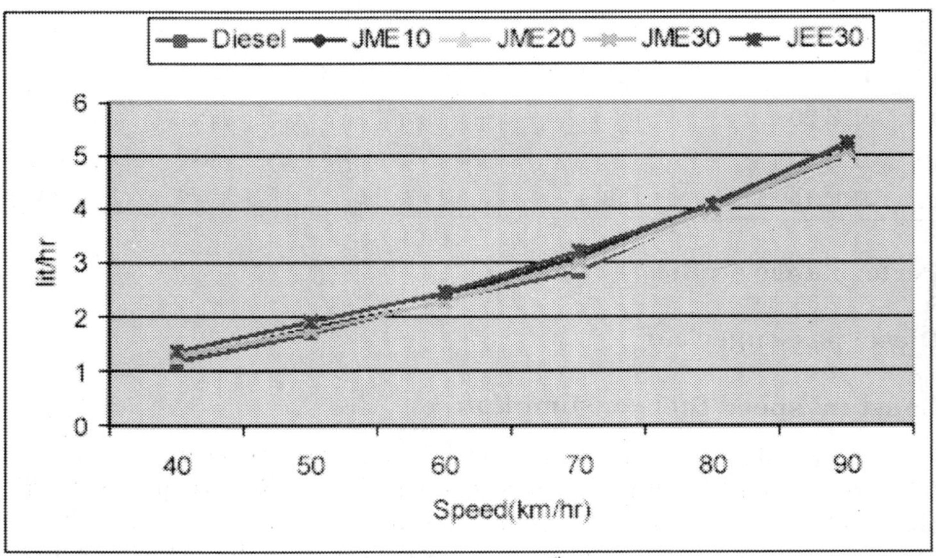

Fig. 2. Constant speed fuel consumption (IVth Gear)

Driving cycle fuel consumption

The results of the driving cycle fuel consumption are given in Table 4. It is observed from the results that the fuel economy of the passenger car was around 16 km/hr on diesel and diesel-biodiesel blends. The difference between the fuel economy data was not significant.

Table 4. Driving cycle fuel consumption

Fuel	Distance	Fuel Consumption (lit)	Fuel Economy Km/lit	Avg. Fuel Economy (Km/lit)
Diesel	10.676	0.6645	16.07	16.32
	10.685	0.6445	16.58	
JME10	10.699	0.66	16.21	16.27
	10.694	0.6551	16.32	
JME20	10.707	0.6306	16.98	16.73
	10.651	0.6464	16.48	
JME30	10.734	0.647	16.59	16.54
	10.707	0.649	16.50	
JEE30	10.715	0.6527	16.42	16.32
	10.637	0.6556	16.22	

Wide open throttle (WOT) power and brake specific fuel consumption

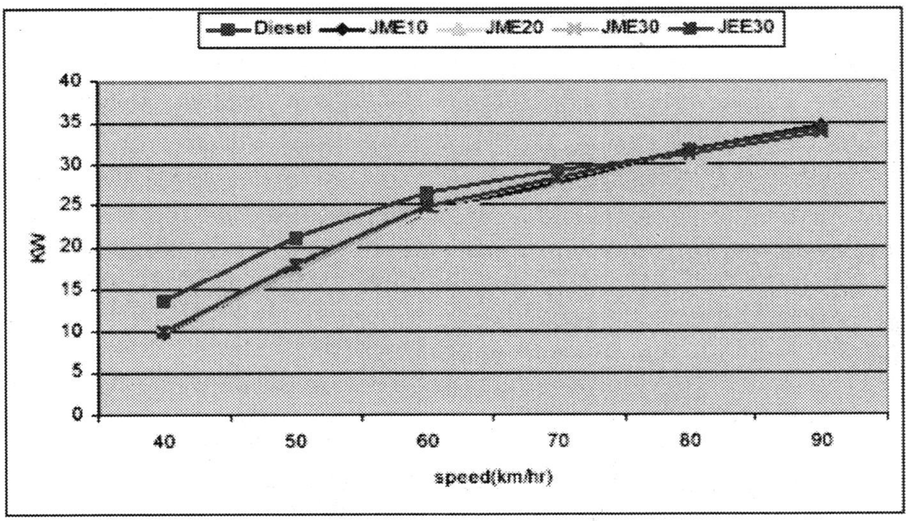

Fig. 3. Power characteristic with speed at WOT

The characteristics of the WOT power and BSFC with speed have been shown in Fig. 3 & 4. It was observed that the WOT power was increasing with speed. The WOT power was around 35kW at 80 km/hr. The power observed using diesel fuel was higher as compared to blends (Raheman and Phadatare, 2004). It was observed that the brake specific fuel consumption was lower at low speed, increases with speed, reaches to a maximum value and then drops down. The value observed was around 240~340 gm/kW-hr for all speeds and different fuel matrix. At lower speeds, the BSFC was lower for diesel fuel and increases with the percentage of blending. This may be due to lower power (due to lower energy content) generated by biodiesel at low speeds. At higher speeds, no specific trend was observed, because at higher speeds the power was almost same and biodiesel blends may improve the combustion characteristics (Ramadhas *et al.*, 2005) and the thermal efficiency resulting in lower BSFC.

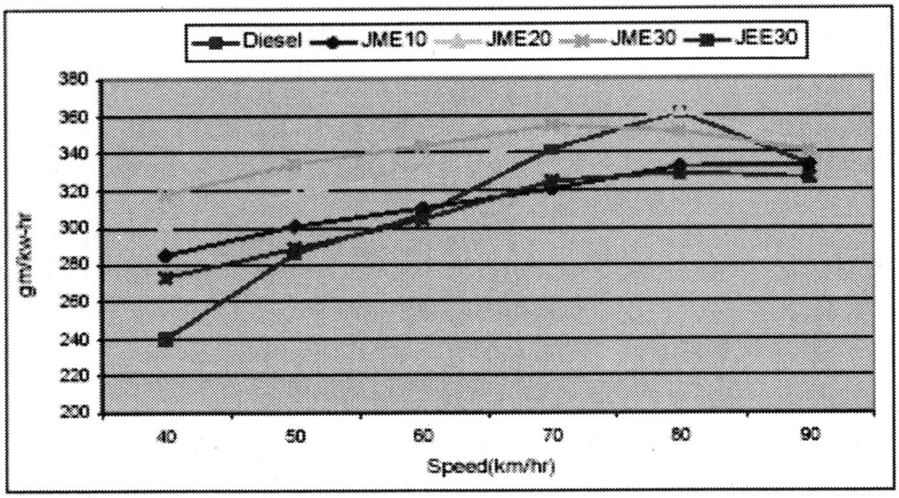

Fig. 4. Wide open throttle brake specific fuel consumption (BSFC)

Estimated CO_2 emission

The values of the CO_2 emissions were estimated on the basis of the driving cycle fuel consumption. These values are given in Table 5. It was observed that CO_2 emission for all the fuels studied is same within the range of 2%. These values are higher in comparison to European Union target of 125 gm/km emissions from passenger cars.

Table 5. Estimated CO_2 emission

Fuel	CO_2
Diesel	167.50
JME10	168.02
JME20	163.40
JME30	165.27
JEE30	167.50

Acceleration test

The results of the acceleration test on actual road conditions are given in Table 6-9. For the study, two different drivers were used for performing the test. It was observed from the results that for diesel fuel, the average maximum speed achieved during the test was around 123~125 km/hr in comparison to about 125 km/hr for JME30 fuel. The time taken for completing the test (1 km distance) was around 42~43 seconds in comparison to 41~42 seconds for the JME 30fuel. Similarly, the time elapsed for achieving the 60 km/hr and 90km/hr speed was also nearly the same.

Table 6. Acceleration test results on indigo diesel car using diesel fuel (Driver A)

Run Number	Driver	Direction	Max Speed (km/hr)	Time (seconds)		
				0-Max Speed	0-60	0-90
1	A	Up	123.49	44.03	12.74	20.95
2	A	Down	122.56	43.82	12.36	20.35
3	A	Up	123.46	44.69	13.27	21.29
4	A	Down	124.28	42.45	10.65	18.99
5	A	Up	121.69	42.77	10.92	18.86
6	A	Down	122.84	43.01	11.59	19.67
7	A	Up	123.17	42.95	11.35	19.45
8	A	Down	122.98	43.04	11.16	19.33
		Mean	123.06	43.35	11.76	19.86
		SD	0.76	0.76	0.93	0.90

Table 7. Acceleration test results or indigo diesel car using diesel fuel (Driver B)

Run Number	Driver	Direction	Max Speed (km/hr)	Time (seconds)		
				0-Max Speed	0-60	0-90
1	B	Down	126.38	40.57	8.66	16.21
2	B	Up	125.33	42.40	11.28	18.70
3	B	Down	125.81	42.00	10.69	18.17
4	B	Up	124.83	42.30	10.72	18.63
5	B	Down	124.80	42.73	11.17	18.94
6	B	Up	123.34	42.87	11.21	18.94
7	B	Down	125.12	42.51	10.98	18.77
8	B	Up	124.40	42.20	10.51	18.63
9	B	Down	124.55	42.87	11.19	19.31
10	B	Up	124.37	42.78	11.12	19.35
		Mean	124.89	42.32	10.75	18.53
		SD	0.84	0.68	0.78	0.90

Table 8. Acceleration test results on indigo diesel car using 30% Bio-diesel fuel (Driver A)

Run Number	Driver	Direction	Max Speed (km/hr)	Time (seconds)		
				0-Max Speed	0-60	0-90
1	B	Down	128.98	40.88	9.53	17.05
2	B	Up	122.64	42.10	9.85	17.96
3	B	Down	128.79	40.68	9.18	16.70
4	B	Up	123.27	41.79	9.53	17.05
5	B	Down	125.47	41.15	9.36	17.03
6	B	Up	123.04	41.42	9.60	17.10
7	B	Down	125.45	40.96	9.09	16.81
8	B	Up	121.55	41.75	9.65	17.56
9	B	Down	124.55	41.38	9.32	17.48
10	B	Up	123.42	42.29	10.22	18.35
		Mean	124.72	41.44	9.53	17.31
		SD	2.51	0.54	0.33	0.52

Table 9. Acceleration test results on indigo diesel car using 30% Bio-diesel fuel (Driver B)

Run Number	Driver	Direction	Max Speed (km/hr)	Time (seconds)		
				0-Max Speed	0-60	0-90
1	B	Down	124.02	41.61	9.34	18.54
2	B	Up	123.41	42.12	9.98	18.19
3	B	Down	124.49	41.66	9.67	17.67
4	B	Up	123.93	41.69	9.45	18.45
5	B	Down	124.07	41.82	9.82	17.91
6	B	Up	124.31	41.19	9.11	17.17
7	B	Down	124.41	41.93	10.01	17.97
8	B	Up	124.34	41.87	9.65	17.85
9	B	Down	124.56	41.35	9.29	16.95
10	B	Up	128.52	40.73	9.12	16.48
		Mean	124.61	41.60	9.54	17.72
		SD	1.42	0.41	0.33	0.66

Gasoline passenger car

Mass emission test

The results of the mass emission tests on Gasoline and E 10 fuels are shown in Figure 5.

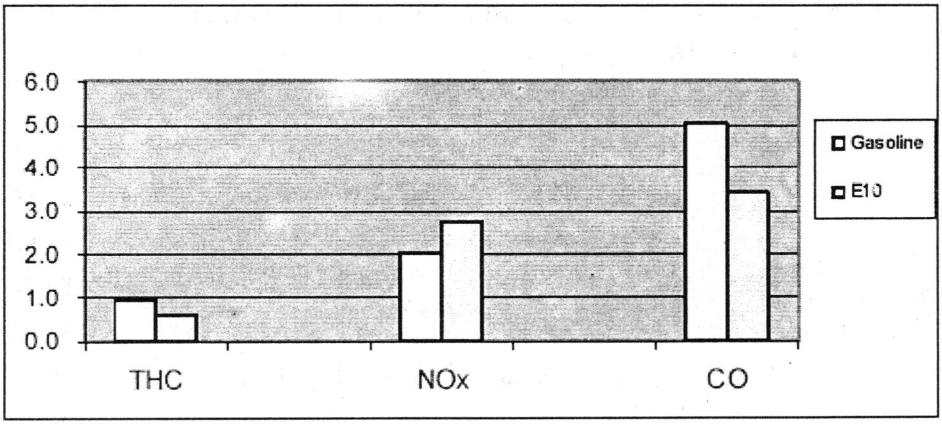

Fig. 5. Mass emission results on gasoline and E10 fuel

It was observed that CO and HC emission were reduced with E10 blend while the NOx emission increased with ethanol blending.

Acceleration test

The results of the acceleration test on actual road conditions are given in Table 10-16. For the study, two different drivers were used for performing the test. It was observed from the results that the acceleration performances on gasoline and ethanol blend up to 30% level were not having any significant difference.

Table 10. Acceleration test results on Hyundai santro car using Gasoline (Driver B)

Run Number	Direction	Max Speed (km/hr)	Time (Second)		
			0-Max speed	0-60	0-90
1	Up	110.77	44.75	9.46	20.98
2	Down	111.61	43.42	8.56	19.1
3	Up	113.32	48.70	8.83	19.64
4	Down	111.1	43.77	8.83	19.29
5	Up	114.1	43.23	8.61	14.17
6	Down	111.79	43.57	8.92	19.79
7	Up	114.82	43.32	8.88	19.45
8	Down	114.82	43.6	9.18	19.57
9	Up	110.5	44.98	11.14	20.95
Mean		112.54	44.37	9.16	19.22
SD		1.74	1.74	0.79	2.01

Table 11. Acceleration test results on Hyundai santro car using 10% Ethanol (Driver A)

Run Number	Direction	Max Speed (km/hr)	Time (Second)		
			0-Max speed	0-60	0-90
1	Up	109.40	44.75	10.39	21.50
2	Down	113.70	43.66	9.71	20.02
3	Up	108.97	44.63	10.02	20.96
4	Down	112.55	43.85	8.18	22.16
5	Up	109.56	43.98	8.64	20.60
6	Down	114.63	42.64	8.59	18.18
7	Up	108.22	44.30	9.74	20.13
8	Down	113.71	43.08	8.52	19.09
9	Up	109.69	44.32	10.00	21.49
10	Down	114.33	42.77	8.61	18.64
Mean		111.48	43.80	9.24	20.28
SD		2.52	0.75	0.80	1.32

Table 12. Acceleration test results on Hyundai santro car using 10% Ethanol (Driver B)

Run Number	Direction	Max Speed (km/hr)	Time (Second)		
			0-Max speed	0-60	0-90
1	Up	111.55	44.05	9.34	20.36
2	Down	114.08	43.22	8.69	19.22
3	Up	111.85	43.63	8.98	19.25
4	Down	113.99	43.89	9.46	19.93
5	Up	111.10	43.26	8.79	18.82
6	Down	114.63	43.35	8.96	19.43
7	Up	114.65	42.95	8.78	19.23
8	Down	110.50	44.42	9.43	20.61
9	Up	113.99	44.31	9.57	19.98
10	Down				
Mean		112.93	43.68	9.11	19.65
SD		1.65	0.52	0.33	0.60

Table 13. Acceleration test results on Hyundai santro car using 20% Ethanol (Driver A)

Run Number	Direction	Max Speed (km/hr)	Time (Second)		
			0-Max speed	0-60	0-90
1	Up	111.50	44.18	9.37	20.25
2	Down	113.25	43.14	8.58	18.97
3	Up	112.52	43.65	9.03	19.47
4	Down	111.97	43.17	8.70	18.75
5	Up	112.69	43.33	8.65	18.90
6	Down	114.00	42.85	8.28	18.38
7	Up	112.93	43.23	8.62	18.69
8	Down	113.44	43.22	8.76	18.67
9	Up	113.04	43.37	8.79	19.27
10	Down	114.99	42.59	8.52	18.45
Mean		113.03	43.27	8.73	18.98
SD		0.99	0.43	0.30	0.56

Table 14. Acceleration test results on Hyundai santro car using Ethanol 20% (Driver B)

Run Number	Driver	Max Speed (km/hr)	Time (Second)		
			0-Max speed	0-60	0-90
1	B	111.86	43.71	8.76	19.55
2	B	113.02	43.38	8.23	20.85
3	B	113.16	42.84	8.2	18.08
4	B	113.05	43.36	8.55	18.10
5	B	114.33	42.9	8.73	18.51
6	B	115.03	42.81	8.7	18.17
Mean		113.41	43.17	8.53	18.88
SD		1.12	0.37	0.25	1.12

Table 15. Acceleration test results on Hyundai santro car using Ethanol 30% (Driver A)

Run Number	Direction	Max Speed (km/hr)	Time (Second)		
			0-Max speed	0-60	0-90
1	Up	107.53	45.14	9.42	22.68
2	Down	110.41	43.85	9.04	19.50

3	Up	111.77	43.84	9.09	20.23
4	Down	109.89	44.41	10.19	20.39
5	Up	111.80	43.77	8.91	19.87
6	Down	111.63	43.76	8.79	20.02
7	Up	112.36	43.29	8.90	19.34
8	Down	106.72	44.26	8.71	19.40
9	Up	112.56	43.21	8.52	19.52
10	Down	111.93	43.44	8.59	19.23
Mean		110.66	43.90	9.02	20.02
SD		2.04	0.58	0.49	1.01

Table 16. Acceleration test results on Hyundai santro car using 30% Ethanol (Driver B)

Run Number	Direction	Max Speed (km/hr)	Time (Second)		
			0-Max speed	0-60	0-90
1	Up	113.39	43.68	9.37	19.86
2	Down	112.45	42.95	8.46	18.93
3	Up	112.98	42.93	8.53	18.77
4	Down	112.15	43.23	8.42	18.91
5	Up	112.66	43.21	8.75	19.18
6	Down	113.53	43.04	8.66	19.18
7	Up	112.70	43.21	8.67	19.26
8	Down	113.36	43.00	8.41	18.91
9	Up	113.55	42.65	8.36	18.34
10	Down	113.46	42.85	8.45	18.91
Mean		113.02	43.08	8.61	19.03
SD		0.51	0.28	0.30	0.39

CONCLUSION

Based on the results of this study it can be observed that the blends of Jatropha methyl ester and ethyl ester up to 30% with diesel gave satisfactory performance in terms of the vehicle driving characteristics, fuel consumption and power requirement. These fuels provide satisfactory acceleration performance in actual road conditions. Ethanol blending at higher level affects the emission characteristic but the acceleration performance up to 30% blending is satisfactory.

REFERENCES

Raheman, H. and Phadatare, A.G. 2004. Diesel engine emissions and performance from blends of Karanja methyl ester and diesel. *Biomass and Bioenergy* 27:393-397.

Ramadhas, A.S., Muraleedharan, C. and Jayaraj, S. 2005. Performance and emission evaluation of a diesel engine fuelled with methyl esters of rubber seed oil. *Renewable Energy* 12: 1789-1800.

203-204

Biofuels plantation development and management

Chapter - 1

FACTORS TO CONSIDER IN MANAGING LARGE SCALE PLANTATIONS

Satish Lele

J-22, Sector 7, Vashi, Navi Mumbai, 400 703.
E-mail: satish.lele@gmail.com / lele@vsnl.com;
Web Site: http://www.svlele.com

ABSTRACT

Jatropha is a most promising and potential plant for biodiesel production because of its seed oil characteristics, hardy and easy to establish nature. For commercial biodiesel it is necessary to have Jatropha seeds in large quantities and that can only be achieved through establishing large scale plantations of Jatropha. The ideal size of large scale manageable plantation is 2500 hectares (25 square km). Boundary and contours of plantation area, water bodies, labour and labour facilities, nursery, pruning and maintenance etc. are some factors that should be considered for efficient management of large scale plantations. This paper provides a brief view on the factors to be considered in managing plantations for biodiesel production, as this information would be very useful for long term sustainable output from the new and existing biodiesl plantations.

Key words: Factors, Jatropha, manageable plantations

INTRODUCTION

For producing 1,000 litres of biodiesel/day for 330 days in a year, a 200 hectares of Jatropha plantation is needed, whereas a 10,000 hectares plantation can support a 50,000 litres/day or 4.4 million gallons/ year. According to the size of plantation area, plantations can be classified into following types-

(a) Extra large plantations- The plantation area is 10,000 hectares ($100\ km^2$) or more.

(b) Large plantations- The plantation area is of about 5,000 hectares ($50\ km^2$) or more.

(c) Medium plantation- The plantation area is of about 2,500 hectares (25 km^2) or more.

(d) Small plantation- The plantation area is of about 1,000 hectares (10 km^2) or less but not more than 2,000 hectares.

Medium size plantation of about 2,500 hectares is most manageable plantation, therefore, large scale plantations should be divided into segments of 2,500 hectares for efficient management.

Apart from quality planting material, following factors should also to be considered for managing and maximizing the production from the plantations.

1. **Boundary of area**: The plot of land can be square, elliptical, rectangular or zigzag in shape. The entire area can be in patches of number of plots spread over hundreds of kilometers and these need to be integrated by roads and other services like water supply, electrical power as well as housing and transport. The boundary should be fenced before starting any activity. Plot may be-

- Square, circular or elliptical: These are easier to manage as the total distance of road is smaller.
- Rectangular: This type of area requires longer roads. Roads are expensive compared to cost of plantation.
- Irregular: It is difficult to map initially.
- Sloping: These soils are better for water drainage and irrigation.
- Rocky patches: These are useful for housing or storage sheds.

2. Contours of area: The area can be flat land and can have a number of hills or even mountains. Such slopy areas are ideal for Jatropha plantations as these drain out water. Areas having similar contours should be considered as one unit for plantation purposes. Following points are important for considering the contours of area:

- Mapping to be done: Entire area should be mapped, and all elevations to be mentioned on map.
- Leveling: Generally the soil is not leveled smooth, but some small heaps are leveled to the contour elevations.
- Water channels: If there are irregular water channels, then these are diverted to make smooth flow of water to the reservoirs.
- Localized water bodies: Small earthen dams are put up to hold water at suitable locations, so that water will accumulate there and then percolate in the soil.

3. Water Bodies: There can be water bodies like rivers, channels or reservoirs inside the boundary area and that should be used for irrigation purposes. Additional reservoirs should also be created for holding the water on the plantation area. Following water bodies should be utilized:

- Existing river: Water from such ready water sources should always be utilized for irrigation as these rivers may flow throughout the year.
- Ponds or percolation tanks: This can be created in the flow of rivers, which flow only during monsoon.
- Sink tube wells: Deep tube wells can be dug around such ponds, to irrigate the land in dry seasons.
- Sprinkling network: This should be set up underground, to irrigate the land.
- Pumps: Sufficient number of pumps must be installed on the wells to supply water to main sprinkler line. The pumping costs should be kept to minimum.

4. Clearing the area: The land is not ploughed. Only pits are dug as per required matrix. Following points are kept in mind while clearing the area:

- Large trees should not be cut and left as they are.
- Small bushes and grasses are cut 50-100 mm above the ground and the biomass is left there on soil as cover to avoid evaporation of groundwater.
- The bushes and grasses should not be burnt as it may also burn the subsoil peat and also pollutes the environment.
- The organic system of soil should not be disturbed.

5. Internal roads: A number of internal roads should be laid out in the plantation. These mud roads can not be used for movement of heavy equipment and only light tractors and trailers can run on these roads. These are used during plantation, maintenance and during harvesting. Top surface of the road should be decided depending on rainfall in that area. Following points should be considered for planning road layout:

- The existing straight roads should be used as they exists and any sharp bends on the road or on steep hills should be removed.
- Main road should be laid out at every kilometer. These roads will be used for plying of trucks and heavy equipments.
- Feeder roads should be created at every 200 meters. Generally, one row and column is not planted after every 10 rows and columns forming a square grid of roads at every 20 or 30 meters. This can also be done by leaving one line of plantation blank. These feeder roads should be suitable for tractors, trailers.

- The road should not be steep and should rise gradually at an average angle of 3°(degrees).
- Support sheds should be set up along roads, for storage of farm equipment, produced material and can be also used as shelter for workers.
- Water channels should be away from road so that the road will not get washed away during heavy rain.

6. Housing, food, water, electricity for labour: A large contingent of labour is required during plantation work. Later smaller labour force is required for pruning and harvesting activities. This essential labour should be located at convenient place, to work in fields. For considering the housing, food, water, electricity facilities, following points should be kept in mind:

- Housing for workers should be close to main road for easy transport of labour during plantation activity. However, in later years the labour requirement will be lesser.
- Relocatable houses are preferable as the plantation activity keeps on shifting from place to place.
- Rocky land can not be used for plantation. Therfore, this area can be used for housing as it provides strong foundation.
- Water for drinking and sanitation can be supplied through tube wells, and this water can be used for irrigation after proper treatment.
- The effluent should be disposed close to the point where it is used. The treated effluent should not join the main water source and should be used directly for irrigation.
- Food for the workers can be supplied through mobile kitchens.
- Mobile electric supply units should be installed for supply of electric power to housing complex.

7. Transport of labour and goods: A large transport fleet is required for these activities. Initially a lot of vehicles are required to move the labour and later vehicles are required to transport the Jatropha seeds. While transporting labour and goods following factors should be considered:

- The goods of farm produce should be preferably transported in trucks.
- Labour from camps to farm should be transported in buses/ tractors.
- Distributed storage bins should be located for seeds during harvesting, that can be emptied in trucks for transport to factories.

- A central godown for collection and storage of seeds should be set up.

8. Nursery and plantation: Nursery is set up in the same area as plantation. Generally ratio of nursery area to plantation area is 1:100. If the nursery is developed in soil bed, then the soil is made fertile by adding cow dung or some such manure. About 50 cubic meters of water per acre of nursery per month for 3 months is required, and for this there should be source of water nearby. Additional facilities for pumping and delivery of water should be created. Arrangements of water sprinkler systems should be planed for plantation area. The following are the requirements for establishment of nursery and plantation:

- There should be one nursery for the area of 2,500 hectares. If the location has smaller area, and it is away from other adjoining area, separate nurseries can be set up for each area.
- Nursery size will be 1% of the total area to be catered. It should be at least 10 hectares.
- Jatropha seed requirement for plantation in 2,500 hectares will be around 15 tons.
- Organic fertilizer requirement for plantation in 2,500 hectares will be 15 tons.

9. Man power: Man power requirements for establishing and maintaining nursery and plantation is as follows:

A. Nursery

- For mixing of manure and seed sowing about 50 man hours are required. Manure is spread all over nursery and mixed with soil using tractor. Sowing is also done using tractor attachment.
- Mixing of fertilizer and application requires 10 man hours. This is done few days after the seedling rises above the ground.
- 12 man hours are required for nursery raising. Weeding and removing unhealthy plants is done during the 3 months period.
- Transportation of plants requires 3 man hours. Seedlings are plucked from soil, and transported bare root for plantation.
- For a nursery to cater to 2,500 hectares of plantation, total man hours required are 75.

B. Plantation

Plantation can be done in following matrices:

(i) 2 by 2 meter matrix is used for manual plantation and harvesting.

(ii) 3 x 2 meter matrix is used for mechanized plantation and harvesting.

- Digging of pits requires 250 man hours. These are dug manually or using post hole diggers. Digging time can be reduced by using post hole diggers attached to tractors.
- Pitting requires 200 man hours. The pits are dug manually or using plantation machines.
- About 10 man hours are required for miscellaneous works. This is for maintenance of soil during the plantation.
- Total of about 460 man hours are required for plantation activities.

10. Pruning and maintenance: Pruning is done manually by using large knives or secateurs, Branches are pruned by hand cutters or large knives for initial 2-3 years of plantation. This is done after rest period (winter). Maintenance such as weeding and fertilizer application should be done regularly for initial 2-3 years of plantation.

11. Harvesting: So far machines for harvesting are not available, hence it is done manually. All the fruits are not ready for harvesting at the same time, the fruits ripen gradually over a period of 4 months. There are green, yellow, golden yellow as well as dry fruits in the same bunch. Only golden yellow or dry fruits are ready for harvesting and if the golden yellow fruits are not harvested, these fruits dry up on bush and then it can be harvested mechanically using harvesters similar to Olive. In this process, harvesting gets delayed by 1 or 2 months.

12. Cost of plantation per hectare

S.No	Particular	Amount in Rs (INR)		
		I year	II year	III year
1.	Saplings	6000(2500 nos)	1200(500 nos)	Nil
2.	Fertilizer/manure	2000	400	500
3.	Labour for plantation	6000	1200	Nil
4.	Irrigation/ plant protection	3000	600	500
5.	Labour for fruits collection	Nil	Nil	500
	Total	17000	3400	1500

13. Income from plantation: Income from plantation starts from IIIrd year of plantation. After 3 years, the annual seed production is expected to be around 2000 kgs from one hectare plantation. With the expected price of seeds of Rs 6/kg, the expected income will be Rs 12000 per hectare. After four years the annual seed production is expected to be around 3000 kg/hectare and with the expected price of Rs 6/kg the expected income will be Rs 18000/hectare. The annual seed production from Vth year onwards is expected to be around 4000 kg/ hectare and with the expected price of Rs 6/kg, the expected income will be Rs 24,000/hectare. This annual income is expected to rise with further years of plantation.

14. Establishment of seed procurement cum oil extraction centers: A collection center is required for a 2500 hectare plantation. About 7500 MT of seed will arrive at the procurement center and suitable facilities to store the seed and to extract oil should also be created. The unit cost of seed procurement and expelling center are provided in following table:

S.No.	Particulars	Amount in INR (Rs)
1.	Seed godown	3,100,000
2.	Weighing bridge	700,000
3.	Civil construction	500,000
4.	Cost of land	200,000
5.	Cleaner & grader	100,000
6.	Drier	100,000
7.	DG Set	100,000
8.	Storage tank	200,000
9.	5 sub-centers	1,800,000
10.	Misc	700,000
	Total	7,500,000 (7.5 million)

Expellers: Modern oil expellers with a capacity to crush Jatropha seeds are available. The size of expeller unit may be:

S.No.	Capacity	Capital cost (in Rs)
1.	1 MT/day	7,00,000
2.	1 MT/hour	30,00,000
3.	2 MT/hour	50,00,000

Depending on the capacity of the plant the cost of expelling oil will be between 0.5 cent to 2 cents or Rs. 0.25 per liter to Rs. 0.90 per liter of oil.

Chapter - 2

NURSERY TECHNIQUES FOR HEALTHY JATROPHA SEEDLINGS

Gohil Rajendra Sinh

R & D division (Bio-fuel), Hash Biotech Lab Pvt. Ltd.,
Hash Group of Company, SCO-311-312, Sector-40D, Chandigarh-160036

ABSTRACT

Jatropha curcas is a 3-5 m tall deciduous shrub found in tropical and sub-tropical regions of the world. For getting optimum and healthy seedling of the species, an experiment on seed germination was conducted to know the effect of nursery bed on optimum and speedy development of seedling of *J. curcas* L. The results indicate that the large scale raising of *Jatropha curcas* L. through seeds can provide good opportunity to the people involved in commercial nurseries.

Keywords: Jatropha, seedling, nursery

INTRODUCTION

Jatropha (Euphorbiaceae), a morphologically diverse genus encompassing around 170 species of trees, shrubs, rhizomatous sub-shrubs, herbs, annuals and geophytes, distributed primarily in the seasonally dry tropics (Dehgan, 1984). *Jatropha curcas* is a 3-5 m tall deciduous shrub found in tropical and sub-tropical regions of the world. The genus Jatropha has attracted the attention of several workers, particularly the species *J. curcas*, for the use of its seed oil as a commercially viable alternative source of fuel (Banerji *et al.*, 1985, Martin and Mayeux, 1985, Openshaw, 2000). It can be cultivated successfully in the regions having scanty to heavy rainfall, except water logged condition. The plant produces flowers in racemose inflorescences, with dichasial cyme. The flowers are yellowish green in colour with glabrous or pubescent cymes at the end of the branches. The seed of Jatropha have 30-40% oil (Heller, 1996) of high value, which can be used in locomotive industry as an alternative for diesel and is called as biodiesel.

For getting optimum and healthy seedling of the species, an experiment on seed germination was conducted to know the effect of nursery bed on optimum and speedy development of seedling of *J. curcas* L.

MATERIALS AND METHODS

Seeds of *J. curcas* were collected through the field surveys from natural population. Seeds from each plant (at least 10%) were collected and labelled to maintain the identity of seed parent. An uniform pre-treatment was given to the seeds prior to sowing them in LSF (Liquid Sea Weed Fertilizer) for two hours, and then soaked for 12 hours. Sixty seeds were sown directly in each treatment blocks containing potting mixture of sand, soil and farm yard manure in the ratio of 1:1:1 (by volume) in three replication using randomized block design as per ISTA rules (1976) at R&D centre of Hash Biotech lab Pvt. Ltd. at Mohali (Chandigarh, India) on 4th April, 2008. Prior to sowing, the block containing potting mixture were given fungicide treatment (0.2% Bavistin) and kept covered with plastic sheet for 24 hours. Seed sowing was done in three types of nursery beds viz. raised beds, sunken beds and flat bed. The raised bed was prepared 8.0 cm above the ground level while sunken bed was kept 8 cm below the ground level. Daily germination observation was taken till the germination was completed (upto 20 days).

RESULTS

Results of various treatments are presented in Table 1. In case of sunken bed and flat bed, the germination percentage turned out to be 35 % and 45%, respectively. Sunken and flat beds gave poor germination when compared to raised bed which gave 75% germination. In all three treatments, time taken for germination was same i.e. 11 to 20 days. The possible reason for better germination in raised bed could be that the Jatropha seeds prefer well drained soils, with optimum levels of moisture.

Table 1. Types of bed and its effect on germination

Bed type	Bed size (cm)	Sowing depth	Time taken to germinate (days)	Germination (%)
Flat bed	60 x 60	Ground level	11-20	35
Sunken bed	50 x 50	8 cm below ground level	11-14	45
Raised bed	60 x 60	8 cm above ground level	11-15	75

The results are quite encouraging and indicate that the large scale raising of *Jatropha curcas* L. through seeds can provide good opportunity to the people involved in commercial nurseries and also help to meet the heavy demand of seedling of this important oil bearing plant for the production of biofuels.

REFERENCES

Banerji, R., Chowdhury, A.R., Mishra, G., Sudarshanam, G., Verma, S.C. and Srivastava, G.S. 1985. Jatropha seed oils for energy. *Biomass* 8: 277-282.

Dehgan, B. 1984. Phylogenetic significance of interspecific hybridization in Jatropha (Euphorbiaceae). *Syst. Bot.* 9: 467-468.

Heller, J. 1996. Physic nut, (*Jatropha curcas*): Promoting the conservation and use of underutilized and neglected crops. International Plant Genetic Resources Institute, Rome.

ISTA 1976. International rules for seed testing. Proc. *Int. Seed Testing Assoc.* 31: 1-152.

Martin, G. and Mayeux, A. 1985. Curcas oil (*Jatropha curcas* L.) - A possible fuel. *Agr. Trop.* 9: 73-75.

Openshaw, K. 2000. A review of *Jatropha curcas*: An oil plant of unfulfilled promise. *Biomass and Bioenergy* 19: 1-15.

Chapter - 3

PROPAGATION OF *JATROPHA CURCAS* THROUGH SHOOT SPROUTS

Dinesh Kumar, Sanjay Singh, Vishavjeet Kumar, Ritesh Sharma and Kshitij Malhotra

Silviculture Division, Forest Research Institute, Dehradun

ABSTRACT

Jatropha curcas has been identified as one of the most suitable species under the biodiesel programme. The Planning Commission of India seeks to finally blend diesel with 20% biodiesel by 2011-12 which will substitute for 13.38 million tonne diesel. This is estimated to require plantations on nearly 11 m ha land. In order to allow planting on such extensive areas, it is necessary to have reliable techniques for rapid propagation of this species through seeds and cuttings. The present paper reports the results of a successful attempt to multiply *J. curcas* through branch sprouts.

Key words: *Jatropha curcas*, gestation period, rooting of cutting, clonal propogation

INTRODUCTION

Jatropha curcas has been identified as one of the most suitable species under the biodiesel programme of the Government of India. It is a cheap source of renewable liquid fuel which is extremely important for a developing country like India. Its gestation period is short and yield seeds profusely with a high oil content. It gives highest oil yield per hectare among all tree borne oilseeds (Panwar, 2004). The plant has the ability to thrive well under a wide variety of ecological conditions in greater part of India. Massive efforts are on to plant this species in wastelands all over the country. The Planning Commission seeks to finally blend diesel with 20% biodiesel by 2011-12 which will substitute for 13.38 million tonne diesel. This is estimated to require plantations on nearly 11 m ha land. In order to allow planting on such extensive areas, it is necessary to have reliable techniques for rapid propagation of this species through seeds and cuttings.

Clonal planting of *J. curcas* is likely to gain popularity in future as cloning allows capturing additive as well as non-additive components of genetic variance, and it results in greater uniformity in plantation. Cloning also results in seed production at a younger age than seedling crop. Non-availability of sufficient number of cuttings is one of the major handicaps in clonal planting of *Jatropha* in operational planting programme. The prevalent clonal technology involves planting of 30 cm to 45 cm-long cuttings in nursery beds. Rooting percentage is variable (Kaushik and Kumar, 2005) and finally one rooted cutting results into production of one plant (Kumar *et al.*, 2004). Shortage of cuttings is a major handicap in large scale clonal planting of this species (Srivastava *et al.*, 2004). Dhiman (1999) has multiplied *Paulownia* through root sprouts, thereby producing about twenty plants from one root cutting instead of producing only one plant by the standard nursery practice of that species. Shoot sprouts originating after coppicing of the stems have been successfully used in producing nursery stock of some woody species (Gurumurthi, 2000; Zobel and Talbert, 1984). Sprouts from branches, without coppicing of trees, have been successfully used in micropropagation since long (Nanda *et al.*, 2004, Quairashi *et al.*, 2004, Hernandez *et al.*, 2003). However, reports of use of sprouts from mature branches without cutting back of stems or branches in macropropagation are not reported in the literature for any tree species. Use of sprouts in *J. curcas*, if successful, would increase clonal multiplication rate several times as this species grows fast producing flushes of sprouts all along the branches.

The present paper reports the results of a successful attempt to multiply *J. curcas* through branch sprouts. In view of this success, experiments were also carried out to standardize sprout length and frequency of collecting sprouts. Findings of these studies are summarised in this paper.

MATERIALS AND METHODS

The studies were carried out in Central Nursery, Forest Research Institute; Dehradun located at 30°19'N latitude and 78°03' E longitude. The studies are described as under:

Clonal multiplication of selected clones through branch sprouts

Branches were collected during second and third weeks of January 2006 from eight phenotypically superior mature plants of *Jatropha curcas*. The branches were packed in moist gunny bags and brought to Dehradun. The material was kept moist and placed in shade throughout transportation and storage period. Cuttings of 30 cm length were made within three days of bringing the branches to Dehradun. The cuttings were planted in a medium consisting of loamy soil and sand in 1: 1 ratio inside a polyhouse and irrigated immediately after planting. Average noon temperature inside the polyhouse was 18.1°C, 25.3°C, 33.7°C and 36.7°C during January, February, March and April months, respectively.

Lateral buds on the cuttings started sprouting in mid-March. In cuttings with multiple sprouts, 5 to 10 cm-long sprouts were removed with a sharp blade on April 10, 2006. The longest sprout on the cutting was not removed so as to allow normal growth and development of the cutting into plant. The sprouts were planted in a non-mist propagation chamber constructed as per design described by Dhiman (1999). The medium in contact with sprout was river sand. Average temperature and humidity during noon were 35.8°C and 96 percent during propagation of sprouts in the chamber. Rooting was recorded after 14 days of planting. The rooted sprouts were removed from sand and planted in perforated polythene bags having sand: soil : FYM in 1: 1: 1 ratio. Polythene bags were kept in propagation chamber for 7 days. The plants in polythene bags were removed from propagation chamber on the evening of seventh day and kept in shade for one week. Finally these plants were shifted to open sun. Plant per cent i.e. percentage of sprouts that successfully developed into plants after hardening, was recorded after two weeks.

Standardization of sprout length

Cuttings of mature plants of *Jatropha curcas* were brought from Premnagar, Dehradun and planted in polyhouse during fourth week of January. Sprouts developing from lateral buds were removed from the cuttings on April 26, 2006. The sprouts were graded into four classes on the basis of their length: 3 cm - 5 cm, 6 cm -10 cm, 11 cm - 15 cm and 16 cm - 20 cm. The average diameter of sprout's base i.e. diameter of base of the sprouted shoot for the four length classes was 3 mm - 5 mm, 4 mm - 5 mm, 5 mm - 7 mm and 7 mm - 9 mm. The sprouts were planted in non-mist propagation chambers. The experiment was laid out in randomized block design with 3 replications and 10 sprouts per plot. The resulting plants were hardened as in the previous experiment. Data was recorded as in the above experiment.

Study of performance of sprouts collected on different dates

Cuttings of mature plants of *J. curcas* were brought from Saknidhar, Tehri Garhwal and planted in nursery beds during the third week of February. Sprouts of 7.5 cm to 10 cm length were collected from cuttings on May 10, 2006 (i.e. second week of May) and planted in non-mist propagation chambers. Ten sprouts were planted per treatment with three replications. The sprouts were further collected and planted at 20-day intervals on May 30 (i.e. last week of May) and June 19 (third week of June). Observations on rooting were recorded after 10, 15 and 20 days of planting for each planting date. Rooted sprouts were planted in polythene bags after 20 days and subjected to hardening procedure.

RESULTS

Clonal multiplication of selected clones through branch sprouts

Table 1 reveals that *Jatropha curcas* can be multiplied by vegetative propagation through sprouts with a high degree of success. 85.14 per cent of the sprouts developed roots and all the rooted sprouts survived hardening process and developed into plants. Variation was observed among different clones; the percentage of sprouts that developed into rooted plants ranged from 30 per cent to 100 per cent.

Table 1. Propagation of clones of *Jatropha curcas* through shoot sprouts

Name of clone	Location of ortet	No. of sprouts planted	No. of sprouts rooted	Per cent rooting (%)	No. of plants survived after hardening	Plant per cent obtained from sprouts (%)
		(a)	(b)	(c)= (b) x 100/(a)	(d)	(e)=(d)x100/(a)
Clone-1	Jhajjar	35	32	91.43	32	91.43
Clone-2	Varanasi	20	20	100.00	20	100.00
Clone-3	Dehradun	15	14	93.33	14	93.33
Clone-4	Gorakhpur	25	24	96.00	24	96.00
Clone-5	Rewari	10	3	30.00	3	30.00
Clone-6	Jhajjar	25	22	88.00	22	88.00
Clone-7	Dehradun	5	4	80.00	4	80.00
Clone-8	Hathras	20	11	55.00	11	55.00
Clone-9	Dehradun	20	19	95.00	19	95.00
Total		175	149	85.14	149	85.14

Standardization of sprout length

Sprouts of 6 cm to 15 cm length gave best rooting response. 80 to 86.7 per cent rooting was recorded with this range of sprout length. Sprouts shorter than 5 cm length frequently rotted before rooting could take place, thereby giving 40 per cent rooting. Sprouts longer than 15 cm were not able to stand erect, their apices tended to touch the sand; they wilted and died without rooting in 36.7 percent of the cases. There was no significant mortality in rooted plants during hardening process (Table 2).

Table 2. Effect of sprout length on rooting and plant development

Sprout length (cm)	Basal diameter of sprouts (mm)	Percentage rooting in sprouts (%)	Plant per cent obtained from sprouts (%)
3 to 5	3 to 4	40.0 ± 10.0	36.7 ± 11.6
6 to 10	4 to 5	86.7 ± 5.8	86.7 ± 5.8
11 to 15	5 to 7	80.0 ± 10.0	80.0 ± 10.0
16 to 20	7 to 9	63.3 ± 5.8	63.3 ± 5.8

Study of performance of sprouts collected on different dates

Table 3 shows that callus formation was most rapid in sprouts collected and planted in second and last week of May, more than 43 percent sprouts had sprouted within 10 days in these planting dates. Root formation and plant survival were also better in these dates as compared to planting in third week of June. Rooting of 66.7 percent and 80 percent had been achieved in 15 days and 20 days, respectively in case of sprouts planted in second week of May while corresponding figures were 76.7 percent and 93.3 percent for planting in last week of May, and 46.7 percent 66.7 percent for planting in third week of June. Significant difference was not observed between percentage of sprouts rooted in 20 days and percentage of plants developed after hardening, thereby indicating negligible mortality during hardening process.

Table 3. Variation in rooting performance of sprouts collected on different dates

Time of coll-ecting and planting of sprouts	Percentage of sprouts with callus after 10 days (%)	Percentage of sprouts with rooting (%)			Plant per cent obtained from sprouts (%)
		After 10 days	After 15 days	After 20days	
Second week of May, 2006	43.3 ± 11.5	0	66.7 ± 15.3	80.0 ± 10.0	80.0 ± 10.0
Last week of May, 2006	46.7 ± 5.8	0	76.7 ± 5.8	93.3 ± 11.5	90.0 ± 10.0
Third week of June, 2006	20.0 ± 10.0	0	46.7 ± 11.5	66.7 ± 15.3	56.7 ± 15.3

DISCUSSION

The results in Table 1 reflect that *J. curcas* can be successfully propagated through branch sprouts. About 85 percent sprouts developed roots within 20 days and grew into healthy plants during the hardening process. These sprouts had been taken from branch cuttings while retaining the longest sprouts with the cutting. Thus plants in these studies were produced through detached sprouts over and above the plants that were produced through the cuttings by the normal procedure of propagation through cuttings. Different workers have so far suggested propagation of this species by the standard procedure of vegetative multiplication whereby one plant is produced per cutting (Kaushik and Kumar, 2005, Kumar *et al.,* 2004). The procedure of propagation through sprouts developed by us produces several plants per cutting and can, therefore, be used for mass propagation of clonal material of this species. Data on number of sprouts produced per cutting was not recorded in this study. Studies will be undertaken in the future to determine the number of plants that can be produced per cutting. Variation was observed among clones for propagation success through sprouts. Studies will be continued to examine the differential behaviour of different genotypes.

Data in Table 2 suggested that sprouts of 6 cm to 15 cm length give better propagation success than shorter or longer sprouts. It emerged that it would be judicious to use sprouts of 6 cm to 10 cm length since significant difference was not observed during its comparison with 11 cm to 15 cm length in the present study. Removal of shorter sprouts may, in fact, lead to development of more sprouts on the same cutting, thereby further increasing the clonal multiplication rate.

As far as suitable period for collection and planting of sprouts is concerned, June 19 appears to give relatively poor performance. At least 80 per cent rooting was observed when sprouts were planted for rooting from second to last week of May (Table 3). Planting of 6 cm to 15 cm long sprouts on April 26 also resulted in more than 80 per cent rooting (Table 2). Sprouts collected before April 26 or after June 19 were not subjected to rooting. Different germplasm used in different experiments may respond differently to different dates. Hence it is not feasible to suggest the best period of the year for propagation through sprouts. However, it can be safely stated that last week of April to last week of May is a suitable period for this purpose.

The technique of vegetative propagation through sprouts can be used when largescale multiplication of superior germplasm is required but the quantity of cuttings available for propagation is not sufficient to meet the requirement by the conventional practice. The findings of this study may also prove helpful for developing practices for management of stools in a vegetative multiplication garden.

ACKNOWLEDGEMENT

The study was carried out under the NMITLI research project on Jatropha financed by Council of Scientific and Industrial Research, Govt.

of India. The authors are thankful to the funding agency for this assistance.

REFERENCES

Dhiman, R.C. 1999. *Paulownia*. No. 61, ICFRE Brochure 9, Forest Research Institute, Dehradun. 27 p.

Gurumurthi, K. 2000. Clonal forestry for tree improvement. In: *Genetic improvement and propagation of forest trees* (eds. Bennet, S.S.R., Subramanian, K. and A.J. Kurian) Institute of Forest Genetics and Tree Breeding, Coimbatore. pp. 37-61.

Hernandez, I., Celestino, C. and Toribio, M. 2003. Vegetative propagation of *Quercus suber* L. by somatic embryogenesis. I. Factors affecting the induction in leaves from mature cork oak trees. *Plant Cell Reports* 21(8): 759-764.

Kaushik, N. and Kumar, S. 2005. *Jatropha curcas* L. silviculture and uses. Agrobios, India, Jodhpur. 53 p.

Kumar, R.V., Ahlawat, S.P. and Gupta, V.K. 2004. Mass propagation of *Jatropha*. In: *Compendium on National Conference on Biodiesel*. Utthan, Centre for Sustainable Development & Poverty Alleviation. Allahabad. pp. 31-36.

Nanda, R.M., Premananda, D. and Rout, G.R. 2004. *In vitro* clonal propagation of *Acacia mangium* Willd. and its evaluation of genetic stability through RAPD marker. *Annals of Forest Science* 61(4): 381-386.

Panwar, K.S. 2004. Sustaniable development of degraded and wasteland areas through Jatropha cultivation. In: *Proceedings of International Conference on Biofuels: Perspectives and Prospects*, New Delhi. Sept 16-17, 2004. Winrock International, India. New Delhi. pp. 229-232.

Quraishi, A., Koche,V., Sharma, P. and Mishra, S.K. 2004. *In vitro* clonal propagation of neem (*Azadirachta indica*). *Plant Cell, Tissue and Organ Culture* 78(3): 281-284.

Srivastava, R.K., Kumar, D. and Kumar, S. 2004. Clonal planting of *Jatropha curcas*: need, challenges and prospects. In: *Compendium on National Conference on Biodiesel*. Utthan, Centre for Sustainable Development and Poverty Alleviation. Allahabad. pp. 7-12.

Zobel, B and Talbert, J. 1984. Applied forest tree improvement. John Wiley and Sons, New York. pp. 309-344.

Chapter - 4

CANOPY DEVELOPMENT AND MANAGEMENT OF *JATROPHA CURCAS* FOR HIGHER SEED YIELD

K. Chalapathy Reddy, Vishal V. Naole, Tarun Thakur, V.K. Kiran Kumar, B.H. Raghu and Satyavir Singh

Mission Biofuels India Pvt. Ltd., 608 Powai Plaza,
Hiranandani Business Park, Powai, Mumbai 400076
E-mail: dr.chalapathy@missionnewenergy.com

ABSTRACT

Jatropha seed and oil yield is the function of planting material used, growing conditions, plant architecture development and maintenance. Building ideal canopy by pruning during first growing season is very important. It is an important process to increase the productivity of the plant and to reduce the gestation period. Pruning induces more number of lateral bran-ches and these branches produce more number of fruits. The pruning helps in improving the plant architecture and maintaining the balance between vegetative and reproductive growth of the plants. For developing the plant into a desired architecture including height atleast more than two pruning are essential in three years.

Keywords: *Jatropha curcas*, pruning, canopy

INTRODUCTION

Jatropha seed and oil yield is the function of planting material used, growing conditions, plant architecture development and maintenance. Jatropha is hardy plant; thrives well in arid and semi arid regions with an average rainfall ranging from 500 to 1200 mm. It has been planted in varied soil types, marginal and waste lands wherein utilization of *in situ* resources by agronomic practices and development and management of canopy by pruning are very crucial for harvesting higher seed yield for biodiesel production. Building ideal canopy by pruning during first growing season is very important.

The pruning is a simple cutting off the main stem or the top branches at predetermined height. This pruning induces more number of lateral branches and these branches produce more number of fruits. The number of branches and the number fruits are directly related to the quality and success of pruning. It is an important process to increase the productivity of the plant and to reduce the gestation period.

Importance of pruning in Jatropha

- Induces more number of new lateral shoots (Fig. 2 and 5).
- Increases flower and fruit production per plant.
- Induces bushy architecture (Fig. 2).
- Increases light penetration into the bush and improves overall fruit quality.
- Helps in better management of pests and diseases.
- Facilitates easy harvesting of matured fruits.
- Helps in removing diseased, whippy, defunct and or moribund branches.
- Reduces the size of the above-ground portion of the plant, which allow the farmer to take inter cultural operations in inter-space available between the Jatropha plants.

Fig. 1. One year old non pruned plants

Fig. 2. First year pruned plant

Number of pruning needed for *Jatropha curcas*

It is ideal to keep the plant height around 2 m for easy harvesting and accommodating more plants per unit area. For developing the plant n to a desired design and height atleast two pruning are essential in hree years. First pruning should be done within 6 months of ansplanting and second pruning should be completed during winter ason of second year of planting. Comparison between pruned and non-ned plants are given in Table 1.

Table 1. Comparison between pruned and non pruned 2 year old plants

Character	Non Pruned	Pruned
Branches number	1. Less than 2 branches after first year of planting 2. Less than 4 branches after second year of planting (Fig. 1)	1. 4 to 8 branches after first year of pruning 2. 20 to 32 branches after second year pruning (Fig. 2)
Fruit number	1. 2 to 3 fruit bunches per branch 2. 6 to 8 fruits per bunch 3. 48 to 96 fruits per plant	1. 2 to 3 fruit bunches per branch 2. 6 to 8 fruits per bunch 3. 240 to 768 fruits per plant
Seed yield (kg/plant)	1. No. of seeds: 144 – 288 2. Seed yield: 0.10 to 0.25 kg	1. No. of seeds: 720 – 2304 2. Seed yield: 0.75 to 1.50 kg

The following criteria for pruning are important:

- Plants should have attained more than 3.5 cm stem diameter.
- Plant height should be more than 60 cm above ground (Fig. 3).
- The basal stem colour should be silvery whitish at least up to 46 cm height (Fig. 4).
- Pruning should be done at the height of 30 cm to 46 cm from the ground (Fig. 4).
- The basal stem of 46 cm height will have 10 to 15 eye buds for establishment of lateral branches.

Fig. 3. Height of plant 3 feet

Fig. 4. 1.5 Feet tall pruned stump

Fig. 5. Growth of new lateral shoots

Pruning season: Pruning should be done during winter season when plants are in dormant condition. The basic principle is that during dormant condition, mobilization of photosynthates takes place from leaves to the root and shoots system and all the resources are stored in the main root system. After pruning the new growth utilizes the stored resources from the root system; while new branches start synthesizing photosynthates. Pruning should not be done when the plant is having more new vegetative growth and it is ideal to prune when the plant is shedding its leaves during winter season.

Pruning of lateral branches

The lateral branches of the plant should be pruned along with main stem and pruned stem should have atleast 6 to 8 buds (Fig. 6, 7).

Fig. 6. Plant with lateral branch

Fig. 7. Pruned stump with lateral branch pruned

Pruning instrument

Sharp knife: Pruning is done with the help of sharp knife enough to make clean cut ends of stumps (Fig. 8). Sharpening of cutting tool should be repeated after pruning of every 100 plants. Use of secateurs or hedge scissors should be avoided, as they may result in injury to bark and invite entry of fungus/insects into branch.

Specifications for pruning cuts

- A pruning cut can be a point of infection for various diseases and this risk must be minimized. For a good clean cut, slant and smooth edge is essential (Fig. 11). This will also avoid the risk of damaging the bark on the main stem below the pruned branch.
- There should be no splitting of branches or bark during pruning operation (Fig. 12).
- Improper pruning technique will seriously damage or kill the plants.

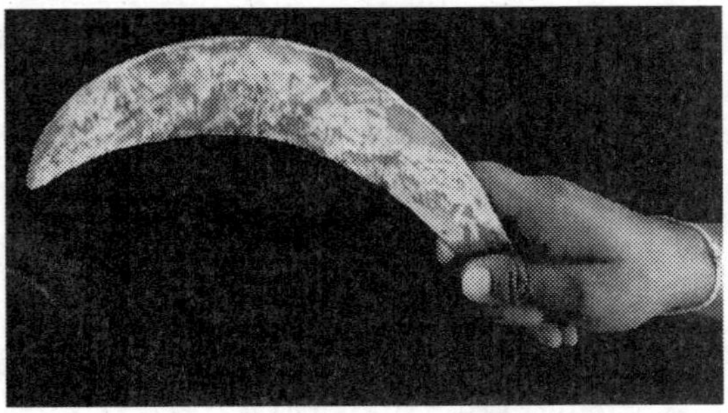

Fig. 8. Sharp knife for pruning

Fig. 9. Position taken for ideal pruning

Precautions to be taken during pruning

- Remove old leaves from pruned stem (Fig. 4 and 7).
- Do not hold the knife loosely as this may result in the knife falling from the hand and cutting oneself. The sharp knife is handled in such way that the angle between plant and knife is almost 45 degree (Fig.11).
- Sharpen the knife after pruning every 100 plants.
- In any circumstances do not use blunt knives as this will result in uneven cut surface.

Precautions to be taken after pruning

- The cut ends of the stumps should be covered with cow dung or handful of soil; to prevent loss of moisture, fungal attack and drying of the stumps.
- Do not leave broken and split branches hanging in the field as it is an open invitation for disease and insect pests.
- The pruned branches should be kept around the pruned stump, as they will act as a mulch for the plant (Fig. 10).

Fig. 10. Pruned materials used as mulch

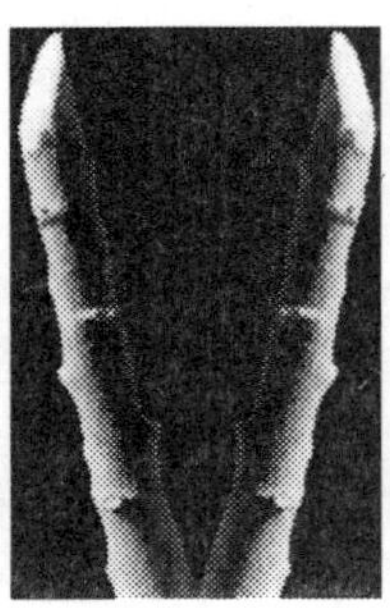

Fig. 11. Stumps with clean cuts

Fig. 12. Stumps with damaged cuts

CONCLUSION

The pruning helps in improving the plant architecture and maintaining the balance between vegetative and reproductive growth of the plants. It is ideal to keep the plant height of around 2 m for easy harvesting and accommodating more plants per unit area. For developing the plant into a desired architecture and height; more than two pruning are essential in three years.

Chapter - 5

COMPARATIVE EVALUATION OF JATROPHA CLONES (GRAFTED) AND SEED SOURCE

P. Masilamani , K.T. Parthiban[1], M. Paramathma[2] and K. Annadurai

Agricultural Engineering College & Research Institute,
Tamil Nadu Agricultural University, Kumulur – 621 712, Tamil Nadu.
[1]Forest College and Research Institute, Mettupalayam
[2]Centre for Excellence in Bio-fuels, AEC&RI, TNAU, Coimbatore-3
E-mail: masil_mahesh @ yahoo.com

ABSTRACT

Experiment was conducted to develop a drought tolerant *Jatropha curcas* through cleft (or) wedge grafting with *Jatropha glandulifera* and *Jatropha gossypifolia* (drought tolerant species) and to find out the grafting compatibility, survival and growth attributes of grafted plant at weekly intervals up to seven weeks. Another study was conducted to find out the growth and yield potential of grafted Jatropha clones along with check. On seven weeks after grafting, the results revealed that survival percentage (80.3 per cent) and number of sprouts per plant (3.2) was higher in plants grafted with *Jatropha gossypifolia* than the plants grafted with *Jatropha glandulifera*. It is concluded that considering the survival percentage and number of sprouts per grafted plant, cleft or wedge grafting of *Jatropha gossypifolia* with *Jatropha curcas* was found to have good compatibility for grafting. Sixteen months after planting under field condition the results revealed that grafted Jatropha clones produced more number of fruits (32) per plant, fruits weight (270g.), 100 seed weight (67.2g.) and seed yield per plant (75.68g) when compared to check (6.9 fruits per plant, fruit weight 63g. per plant and seed yield of 15.3 g. per plant).

Keywords: Jatropha, cleft grafting, compatibility, survival percentage

INTRODUCTION

Jatropha curcas L. is a multipurpose tropical shrub native to Mexico and Central America, but it is widely distributed in wild or semi cultivated lands in Latin America, Africa, India and South East Asia (Swamy and Singh, 2006). The economic importance of *Jatropha curcas*

has been increasing because of suitability of its oil as an alternative for diesel fuel, and for soap production in rural areas, which would improve the standard of living conditions of the people and would offer additional income (Ghosh *et al.*, 2008). In addition to *Jatropha curcas*, 13 species were reported in South India (Paramathma *et al.*, 2004). Among them, six species viz., *Jatropha curcas, Jatropha gossypifolia, Jatropha glandulifera* (widely distributed in wild), *Jatropha tanjoransis* (natural hybrid – sterile plant), *Jatropha multifida* and *Jatropha podagrica* (ornamental type) were distributed in Tiruchirappalli, Perambalur, Cuddalore, Pudukottai districts and Cauvery delta regions of Tamil Nadu (Paramathma *et al.*, 2007). Based on observations, among the above six species, two species viz., *Jatropha glandulifera* and *Jatropha gossypifolia* (highly drought tolerant species) were found predominantly distributed, but produced lesser seed and oil yield. On the other hand *Jatropha curcas* produce comparatively higher seed and oil yield, but the Jatropha planters (promoters, private industry personnels, researchers and farmers) experienced that *Jatropha curcas* could not tolerate extreme drought. Against the backdrop, efforts were initiated to develop a drought tolerant *Jatropha curcas* through grafting with *Jatropha glandulifera* and *Jatropha gossypifolia* (drought tolerant species) and to find out the grafting compatibility, survival and growth attributes of grafted plant and also to find out the comparative evaluation of Jatropha (grafted) clones and seed sources planted in the main field.

MATERIALS AND METHODS

Jatropha glandulifera and Jatropha gossypifolia seeds were collected from wild stand in the wastelands of Lalgudi block, Tiruchirappalli District, Tamil Nadu. The seeds were subjected to 12 hours of water soaking (Kathiravan, 2004) and then placed for germination in earthen pots containing red earth, sand and farm yard manure at 2:1:1 ratio. Germination was initiated 7 days after sowing, the seedlings were allowed for establishment up to 3 months. Cleft or wedge grafting of above two drought tolerant species was made with *Jatropha curcas*. By this method, the three-month-old seedlings (1.5 cm diameter) were prepared as a root stock (the shoot tip of the seedling was removed and a vertical cut was made through the centre pith). Same thickness of scion (shoot portion) was prepared (by making tapering cuts in the form of wedge) from two year old *J. curcas*. Scions were inserted to the rootstock of *Jatropha glandulifera* and *Jatropha gossypifolia,* and then the union was tied with polythene tape to avoid water entry (Plate 1 & 2). Replicated trial was conducted with five replications and fifty seedlings per replication. The observations on survival percentage and number of sprouts per grafted plant were made at weekly intervals up to seven weeks for graft union. The mean values of five replications on survival percentage and number of sprouts per grafted plant are given in Table 1. The compatible grafted plants of 380 days old were planted in the main field along with check to find out the survival percentage (Plate 3). Another study was conducted to find out the growth and yield potential of grafted Jatropha clones (plants grafted with *Jatropha gossypifolia*) along with check. The trial was established in 0.40 acre adapting a spacing of 2.5 x 2.5 m at Agricultural Engineering College & Research Institute Farm, Kumulur, Tamil Nadu. Each treatment was replicated six times with nineteen seedlings along with border rows. The observations on survival percentage, plant height (cm), number of leaves per plant, total number of fruits per plant, fruits weight per plant (g), seed yield per plant (g) and 100 seed weight (g) were estimated. The germination test was carried out in sand medium in quadruplicate using 200 seeds for each treatment with four replications of 50 seeds (ISTA, 1999) in a germination room maintained at a temperature of 25 ± 2°C with diffused light (approx. 10 hr) during the day. Final count on normal seedlings was recorded on fourteenth day (Kathiravan *et al.,* 2008) and germination per cent was computed. The seedling dry weight was determined using normal seedlings after drying at 85°C for 24hr and expressed as g seedlings^{-10}. Vigour index was calculated as the product of germination percentage and seedling length (Baki and Anderson, 1973). Oil content of the seeds was estimated and expressed in percent.

Jatropha Grafting

Plate 1. Cleft grafting *Jatropha glandulifera* with *Jatropha curcas*

Plate 2. Cleft grafting *Jatropha glandulifera* with *Jatropha curcas*

Plate 3. Grafted plant (*Jatropha gossypifolia* - stock with *Jatropha curcas* - scion) ready for field planting

RESULTS AND DISCUSSION

The results on influence of grafting on survival percentage of grafted plant revealed that survival percentage was high in plants grafted with *Jatropha gossypifolia* than grafted with *Jatropha glandulifera* irrespective of week's after grafting. In the plants grafted with *Jatropha gossypifolia, the* survival percentage was high from 1st week to 3rd week (72.0% to 86.3%). And then, there was a gradual reduction in survival percentage (i.e.) 84.2 to 80.3%.Where as in the case of plants grafted with *Jatropha glandulifera,* up to 2 weeks, there was an increase in survival percentage (i.e.) 19.0% to 47.2%. After then, a steady reduction in survival percentage was observed (i.e.) 46.1 to 29.3%. Between these two species, the plants grafted with *Jatropha gossypifolia* showed good compatibility for grafting in terms of survival percentage. The results on influence of grafting on number of sprouts per plant revealed that among the two species tried for grafting, the number of sprouts per plant was high in the plants grafted with *Jatropha gossypifolia* than the plants grafted with *Jatropha glandulifera* irrespective of weeks (time) after grafting. The number of sprouts per grafted plants varied from 3.1-4.2 and 1.6-3.0 for the plants grafted with *Jatropha gossypifolia* and *Jatropha glandulifera* respectively (Table 1).

Table 1. Influence of grafting on survival percentage and number of sprouts per grafted plant

Weeks after grafting	Survival Percentage		Number of sprouts per grafted plant	
	Jatropha gossypifolia	*Jatropha glandulifera*	*Jatropha gossypifolia*	*Jatropha glandulifera*
1	72.0 (58.05)	19.0 (25.84)	3.4	1.6
2	86.0 (68.03)	47.2 (43.39)	3.8	2.2
3	86.3 (68.28)	46.1 (42.76)	4.2	3.0
4	84.2 (66.58)	41.4 (40.05)	3.5	2.7
5	82.7 (65.42)	37.0 (37.46)	3.2	2.0
6	81.0 (64.16)	32.7 (34.88)	3.1	1.9
7	80.3 (63.65)	29.3 (32.77)	3.2	1.9

Figures in parenthesis indicate *Arc sine* values

Note: Pooled mean of five replication, data not statistically analyzed.

Plate 4. Grafted plant (*Jatropha gossypifolia* stock with *Jatropha curcas* - scion) planted in the field.

Sixteen months after planting, the results under field conditions revealed that even though there was a reduction in plant height and number of leaves per plant, grafted Jatropha clones produced more number of fruit per plant (32), higher fruit weight (270 g), 100 seed weight (67.2 g) and seed yield per plant (75.68g) as compared to check (6.9 fruits per plant, fruit weight per plant (63 g) and seed yield of 15.3 g. per plant) (Table 2). The oil content was slightly high with the grafted clone than check. The seeds collected from the grafted clones showed higher germination percentage than the check. From this experiment, it could be concluded that considering the survival percentage and number of sprouts per grafted plant, cleft or wedge grafting *Jatropha gossypifolia* with *Jatropha curcas* was found to have good compatibility for grafting (Plate 3). The results on the performance of grafted plant and its survival under field conditions were promising (Plate 4 & 5). Hence, it can be concluded that grafted Jatropha clones have good potential to tolerate drought conditions. However, it needs further investigation to stabilize its yield potential at least for a minimum period of 5 years.

Table 2. Comparison of biometric observations between grafted plant and checks (16 months old)

Biometric observations	Check	Grafted plant
Survival (%)	98.3	92.1
Plant height (cm)	161.0	134.5
Number of leaves / plant	136	84.7
Total No. of fruits / plant	6.9	32.3
Fruit weight / plant (g)	63.0	270.0
Seed yield / plant (g)	15.35	75.68
100 seed weight (g)	66.20	67.20
Germination (%)	76.0	78.0
Dry matter production(g^{-10})	2.7	3.0
Seedling vigour	3044	3140
Oil content (%)	26	28

Note : Pooled mean of five replication, data not statistically analyzed.

Plate 5. Grafted plant (*Jatropha gossypifolia* - stock with *Jatropha curcas* - scion) initiated for bearing.

REFERENCES

Abdul Baki, A.A. and Anderson, T.D. 1973. Vigour determination in soybean by multiple criteria. *Crop Science* 13: 630-633.

Ghosh, L., Singh, L. and Singh, A.K. 2008. Variation in rooting response in cuttings of *Jatropha curcas* L. *Indian Journal of Agroforestry* 2:49-53.

International Seed Testing Association. 1999. International rules for seed testing. *Seed Science and Technology* 21 (Suppl.) 27: 25-30.

Kathiravan, M. 2004. Seed Technological studies in *Jatropha curcas.* Ph.D. Thesis, Department of Seed Science and Technology, Tamil Nadu Agricultural University, Coimbatore, Tamil Nadu. 220 p.

Kathiravan, M., Ponnuswamy, A.S., Vanitha, C. and Balamurugan, P. 2008. Standardization of seed storage techniques to minimize loss of vigour and viability in Jatropha (*Jatropha curcas* L.). *Seed Research* 36: 84-89.

Paramathma, M., Parthiban, K.T. and Neelakandan, E.S. 2004. *Jatropha curcas.* Forest College and Research Institute Publication. Mettupalayam, Tamil Nadu, India. 48 p.

Paramathama, M., Srimathi, P. and Nelson Navamaniraj, K. 2007. Seed standards for quality seedling production. In: *Jatropha Improvement, Management and Production of Biodiesel.* Paramathma, M., Venkatachalam, P and Sampathrajan, A. (ed.) Tamil Nadu Agricultural University Publication, Coimbatore, Tamil Nadu. 190-218 pp.

Swamy, S.L. and Singh, L. 2006. Strategies for development of quality planting stock of *Jatropha curcas* for biofuel plantations. Biodiesel Conference Towards Energy Independence – Focus on Jatropha. Rashtirapathi Bhawan, New Delhi,143-154 pp.

Chapter - 6

CLONING OF *JATROPHA CURCAS* (L.) : AN EXCELLENT SOURCE OF BIODIESEL

Vinod Saharan, Birchand Patel, Rupesh Singh, Rajveer Singh, Anita Sharma* and Vimal Sharma

Plant Tissue Culture Laboratory,
Department of Molecular Biology and Biotechnology
* Department of Plant Breeding and Genetics, Maharana Pratap University of Agriculture and Technology, Udaipur 313 001
E-mail: vinodsaharan@gmail.com

ABSTRACT

A reproducible protocol would be a prime factor for mass multiplication of Jatropha species. *Jatropha curcas* still regarded as highly recalcitrant for tissue culture. Shoot tip necrosis (STN) is the major problem encountered during successful regeneration of Jatropha. Additional amount (80 and 40 mg/l) in addition to standard concentration of Ca and K in MS basal media is imperative for *in vitro* regeneration of Jatropha plants. TDZ, BA, Kin, IAA and GA_3 formulated MS media induced multiple shoots via adventitious regeneration from leaf explant. Enrichment of MS basal medium with Ca and K played significant role in order to overcome STN.

Keywords: Biodiesel, Jatropha, *in vitro* regeneration, shoot tip necrosis

INTRODUCTION

Large-scale plantation of *Jatropha curcas* (L.) is being practiced all over the world for maintaining the supply of energy as a biodiesel. In this connection larger number of private and public organization investing on research and development of Jatropha plants. The document, India vision 2020 presented by the Planning Commission as a framework for policy planning in the coming decades, mentions the potential of biofuels in general and specifically refers to plantations of *Jatropha curcas* to produce large quantities of biodiesel. According to this document, cultivation of 10 million ha of this crop could generate 7.5 million metric tons of fuel annually, while generating year-round employment for 5 million people. India is likely to require about 5-6 million tons of bio-diesel in 2030, if 5% of the diesel used in transport is to be replaced. This amount of bio-diesel can be generated from 10 million ha of Jatropha plantations (Francis *et al.*, 2005). As Jatropha is still a wild plant, improvement of germplasm in respect to insect-pest, viral disease, toxicity and yield are necessary before mass-production can be realised. Therefore, additional genetic tools are required to further exploit the potential resources of *J. curcas*. In present scenario, Jatropha biotechnology could play an important role in improving the existing Jatropha germplams. Mass multiplication of desire clone, production of inbred lines through pollen culture, induction of somaclonal variation, generation of variation via transgenic technology are some of the possible applications of biotechnology in Jatropha. To date, many efforts have been made on *in vitro* regeneration protocol of Jatropha. A number of workers have done research on tissue culture of Jatropha and developed protocols for *in vitro* propagation from various explants (Sujatha and Dhingra, 1993; Sujatha and Mukta, 1996; Sujatha *et al.*, 2005). More reports have been published for achieving *in vitro* culture system in *J. curcas* (Wei *et al.*, 2004; Rajore and Batra, 2005; Datta *et al.*, 2007). Recently, efficient regeneration protocol was standardised to raise plan-ting material from somatic embryos of Jatropha (Jha *et al.*, 2007).

MATERIAL AND METHODS

In present investigation, *J. curcas* leaf explants were collected from 1.5 year old *Jatropha curcas* plantation in Department of Molecular Biology and Biotechnology, Maharana Pratap University of Agriculture and Technology, Udaipur, Rajasthan. Leaf explant was initially submerged in 70% (v/v) ethanol solution for 1 min, then surface disinfested with 0.1% (m/v) $HgCl_2$ for 15 min, washed several times with sterile water and cultured on shoot induction media consist of Murashige and Skoog (1962), Himedia, Mumbai, India with 1.0 mg / L TDZ + 0.5 mg/ L BAP. The pH of MS medium was adjusted to 5.8 prior to sterilization at 120°C for 20 min. Leaf explants were cut into 1.0 cm^2 size before culture into shoot induction media. After two subculture, at an interval of 2 weeks on shoot induction media, all the cultures were transferred to shoot multiplication/elongation media. Shoot elongation /multiplication media consisted of 0.5 mg / L BAP + 0.5 mg / L Kin+ 0.25 mg / L IAA + 0.25 mg / L GA_3 in MS basal media. After two subcultures on multiplication media, microshoots were evaluated to see the effect of Ca and K on survival percentage. Every experiment was repeated twice and every treatments consisted of 30 microshoots.

RESULTS AND DISCUSSION

We have established protocol for direct organogenesis from young leaf of *Jatropha curcas* (L.) using a combination of TDZ and BAP in shoot induction and BAP, Kin, IAA and GA_3 combination for shoot multiplication/elongation. Frequency of shoot induction from leaf explants was recorded around 55% and 5-6 shoots can be harvested from 1.0 cm^2 leaf explant. After two subculture on shoot induction media, small protuberance emerged from surface of explants and in 6-weeks time 0.5-1.0 cm length microshoots developed from leaf explants. Microshoots transferred to shoot multiplication/elongation media for further 4 weeks (6-10 week). During this period microshoots start to elongate and multiplied (Fig. 1). After 10th week, elongated shoots were separated and transferred to rooting media (0.5 mg / L IBA). In period of 2-3 weeks root emerged from the base of growing shoots (Fig. 2). From 5th to 6th week, microshoots started to show symptoms of STN and dieback (Fig. 3). Jatropha microshoots exhibited leaf burn followed by shoot necrosis and death of regenerated microshoots. Browning / blackening of the margin and tip of young leaves were first symptoms of the STN. Reaching to 9th week, 42% of microshoots were dead due to STN. However, some of the microshoots died during well-established root stage. Therefore, almost 50% regenerated shoot died before getting into hardening stage. The occurrence of necrosis after several subcultures has been reported in *Psidium guajava* (Amin and Jaiswal, 1988) and *Cercis canadensis* (Yusnita *et al.*, 1990). Dieback of the shoot tips has been reported in *Gymnema sylvestre* (Komalavalli and Rao, 2000), *Holostemma adakodien* (Martin, 2002) and *Rotula aquatica* (Martin, 2003). In the present study plant regeneration protocol from leaf explants were used for study of effect of Ca and K on STN. Shoot cultured on shoot multiplication media were transferred to various concentration of $CaCl_2$ (0, 20, 40, 60, 80, and100 mg/L) enriched MS basal supplemented with 0.5 mg / L BAP + 0.5 mg / L Kin.+ 0.25 mg / L IAA + 0.25 mg/L GA_3 in 6th week (Table 1). Data on survival of microshoots were recorded up to 9th week with two more subculture. Calcium chloride 60-80 mg/L found effective and showed survival of 77-80% microshoots. Increasing concentration of $CaCl_2$ above 80 mg / L become toxic to microshoots. Further, effect of K was also evaluated for survival percentage of microshoots up to 9th week. Various concentration of $K_2H_2PO_4$ (0, 20, 40, 60, 80 and 100 mg/L) were supplied to media containing 0.5 mg/L BAP + 0.5 mg/L Kin.+ 0.25 mg/L IAA + 0.25 mg / LGA_3 + 60 mg/L $CaCl_2$ in MS basal media. Survival percentage of microshoots further improved, media containing 40-60 mg/L $K_2H_2PO_4$ showed 88-85% survival of micro-shoots (Table 2).

Fig. 1. Multiplication and elongation of Jatropha shoots

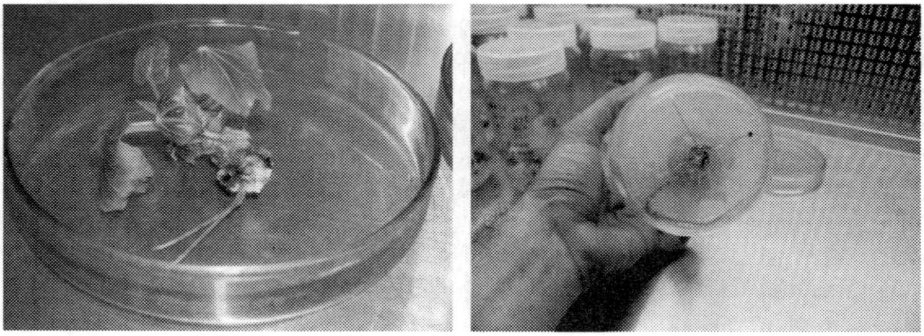

Fig. 2. Induction of root in regenerated shoot of *Jatropha curcas* (L.)

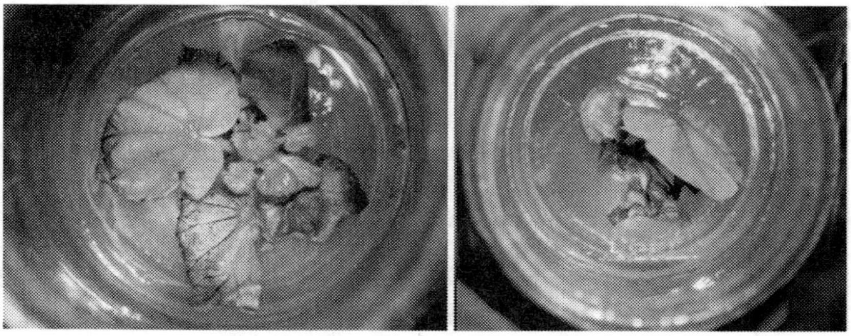

Fig. 3. Initiation of Shoot tip necrosis and dieback of Jatropha plantlets

Table 1. Effect of $CaCl_2$ on survival percentage of microshoots of *Jatropha curcas* culture on MS basal media supplemented with 0.5 mg / L BAP + 0.5 mg / L Kin.+ 0.25 mg/ L IAA + 0.25 mg / L GA_3. Each concentration tested on 30 microshoots, results in percentage are mean of 30 microshoots.

$CaCl_2$ (mg/L)	6th week	7th week	8th week	9th week
0	90	74	62	58
20	90	78	67	60
40	95	83	74	68
60	97	91	85	77
80	97	88	80	80
100	90	80	68	65

CONCLUSION

The results express concern about possible change in the ratio of nutrients in MS basal media. From this protocol, direct multiple shoots can be harvested from leaf explants without callusing phase, which is the key of true-to-type production of Jatropha plants. This protocol could be used for mass multiplication and genetic transformation of *Jatropha curcas*, a highly valuable industrial crop.

Table 2. Effect of $K_2H_2PO_4$ on survival percentage of microshoots of *Jatropha curcas* culture on MS basal media supplemented with 0.5 mg/l BAP + 0.5 mg/l Kin.+ 0.25 mg / L IAA + 0.25 mg / L GA_3 + 60 mg/L $CaCl_2$. Each concentration tested on 30 microshoots, results in percentage are of mean of 30 microshoots.

$K_2H_2PO_4$ (mg/l)	6th	7th	8th	9th
0	96	92	87	80
20	96	90	88	84
40	100	95	92	88
60	95	93	89	85
80	94	88	82	80
100	90	80	68	65

REFERENCES

Amin, M.N. and Jaiswal, V.S. 1988. Micro-propagation as an aid to rapid cloning of a *guava* cultivar 36:89-95.

Datta, M.M., Mukherjee, P., Ghosh, B. and Jha, T.B. 2007. *In vitro* clonal propagation of biodiesel plant (*Jatropha curcas* L.). *Current Science* 93(10):1438-1442.

Francis, G., Edinger, R. and Becker, K. 2005. A concept for simultaneous wasteland reclamation, fuel production, and socio-economic development in degraded areas in India: need, potential and perspectives of Jatropha plantations. *Natural Resources Forum* 29:12-24.

Jha, T.B., Mukherjee, P. and Datta, M.M. 2007. Somatic embryogenesis in *Jatropha curcas* Linn., an important biofuel plant. *Plant Biotechnology Reports* 1:135-140.

Komalavalli, N. and Rao, M.V. 2000. *In vitro* micro-propagation of *Gymnema sylvestre* – a multipurpose medicinal plant. *Plant Cell Tissue and Organ Culture* 61 :97-105.

Martin, K.P. 2002. Rapid propagation of *Holostemma adakodien* Schult., a rare medicinal plant through axillary bud multiplication and indirect organogenesis. *Plant Cell Reports* 21 :112-117.

Martin, K.P. 2003. Rapid *in vitro* multiplication and *ex vitro* rooting of *Rotula aquatica* Lour., a rare rhoeophytic woody medicinal plant. *Plant Cell Reports* 21: 415-420.

Murashige, T. and Skoog, F. 1962. A revised medium of rapid growth and bioassays with tobacco tissue cultures. *Physiology Plantarum* 15: 473-497.

Rajore, S. and Batra, A. 2005. Efficient plant regeneration via shoot tip explant in *Jatropha curcas*. *Journal of Plant Biochemistry and Biotechnology* 14: 73-75.

Sujatha, M. and Dhingra, M. 1993. Rapid plant regeneration from various explants of *Jatropha integerrima*. *Plant Cell Tissue Organ Culture* 35: 293-296.

Sujatha, M. and Mukta, N. 1996. Morphogenesis and plant regeneration from tissue cultures of *Jatropha curcas*. *Plant Cell Tissue and Organ Culture* 44:135-141.

Sujatha, M., Makkar, H.P.S. and Becker, K. 2005. Shoot bud proliferation from axillary nodes and leaf sections of non-toxic *Jatropha curcas* L. *Plant Growth Regulation* 47:83-90.

Wei, Q. Lu, Loao, W.D., Pan, Y., Xu, S.L., Y. Tang, L. and Fang, C. 2004. Plant regeneration from epicotyl explants of *Jatropha curcas*. *Journal of Plant Physiology and Molecular Biology* 30:475-478.

Yusnita, S., Geneve, R.L. and Kester, S.T. 1990. Micro-propagation of white Eastern Redbud (*Cercis canadensis* var. *alba*). *Hort. Science* 25:1091.

Chapter - 7

EVALUATION OF *JATROPHA CURCAS* ACCESSIONS UNDER TROPICAL CLIMATIC CONDITIONS OF CENTRAL INDIA

A.K. Pandey and A.K. Mandal

Tropical Forest Research Institute, P.O. RFRC, Jabalpur - 482 021
E-mail: akpandey10@rediffmail.com

ABSTRACT

Jatropha curcas L. is recognized as most potential species for biodiesel production, since the seeds have high oil content (25–42%) and could be grown under different climatic conditions. Considering its potential as a bioenergy crop, systematic provenance research has been initiated in 2004-2005 at Tropical Forest Research Institute, Jabalpur under NOVOD Board Network programme to find out superior provenance with high seed yield and better quality and quantity of oil. With these objectives multilocational trials (national and zonal) have been established at TFRI campus in the year 2005 and 2006 to evaluate performance of various accessions representing different agroclimatic regions of India. In the study a total of 36 accessions of *J. curcas* received from various networking institutions e.g. IGAU Raipur; TNAU Mettupalayam; SFRI Jabalpur; TFRI Jabalpur; GBPUAT, Pantnagar, MPKV Rauri; PDKV Akola, RRL Bhubhneshwar; CSK HPKV Palampur, were evaluated for their growth performance (height, collar diameter, no. of branches, fruiting pattern, fruit yield, seed characteristics and oil percentage). Significant variation was observed for all growth parameters. The study revealed TNMC-5 and TFRI-2 performed better among all other accessions.

Keywords: *Jatropha curcas,* superior genotypes, accessions, multilocational trials

INTRODUCTION

Jatropha is being grown in almost all states of India up to 1500 meter altitude from coastal areas both in forest and non forest land. There are 176 species of Jatropha in the world, of which *Jatropha curcas* is the predominant one and has very good yield compared to other varieties (Tiwari, 2007). It is a multipurpose and drought resistant large shrub or small tree. It is considered to have originated in Latin America but presently grows throughout the arid, semi arid and tropical regions of the world (Hilwa, 1995; Henning, 1996; Makkar *et al.,* 1997). India is sixth in world in energy demand accounting to 3.5% of world's commercial consumption. In India there is growing demand and it is predicted that consumption would be 5.6 million barrel per day by 2030 out of which more than 94% would be met through imports (Kumar and Mohan, 2005). The distribution of *J. curcas* beyond tropical America was facilitated by the Portuguese and Arab traders who transported the plant to Africa and Asia where it has become naturalized and widely utilized (Jones and Miller, 1992). *J. curcas* has immense potential to make significant contribution to country's growing energy requirements (Srivastava, 1999). The engine performance of *J. curcas* oil has been found to be similar to diesel. In the light of above, present study was taken up to evaluate and identify best provenance for plantation programme.

MATERIALS AND METHODS

In order to evaluate the performance and variation in *Jatropha curcas*, multilocational trials in the form of national and zonal trial have been established at Tropical Forest Research Institute, Jabalpur campus in the year 2005-2006. The study was carried out under the national network programme on research and development of Jatropha funded by National Oilseeds and Vegetable Oil Development Board (NOVOD). Provenances collected by this institute have been exchanged among other members of network for evaluation of their performance. Number of provenances included in the national and zonal trials is given in Table1.

Table 1. Number of accessions included in national and zonal trial

Number of Provenance/Accession			Participating Institute/center
First national trial (July 2005)	Second national trial (July 2006)	Zonal trial (July 2005)	
03		03	IGAU, Raipur
06		0	TNAU, Mettupalayam
0	01	0	Gujarat
02		02	SFRI, Jabalpur
0	02	0	Palampur
02		02	TFRI, Jabalpur
02		0	GBPUAT, Pantnagar
01		01	MPKV, Rahuri
03		02	PDKV, Akola
0		04	RRL, Bhubneshwar

Establishment of multilocational trial

Multilocational trials were established at institute campus in the year 2005-2006. First national trial has been established in July 2005 comprising of 19 accessions received from 8 participating institutes. Second national trial was established in July 2006 comprising of 3 accessions received from 2 participating institutes. Zonal trial was raised from the seed material of 14 accessions received from 6 participating institutes in July, 2005.

Germplasm collection: Superior germplasm were recieved from different participating institutes i.e. from IGAU Raipur, TNAU Mettupalayam, SFRI Jabalpur, TFRI Jabalpur, GBPUAT Pantnagar, MPKV Rahuri, PDKV Akola, RRL Bhubneshwar etc.

Nursery practices: Jatropha seeds were sown in nursery in the first week of April in poly bags filled with soil, sand, FYM in the ratio of 1:1:1. Germination was completed in 10-15 days. Seeds sown in poly bags of 6" x 9" size were found suitable for raising nursery. Seedlings were watered regularly.

Layout and designing: The experimental field was divided into 57 equal size plots for first national trial, 9 plots for second national trial and 42 plots for zonal trial and 16 plants were planted in each plot at the distance of 2 x 2 meter. The pits were digged and filled with the mixture of soil and FYM (2:1) before plantation as the experimental site has very poor soil texture, low nutritional status and water holding capacity. Plantation was done in Randomized Block Design (RBD) with three replications.

Observation and growth data recording: Regular growth data was recorded to find out variation among different provenances on the basis of various growth parameters like height, collar diameter, no. of branches, total no. of fruits, no. of seeds in fruits and oil content etc. A comparative study was done to find out best performing provenance on the basis of observed growth data.

Collection of fruits: The national and zonal trials laid out earlier in the year 2005 and 2006 yielded fruits. The fruits were carefully collected and brought to the laboratory for further analysis. However, only few genotypes yielded good quantity of fruits.

Data analysis: The data were subjected to statistical analysis using mean values (Panse and Sukhatme, 1978). Individual means were used to compute ANOVA while provenance means were used to carry out correlation. Table 2 and 3 represents the statistical data.

RESULT AND DISCUSSION

The growth performances of trials as presented in Table 2 and 3 is described and discussed below. Analysis of various indicated significant variation for all the characters studied in both national and zonal trial.

1. **Plant height:** In national trial-I, the recorded value for height was highest in TNMC-5 (203.46 cm) followed by TNMC-7 (176.8 cm). The height of TNMC-22 (92.0 cm) is found to be lowest (Table 2). The plant height ranged 92.0 – 203.46 cm with a mean value of 136.28 cm ± 25.08. However in zonal trial, it was found to be highest in IGAU-1 (152.48 cm) followed by TFRI-1 (152.36 cm) (Table 3). The height of RRL-1 (87.53 cm) accession was found to be lowest. The plant height ranged 87.53 – 152.48 cm with a mean value of 130.10 cm ± 25.36.
2. **Collar diameter:** In national trial-I, maximum diameter was recorded in TNMC- 5 (10.3cm). TNMC-22 (4.5cm) recorded minimum value for this trait. The collar diameter ranged 4.5-10.3 cm with a mean value of 7.08 ± 1.46. However, in zonal trial SFRI-1 (8.29 cm) and RRL-1 (5.97 cm) recorded maximum and minimum collar diameter respectively. The collar diameter ranged 5.89-8.29 cm with a mean value of 6.70 ± 8.92.
3. **Number of branches:** In national trial-I, PDKV-2 (21.10) registerd maximum number of branches per plant and minimum in 4 accessions i.e. TNMC-2, PDKV-1, TNMC-22 and GBPUAT-2. Number of branches in different accessions in national trial ranged from 12.0-21.10, with a mean of 14.87±3.72. However, in zonal trial number of branches were recorded maximum in RRL-3 (16) followed by TFRI-2 (14.66) and minimum in RRL-1 (8.33). Number of branches ranged from 8.33-15 with a mean value of 13 ±2.89.

Table 2. Growth performance of provenances of national trial –I (2005)

Provenance	Height (cm)	Collar Dia (cm)	Branch number
IGAU-1	116.66	5.61	12.66
TNMC-2	112.90	6.86	12.00
TFRI-1	121.00	7.24	14.20
PDKV-1	122.60	7.30	12.00
TNMC-3	121.53	7.26	15.00
SFRI-1	122.60	7.73	17.00
NAGPUR	108.86	7.54	14.00
IGAU-2	134.40	7.20	14.30
TNMC-4	142.53	6.76	15.33
TFRI-2	154.83	7.46	14.33

IGAU-3	149.23	6.60	13.11
TNMC-5	203.46	10.30	13.66
TNMC-7	176.80	8.33	20.66
CBPUAT-1	143.00	5.56	19.33
PDKV-2	137.83	8.86	21.10
TNMC-22	92.00	4.50	12.00
MPKV-1	108.00	6.03	15.00
SFRI-2	150.53	6.01	13.33
GBPUAT-2	122.60	7.30	12.00
Mean	136.28	7.08	14.87
CD	16.94	1.50	5.09
SE	8.35	7.39	2.51

Table 3. Growth performance of provenances in zonal trial –I (2005)

Provenance	Height (cm)	Collar diameter (cm)	Branch number
IGAU-1	152.48	7.41	12.00
TFRI-1	152.36	6.39	10.00
RRL-1	87.53	5.97	8.33
IGAU-2	119.53	6.25	13.33
TFRI-2	134.23	6.85	14.66
RRL-2	123.66	7.39	13.00
PDKV-2	140.86	6.69	12.33
SFRI-1	143.46	8.29	15.00
IGAU-3	118.03	7.33	14.66
RRL-3	146.13	7.63	16.00
MPKV-1	144.94	6.03	12.00
SFRI-2	141.20	6.48	13.66
PDKV-2	128.60	5.89	12.00
RRL-4	87.66	6.17	12.33
Mean	130.1	6.70	13.0
CD	30.50	0.99	4.19
SE	14.83	4.84	2.04

4. **Number of fruits:** In national trial-I, TNMC-5 yielded maximum number of fruits (892 fruits) followed by TNMC-4 (695 fruits). SFRI-1 yielded minimum number of fruits (452 fruits). However, in zonal trial IGAU-1 yielded maximum number of fruits (583 fruits) followed by TFRI-2 (580 fruits) and RRl-3 yielded minimum numbers (320) of fruits (Table 4).

5. **Fruit weight:** In national trial, the weight of 100 fruits was found highest in TNMC-5 (315.50 gm) followed by TFRI-2 (302.25 gm) and minimum fruit weight was recorded in TFRI-1 (230.40 gm). However, in zonal trial the weight of 100 fruits was

recorded maximum in TFRI-2 (298.00 gm) and minimum in IGAU-3 (205.50 gm). The fruit weight ranged 205.50-315.50 gm.

6. **Seed weight:** In national trial, the maximum seed weight was recorded in TNMC-5 (240.30 gm) and minimum in IGAU-1 (172.60 gm). However, in zonal trial seed weight was recorded maximum in TFRI-2 (204.63 gm) followed by SFRI-2 (198.20gm) and minimum in IGAU-2 (172.50 gm). The weight of 100 seeds ranged 172.50 – 240.30 gm.
7. **Seed length:** In national trial, the maximum seed length was recorded for the seeds of TNMC-5 (18.32 mm) followed by three accessions i.e. TNMC-2, TNMC-3, and TFRI-2 (16.37 mm). The minimum seed length was recorded in IGAU-1 (14.30 mm). However, in zonal trial, the maximum length was recorded for the seeds of IGAU-3 (18.32 mm) followed by IGAU-2 (16.26 mm) and it was recorded minimum in seeds of TFRI-1 (14.20 mm). The seed length ranged 14.20 – 18.32mm.
8. **Seed width:** In national trial, the maximum seed width was recorded in IGAU-1 (9.80 mm) and minimum in TNMC-2, TNMC-3 and TFRI-2 (8.37 mm). However, in zonal trial the maximum seed width was recorded in the seeds of IGAU-3 (10.20 mm) and minimum in two accessions i.e. RRL-3 and SFRI-2 (8.20 mm). The seed width ranged 8.20 – 9.80 mm. Variation in seed characteristics of *Jatropha curcas* has also been reported by Pandey *et al.,* 2007.
9. **Oil percentage:** In national trial, the maximum oil content was recorded in TNMC-5 (36.40%) followed by IGAU-1(34.52%) and minimum in SFRI-1(29.32%). In zonal trial the maximum oil content was recorded TFRI-2 (35.32%) followed by IGAU-2 (35.30%) and minimum oil content was recorded in the seeds of TFRI-1 (30.50%). The oil content in the seeds ranged 29.32 – 36.40%.

Table 4. Fruits characteristics and oil percentage in national and zonal trial.

S.No.	Provenance	Origin	Total No. of Fruits	Wt. of 100 Fruits (g)	Wt of seeds obtained in 100 fruits(g)	Seed length (mm)	Seed width (mm)	Oil percentage (%)
National								
1	IGAU-1	seed	492	250.50	172.60	14.3	9.80	34.52
2	TNMC-2	seed	480	252.16	183.21	16.37	8.37	33.00
3	TFRI-1	seed	508	230.40	198.30	15.30	9.23	35.24
4	PDKV-1	seed	....	....	....	....	....	.....

5	TNMC-3	seed	560	265.50	192.50	16.37	8.37	32.54
6	SFRI-1	seed	452	282.50	175.60	15.30	9.23	29.32
7	NAGPUR	seed	463	264.52	178.52	15.85	8.85	30.00
8	IGAU-2	seed	....	....	.....	....	.....	....
9	TNMC-4	seed	685	287.32	186.50	16.35	9.35	32.50
10	TFRI-2	seed	573	302.25	198.60	16.37	8.37	32.60
11	IGAU-3	seed	....	....	.....	....	.....	....
12	TNMC-5	seed	892	315.50	240.30	18.32	9.65	36.40
Zonal								
13	IGAU-1	seed	583	238.64	176.50	15.20	8.32	31.52
14	TFRI-1	seed	472	247.65	186.30	14.20	9.30	30.50
15	RRL-1	seed	......	....	.....	......	....	....
16	IGAU-2	seed	470	215.40	172.50	16.26	9.32	35.30
17	TFRI-2	seed	580	298.00	204.63	15.20	8.32	35.32
18	RRL-2	seed	.....	.....	.....	.....	....	....
19	PDKV-2	seed	....	.....	.....	....	....	....
20	SFRI-1	seed	.....	....	....	.....	......	.....
21	IGAU-3	seed	553	205.50	184.50	18.32	10.2	35.20
22	RRL-3	seed	320	265.30	186.54	15.2	8.20	33.20
23	MPKV-1	seed	482	215.40	187.65	14.32	9.50	34.60
24	SFRI-2	seed	420	240.30	198.65	14.32	8.20	30.52
25	PDKV-2	seed	.....	.....	.....	.....	....	....
26	RRL-4	seed	.....	.....	.....	.....	.....	.....

Correlation study: Correlation study (Table 6) revealed that of the different growth parameters plant height, collar diameter and number of branches are positively correlated with each other.

Table 5. Correlations of different characters in trials of *J. curcas*

Trial	Variable	Height	Collar diameter	Branch number
National trial	Height (cm)	1.0000		
	Collar diameter (cm)	0.4076	1.0000	
	Branch number	0.3468	0.2521	1.0000
Zonal trial	Height (cm)	1.0000		
	Collar diameter (cm)	0.3699	1.0000	
	Branch number	0.2605	0.2601	1.0000

CONCLUSION

The present study revealed considerable variation with respect to different characters. The findings will provide baseline information for planning *Jatropha curcas* improvement programme.

REFERENCES

Heller, J. 1996. Physic nut (*Jatropha curcas* L.), Promoting the conservation and use of underutilized and neglected crops. Institute of Plant Genetics and Crop Plant Research, Gartersleben, International Plant Genetic Resources Institute, Rome.

Henning, R. 1996. Combating desertification-fuel from Jatropha plants. In: UNIDO symposium on development and utilization of biomass energy in developing countries, Vienna, December 1995.

Hikwa, D. 1995. *Jatropha curcas* L. Agronomy Research Institute, Department of Research and Specialist Services (DR and SS), Harare, Zimbabwe. 4 pp.

Jones, N. and Miller, J.H. 1992. *Jatropha curcas:* a multipurpose species for problematic sites. Land Resources Series No. 1. Asia Technical Department, World Bank, Washington, USA. 12 pp. Annex 1-6.

Makkar, H.P.S., Becker, K. and Schmook, B. 1997. Edible provenances of *Jatropha curcas* from Quintna Roo State of Mexico and effect of roasting on antinutrient and toxic factors in seeds. Institute for Animal Production in the Tropics and Subtropics (480), University of Hohenheim, D-70593 Stuttgart, Germany. 6 pp.

Srivastava, R. 1999. Study in variation in morph-physiological parameters with reference to oil yield and quality in *Jatropha curcas* Linn. Ph.D. Thesis. Forest Research Institute (Deemed University), Dehradun, India.

Kumar, L. and Ram Mohan, M.P. 2005. Biofuel : The key to india's sustainable energy needs. Article published on www.risoe.dk/rispubl/sys/syspdf/ energ-conf05/pp.16.

Tiwari, D.N. 2007. ISBN 81-8430-030-1 Jatropha and Biodiesel pp-1-5.

Chapter - 8

RESPONSE OF DIFFERENT NUTRITIONAL AND PRUNING LEVELS ON GROWTH AND YIELD OF *JATROPHA CURCAS*

S.J. Patil, S.M. Mutanal, H.Y. Patil, Girish Shahapurmath and V. Maheswarappa

AICRP on Agroforestry, University of Agricultural Sciences, Dharwad-580 005 E-mail - sjpatil3@rediffmail.com

ABSTRACT

Self reliance in energy is vital for overall development of our country. Need to search an alternative source of energy which is renewable, safe and non polluting assumes top priority in view of the uncertain supplies and frequent price hikes of fossil fuels in international market. Among many species which can yield oil as source of energy in the form of biodiesel, *Jatropha curcas* has been found to be the most suitable species due to its wild and hardy nature.

With this background, two field experiments were initiated in 2003 and 2004 on silvicultural and nutritional management of *Jatropha curcas* under rainfed conditions at main Agricultural Research Station, Dharwad and Forest Research Station, Prabhunagar, University of Agricultural Sciences, Dharwad, Karnataka on medium black soils. Jatropha seedlings/ seeds were planted/dibbled at 3x2 m apart. Both the experiments were laid out in Factorial Randomized Block Design with three replications. At Dharwad, the experiment consisted of two factors *viz*, pruning levels (1m, 1.5m and no pruning) and sulphur levels (20, 40, 60, kg/ha). Collar diameter, number of branches and seed yield of Jatropha were significantly higher in 1 m pruned treatment as compared to other pruning levels. Seed yield of Jatropha was higher with the application of sulphur @ 40kg/ha as compared to 20kg/ha.. The experiment conducted at Prabhun-agar consisted of two factors *viz*,. fertilizers levels (40:40:40, 60:60:60, 80:80:80 N:P:K g/plant) and pruning levels (1m, 1.5 m 2 m and no pruning). Application of 80:80:80 N:P:K g/plant recorded significantly higher number of branches, collar diameter and seed yield as compared to 40:40:40 N:P:K g/plant. Number of branches and collar diameter of Jatropha were significantly higher in 1m pruning as compared to in other pruning levels.

The silvicultural management *viz*., pruning at 1m height, nutritional management *viz*., application of 80:80:80 NPK g/plant and application of sulphur @ 40 kg/ha proved to be better for higher productivity of jatropha under rainfed conditions. During the winter of 2008-09 looking to biomass produced and branching observed, the pruning levels were modified and implemented as 1m to 0.5m; 1.5m to 1m and 2m to 1.5m.

Keywords: Energy, silvicultural management, nutritional management, jatropha, biofuel, provenance

INTRODUCTION

Research on renewable oils as diesel fuel was conducted at least 100 years ago but the interest lagged because of cheap and plentiful supplies of petroleum fuels. Presently, the known world wide reserves of petroleum are 100 billion barrels and these petroleum resources are predicted to be consumed in about 40 years (Orhan *et al.,* 2004). Feared shortages of petroleum fuels due to rapid depletion, net production of carbon dioxide from combustion sources and stringent emission norms have rekindled interest in renewable oil fuels. In view of these, vegetable oils have been considered as alternative fuels for compression ignition engines.

Vegetable oils are renewable, non toxic, biodegradable and have low emission profiles. The raw materials being used commercially by the biodiesel consuming countries constitute the edible fatty oils derived from rape seed, soybean, palm, sunflower, canola, mustard, coconut and linseed etc. (Korbitz, 1999). Use of such vegetable oils to produce biodiesel in India is not feasible in view of a big gap in demand and supply. Increased pressure to augment production of edible oil has also put limitation on the use of these oils for production of biodiesel. Under Indian conditions, plants which produce non edible oil in appreciable quantities and suitable for large scale cultivation on non cropped marginal and waste lands can be considered for biodiesel production. In this context, the best-suited plant fulfilling the above criteria is *Jatropha curcas*. It is a large shrub or small tree, able to thrive under varied climatic conditions with the rainfall of 250-1200 mm.

India has about 100 M ha of degraded land, which can be effectively utilized for cultivating non-edible biodiesel crops like jatropha. Being an energy plantation, jatropha, may utilize large quantities of nutrients from soil itself and yield satisfactorily but will definitely deplete soil fertility if not properly fertilized. It is therefore, necessary to maintain soil fertility for sustainable production of Jatropha. Deciding the nutrient management options for better yield and quality of oil seems to be the immediate task before going for commercial cultivation of Jatropha.

The Jatropha plant has to be managed into a bushy plant in order to have more branches so that more fruiting points are produced and it also helps in easy harvesting. Hence, experiments were conducted at

Agricultural Research Station, Dharwad and Agricultural Research Station, Prabhunagar in order to optimize the levels of nutrients (NPK & S) and to standardize the pruning level in Jatropha.

Influence of NPK and S is reflected in enhanced growth and productivity. There is very scanty literature on nutrition and silvicultural management of Jatropha. Hence, information available in other oilseed crops is focused. Nitrogen is extremely essential for better growth and healthy appearance in soybean (Starling *et al.*, 1998). It is also known to enhance Relative Growth Rate which is the derived parameter that speaks well about plant growth in oil seed rape (Blake *et al.*, 1998). Phosphorous is a key component in plants, and is responsible to enhance plant growth. Phosphorus deficiency is known to reduce shoot growth and leaf size (Lynch *et al.*, 1991). Shoot development requires the production of leaves by shoot apical meristem and their subsequent expansion. Potassium fertilization influences early plant growth and thereby dry matter accumulation. Sulphur, the constituent part of amino acid is needed in adequate amounts for the normal growth of plants. Removal of sulphur from hydroponic system decreased the Relative Growth Rate of Oilseed rape by 50 percent. While, addition of 1 mM of SO_4 to hydroponic system recorded the RGR of 0.229 per day as compared to 0.100 per day in control (Blake *et al.*, 1998). A positive role of sulphur on chlorophyll content of brassica was observed by Ahmed and Abdin (2000). Application of S @ 40 kg per ha enhanced chlorophyll content from 1.00 to 1.68 mg per g fresh weight as compared to control.

It is well known fact that breaking apical dominance will lead to induction of lateral branches. The heading back of ten month old Jatropha is necessary to enhance side shoot emergence and pruning 2/3rd of the side shoots in the second year will give secondary branches and bushy nature to Jatropha.

MATERIALS AND METHOD

Experiments were conducted at Agricultural Research Station, University of Agricultural Sciences, Dharwad and at Forest Research Station, Prabhunagar on standardization of optimum nutritional and pruning levels in Jatropha.

Locally grown Jatropha plants with minimum of eight years growth were identified for seed collection. The fruits at lime yellow stage were harvested, sun dried and the seeds were sown in polybags with recommended planting mixture. Six months old Jatropha seedlings were planted in July 2004 at 3m x 2m spacing in a pit size of 45 x 45 x 45 cm. Two kg of FYM was added to each pit before planting. The soil type in the experimental area was medium black soil. The experiment consisted of two factors *viz.*,

A. Factor: Pruning levels (P) : P_1 – 1m height; P_2 - 1.5m height; P_3 - No pruning and,

B. Factor : Sulphur levels (S) : S_1- 20kg/ha; S_2- 40kg/ha; S_3- 60kg/ha.

Pruning treatment was imposed in the month of November, 2004 by cutting the upper portion of the plant to required height *viz* 1.00 mt and 1.5 mt. The nutrients, NPK in the form of Urea, Single Super Phosphate (SSP) and Murriate of Potash (MoP) were applied @ 60:60:60 g/plant. Sulphur was applied in the form of gypsum.

The experiment conducted at ARS Prabhunagar on the levels of NPK and pruning is explained in the following lines. At ARS, Prabhunagar, two seeds were dibbled directly into each pit (45 x 45 x 45 cm) filled with red soil and FYM. Dibbling was done in the month of September, 2003. The experiment consisted of two factors *viz.*

A. Factor: Fertilizer levels (F): F_1- 40:40:40
F_2:- 60:60:60
F_3 - 80:80:80 N: P_2O_5 : K_2O_5 g/plant and

B. Factor : Pruning levels (P): P_1- 1m height
P_2- 1.5m height
P_3- 2.0m height
P_4: No pruning

Pruning was imposed in the month of May, 2004 by removing the upper portion as per the treatments. The soil type in the experimental area was medium black.

The experiments were laid out in factorial RBD with three replications. Each treatment had 16 plants. The average rainfall at ARS, Dharwad is 754 mm and at ARS, Prabhunagar it is 1069 mm.

RESULTS AND DISCUSSION

The impact of pruning and sulphur levels on growth parameters and seed yield at the end of third year of growth are presented in Table 1 and 2. The effect of pruning and sulphur levels on height was statistically insignificant. The growth in 1.00 and 1.50 mt pruned treatment was more as indicated by the height. Both these pruned treatments put on secondary branches which grew profusely and came to the level of that of no pruning. The mean height was 2.11m and ranged from 2.00 to 2.22 m. The collar diameter varied significantly among pruned levels and interaction between the treatments was insignificant. The mean collar diameter was 9.36 cm and ranged from 7.81 cm (P_3) to 10.58 (P_1). The pruning had a profound effect on collar diameter by recording 24% and 35% increase respectively in P_2 and P_1 treatments over no pruning. The increase in collar diameter (1.0 and 1.5m pruned treatments) might be due to translocation and storage of assimilates in the lower portion of the stem.

Table 1. Growth of *Jatropha curcas* as influenced by pruning and sulphur levels

Treatment combin-ation	Height (m)				Collar diameter (cm)				Canopy spread (m^2/plant)			
	P_1	P_2	P_3	Mean	P_1	P_2	P_3	Mean	P_1	P_2	P_3	Mean
S_1	2.09	2.32	1.99	2.14	10.07	10.58	11.10	10.16	1.16	1.20	1.14	1.04
S_2	2.19	2.13	2.11	2.14	10.12	9.81	9.16	10.33	1.02	1.00	1.10	1.06
S3	2.06	2.21	1.90	2.06	10.29	10.61	8.53	9.63	0.95	0.96	0.92	1.05
Mean	2.12	2.22	2.00		10.58	9.70	7.81		1.16	1.04	0.94	
For comparing	SEm ±	CD 5%			SEm ±	CD 5%			SEm ±	CD 5%		
Pruning (P)	0.10	NS			0.44	1.32			0.060	0.17		
Sulphur (S)	0.10	NS			0.48	NS			0.092	NS		
Interaction (PxS)	0.17	NS			0.76	NS			0.160	NS		

S_1- 20 kg/ha; S_2 – 40 kg/ha; S_3:- 60 kg/ha. P_1 -1.00 m; P_2 – 1.50 m; P_3 – No pruning

Table 2. Seed yield of *Jatropha curcas* as influenced by pruning and sulphur levels

Treatment combination	No. of branches				Seed yield (kg/plant)			
	P_1	P_2	P_3	Mean	P_1	P_2	P_3	Mean
S_1	5.38	5.21	5.55	5.38	1.040	0.830	0.760	0.876
S_2	5.63	5.13	4.92	5.22	0.960	0.790	0.700	0.816
S3	6.34	5.09	4.75	5.39	0.710	0.720	0.660	0.696
Mean	5.78	5.14	5.07		0.903	0.780	0.706	
For comparing	SEm ± CD at 5%				SEm ± CD at 5%			
Pruning (P)	0.22	0.66					0.06	0.19
Sulphur (S)	0.22	NS					0.06	NS
Interaction (PxS)	0.38	NS					0.11	NS

S_1- 20 kg/ha; S_2 – 40 kg/ha; S_3:- 60 kg/ha. P_1 -1.00 m; P_2 – 1.50 m; P_3 – No pruning

The canopy spread (m^2/plant) varied significantly among pruning levels only. The mean canopy spread (m^2/plant) was 1.05 m^2 and varied from 0.94 (P_3) to 1.16 m^2 (P_1). The sulphur treatment had no effect on number of branches, number of seeds/plant and seed yield (kg/plant). Number of branches (primary) ranged from 5.07 (P_3) to 5.78 (P_1). The seed yield (kg/plant) also showed similar trend with mean value of 0.796 kg/

plant and with range of 0.706 (P_3) to 0.903 (P_1). There was 22% increase noticed in P_1 as compared to P_3. The interspace was used for growing soybean. The grain yield and haulm yield data did not vary statistically. The mean value was 477.4 and 391.9 kg/ha, respectively (Table 3).

Table 3. Soybean grain yield in the interspaces of Jatropha

Treatment combination	Soybean grain yield (kg/ha)				Soybean haulm yield (kg/ha)			
	P_1	P_2	P_3	Mean	P_1	P_2	P_3	Mean
S_1	509.0	525.6	445.4	493.4	359.6	536.7	370.7	422.3
S_2	520.1	456.5	428.8	468.4	462.0	356.9	370.7	396.5
S3	475.1	439.9	495.2	470.3	395.6	326.4	348.6	356.9
Mean	501.6	474.0	456.5		405.7	406.7	363.3	
For comparing	SEm ±		CD 5%		SEm ±		CD 5%	
Pruning (P)	29.04		NS		35.60		NS	
Sulphur (S)	29.04		NS		35.61		NS	
Interaction (PxS)	50.38		NS		61.67		NS	

S_1- 20 kg/ha; S_2 – 40 kg/ha; S_3:- 60 kg/ha. P_1 -1.00 m; P_2 – 1.50 m; P_3 – No pruning

Table 4. Growth and yield of *Jatropha curcas* as influenced by fertilizer and pruning levels (Prabhunagar)

Treatment combination	Height (m)				Collar diameter (cm)				No. of branches/pl				Seed yield (kg/pl)			
	F_1	F_2	F_3	Mean	F_1	F_2	F_3	Mean	F_1	F_2	F_3	Mean	F_1	F_2	F_3	Mean
P_1	2.44	2.45	2.65	2.52	33.66	37.00	35.33	35.33	11.33	12.20	13.66	12.39	0.910	0.920	1.240	1.020
P_2	2.64	2.47	2.56	2.56	38.66	32.66	39.00	36.77	11.33	10.33	8.66	10.11	0.780	0.860	0.980	0.873
P_3-	2.54	2.48	2.56	2.53	38.66	33.33	35.66	35.88	9.00	11.00	9.60	9.88	0.658	0.811	0.780	0.749
P_4 pruning	2.31	2.34	2.35	2.34	33.00	34.33	34.33	33.88	7.00	11.33	11.33	9.88	0.622	0.686	0.763	0.690
Mean	2.48	2.43	2.53		36.00	34.33	36.08		9.66	11.21	10.81		0.742	0.819	0.940	
For comparing	SEm ±		CD 5%		SEm ±		CD 5%		SEm ±		CD 5%		SEm ±		CD 5%	
Fertilizer (F)	0.07		NS		0.45		NS		0.32		0.94		0.06		0.17	
Pruning (P)	0.08		NS		0.86		2.52		0.40		1.17		0.11		0.32	
Interaction (FxP)	0.15		NS		1.86		NS		0.94		NS		0.49		NS	

F_1- 40:40:40; F_2 - 60:60:60; F_3:- 80:80:80 N: P_2O_5 : K_2O_5 g/plant

The seed yield (kg/plant) recorded in the experiment on nutritional and pruning levels at Agricultural Research Station, Prabhunagar indicated that the NPK level and pruning level were statistically superior in mean seed yield (0.834 kg/plant and 0.833 kg/plant, respectively) (Table 4). The treatment F_3 (80:80:80 NPK g/plant) was statistically superior and F_1 and F_2 were on par with each other and the result is in support of Starling *et al.* (1998) who reported the benefits of nitrogen application in soybean. Similarly, Mandal *et al.* (2003) and Dodd and Mallarino (2005) reported positive impact of phosphorus in soybean and Munda *et al.* (2004) reported higher yield (32% and 36%) in soybean by applying 30:26:25 kg NPK/ha. Among pruning levels P_3 and P_4 with yield value of 0.562 and 0.549 were superior over P_1 and P_2.

REFERENCES

Ahmed, A. and Abdin, M.Z. 2000. Effect of sulphur application on lipid, RNA and fatty acid content in developing seeds of rape seed. *Plant Science* 150:71-76.

Blake Kalff, M.M.A., Harrison, K.R., Hawkesford, M.J., Zhao, F.J. and McGrath, S.P. 1998. Distribution of sulfur within oilseed rape leaves in response to sulfur deficiency during vegetative growth. *Plant Physiology* 118:1337-1344.

Dodd, J.R. and Mallarino, A.P. 2005. Soil test phosphorus and crop grain yield responses to long term phosphorus fertilization for corn-soybean rotations. *Soil Sci. Soc. Am. J.* 69: 1118 – 1128.

Korbitz, W. 1999. Biodiesel production in Europe and North America, an encouraging prospect. *Renewable Energy* 16:1078-83.

Lynch, J., Lauchli, A. and Epstein, E. 1991. Vegetative growth of the common bean in response to phosphorus nutrition. *Crop Science* 31: 380-382.

Mandal, B.K., Manda, B.B., Murmu, N.N., Tudu, B.C. and Khanda, C.M. 2003. Growth and yield of sunflower as influenced by dates of sowing and levels of N, P and K. *Indian J. Plant Physiology* 8(1):103-105.

Munda, G.C., Patel, D.P. and Islam, M. 2004. Effect of macro and micronutrients on growth and yield of groundnut. *Ann. Plant Physiology* 18 (1):9-12.

Orhan, A.S., Dulger, Z., Kahraman, N. and Veziroglu, T.N. 2004. Internal combustion engines fueled by natural gas-hydrogen mixtures. *International Journal of Hydrogen Energy* 29: 1527-39.

Starling, M.E., Wood, C.W. and Weaver, D.B. 1998. Starter nitrogen and growth habit effect on late-planted soybean. *Agronomy Journal* 90: 658-662.

Chapter - 9

STUDY OF ASEXUAL PROPAGATION METHOD IN PHYSIC NUT (*JATROPHA CURCAS* L.)

Gohil Rajendra Sinh

R & D division (bio-fuel), Hash Biotech Lab Pvt. Ltd.,
Hash Group of Company, SCO-311-312, Sector-40D, Chandigarh-160036

ABSTRACT

Jatropha curcas L., a highly cross pollinated plant species which have potential for bio-fuels production and adapted to arid/semi-arid conditions. Method of propagation using asexual branch cutting can solve the problem in transmission of mother characters in offspring leading to promote vegetative propagation using superior plant material for multiplication. With the objective to study asexual propagation method and to obtain information about the optimum length and diameter of branch cutting for vigorous growth of planting material, experiments were conducted at Biofuel R&D center, Hash Biotech lab at Mohali (Chandigarh, India). Results show that 8-10 cm thick branch cutting is the most suitable for good plant vigour, owing to maximum number of leaf, largest crown and leaf area, maximum lateral root and total plant weight.

Keywords: Physic nut, cross pollination, asexual propagation

INTRODUCTION

Jatropha curcas L., a highly cross pollinated plant species which have potential for bio-fuels production and adapted to arid/semi-arid conditions (Heller, 1996). Raw oil content being squeezed from dry seeds with the use of a hydraulic pressure after filtering with an intensive sieve can be used in substitution to a diesel fuel oil for diesel tractor, diesel water pumps and other traditional diesel engines with an outstanding performance comparable to diesel fuel oil (Pholsen, 1983).

It is highly cross pollinated in nature, which creates problem in maintaining purity of germplasm. Method of propagation using asexual branch cutting can solve the problem. So objective of the study include obtaining the information about the optimum length and diameter of branch cutting for vigorous growth of planting material.

The experiment was carried out in polyhouse of Biofuel R&D center, Hash Biotech lab at Mohali (Chandigarh, India) with the objective of evaluating the effect of different length and branch cuttings on initial plant vigour under asexual propagation system. It was conducted under high humidity conditions (75% RH, Temperature-35°C) non-mist propagation chamber during March, 2008. Cuttings having three different diameters viz., D_1 (4-6 cm), D_2 (6.5-8 cm) and D_3 (8-10 cm) and different lengths as H_1 (20 cm), H_2 (30 cm) and H_3 (40 cm) were collected from natural/wild plant. The branch cuttings were planted directly in polythene bags (one cutting per bag) containing potting mixture of sand, soil and farm yard manure in the ratio of 1:1:1 (by volume) in three replication of 20 cutting each using randomized block design as per ISTA rules (1996).

Prior to planting the polythene bags containing potting mixture were given fungicide treatment (0.2% Bavistin) and kept covered with plastic sheet for 24 hours. Daily sprouting observation was taken till sprouting in all treatment was completed (up to 20 days). These stocks were raised in nursery in poly bags of size of 9" x 9" in the non-mist chamber and were watered as per requirement. The experiment was laid out in randomized block design with three replications and three treatments. Various data were recorded on five cuttings in each treatment i.e., number of leaf and branches per cutting, height, crown area, leaf area, petiole length, length of tap root, mean length and number of lateral root, root weight, shoot weight, leaf weight and total plant green biomass on 75th day of planting and were statistically analyzed to conclude the treatment effect.

Table 1. Effect of diameter of cutting on their performance

Treat.	No. of leaf	No. of bra-nch	Plant hei-ght (cm)	Crown area (cm)	Leaf area (cm^2)	Pet-iole len-gth	Tap root (cm)	Lat-eral root (cm)	No of late-ral root	Root wt (gm)	Shoot wt (gm)	Leaf wt (gm)	Total plant wt (gm)
D_1	10.00	1.40	30.77	111.93	702.56	80.67	15.50	8.99	12.87	2.81	37.79	13.36	55.45
D_2	15.27	2.40	44.13	147.53	1560.7	208.4	22.47	15.38	20.67	12.91	98.44	32.88	162.54
D_3	21.60	2.13	63.40	168.87	2843.6	270.8	22.20	15.40	22.80	16.42	133.92	53.97	228.73
$CD_{at0.05}$	4.04	0.62	9.39	12.97	588.03	48.82	2.49	3.839	4.45	2.2	15.71	9.23	22.11

The data of Table 1 reveal that the plant raised through cutting having diameter D_3 (8-10 cm) showed best performance, significantly higher over D_1 (4-5.5 cm) in terms of number of leaf per plant (21.60), height (63.40 cm), crown area (168.87 cm), GBH (6.9 cm), leaf area (2843.6 cm), petiole length (270.8 cm), length of lateral root (15.4 cm), number of lateral root (22.8 cm), weight of root, shoot, leaf and total plant biomass (16.42, 133.92, 53.9 and 228.7 gm, respectively). In respect of the number of branches and tap root, D_2 showed the best result (2.4, and 22.47 respectively).

Highest leaf area (2286 cm²) was recorded with 40 cm of cutting length as against 1336 cm² in 20 cm length (Table 2). There was significant difference among the various length for number of leaf, plant height, girth diameter, petiole length and weight of root, shoot, leaf and total plant. The suitable length for cuttings found with this work is 40 cm and then a decline with the shorter lengths of cuttings. The cutting of 40 cm in length provided a greater amount of reserved assimilates in the plant tissues, thus a rapid growth of physic nut should be attained. Similar results were also found by Ratree (2004). The H_1 cutting exhibited higher lateral root. Significantly higher number of branches, girth diameter, number of lateral root and root weight were observed in H_2 length of cutting.

Table 2. Effect of length of cutting on their performance

Treat.	No. of leaf	No. of bra-nch	Plant hei-ght (cm)	Crown area (cm)	Leaf area (cm²)	Petiole length	Tap root (cm)	Lateral root (cm)	No of lateral root	Root wt (gm)	Shoot wt (gm)	Leaf wt (gm)	Total plant wt (gm)
H_1	13.13	1.87	34.30	136.27	1336.27	131.52	20.03	14.23	18.20	8.26	52.59	27.74	98.92
H_2	14.07	2.27	39.53	134.87	1483.77	160.15	17.20	11.51	20.40	12.20	95.66	29.75	147.06
H_3	19.67	1.80	64.47	157.20	2286.95	228.74	22.93	14.02	17.73	11.68	121.9	42.72	200.74
$CD_{at0.05}$	4.56	NS	9.18	NS	NS	57.99	NS	NS	NS	3.29	15.12	12.27	27.21

The combined effect of length and diameter as presented in Table 3 reveals that diameter D_3 (8-10 cm) exhibited maximum number of leaf, crown and leaf area, petiole length, and weight of shoot, root and total plant with height H_3. Also D_3 recorded maximum mean length of lateral root and number of lateral root in H_1 and H_2, respectively. D_2 recorded maximum number of branches, girth diameter, GBH and root weight in H_2. D_2 indicate longest tap root and plant height in H_3.

Table 3. Combined effects of length and thickness of cutting on their performance

Treat.	No. of leaf	No. of bra-nch	Plant height (cm)	Crown area (cm)	Leaf area (cm²)	Petiole length	Tap root (cm)	Lateral root (cm)	No of lateral root	Root wt (gm)	Shoot wt (gm)	Leaf wt (gm)	Total plant wt (gm)
D_1H_1	19.40	3.80	43.40	206.60	1083.1	157.4	30.00	19.53	22.60	4.15	51.98	24.32	80.45
D_1H_2	15.20	2.40	44.60	180.40	1251.1	110.1	24.20	17.20	17.20	4.67	57.80	18.73	83.27
D_1H_3	19.60	1.20	78.80	217.60	1534.8	176.0	28.70	13.70	29.00	7.03	96.38	32.21	141.04

D_2H_1	19.80	3.60	44.40	222.00	2037.6	379.0	34.00	25.20	43.20	8.26	92.02	37.74	152.67
D_2H_2	23.60	6.00	57.00	269.00	2154.0	294.4	35.60	25.05	42.20	38.67	209.88	46.52	323.29
D_2H_3	38.60	3.60	133.20	307.80	3942.1	452.5	51.60	32.07	26.80	24.66	239.97	89.76	410.06
D_3H_1	31.00	3.00	93.00	308.40	4019.8	392.3	45.80	32.73	32.20	33.24	141.12	85.95	299.32
D_3H_2	37.20	3.80	110.80	284.00	4725.5	463.3	30.80	20.23	54.00	24.72	247.31	97.65	388.40
D_3H_3	47.80	4.60	133.00	328.40	6661.3	602.2	43.80	29.53	39.00	30.25	326.95	104.35	530.58
$CD_{at\ 0.05}$	6.61	0.98	12.59	23.35	1234.9	85.99	7.61	5.76	6.61	4.85	20.21	19.22	38.3

Based on the above findings it may be concluded that 8-10 cm thick branch cutting is the most suitable for good plant vigour, owing to maximum number of leaf, largest crown and leaf area, maximum lateral root and total plant weight may lead to subsequent seed production (Ratree, 2004). Where as in regard to length of cutting, H_2 (30 cm) is better suited for number of branches, girth diameter and GBH, while H_3 (40 cm) long root stock is observed best for number of leaf, crown and leaf area and number of lateral roots. For getting optimum initial vigour of *J. curcas* L. branch cutting should be used of 40 cm length and 8-10 cm thick.

REFERENCES

Heller, J. 1996. Physic nut (*Jatropha curcas* L.) : promoting the conservation and use of underutilized and neglected crops. International Plant Genetic Resources Institute, Rome.

ISTA 1976. International rules for seed testing. *Proc. Int. Seed Testing Assoc.* 31: 1-152.

Pholsen, S. 1983. Physic nut (*Jatropha curcas* L.) cultivation and the replacement of its oil to a diesel fuel oil for a small diesel tractor unit. Annual Report, Khon Kaen Agricultural and Technology College, Khon Kaen, Thailand.

Ratree, S. 2004. A Preliminary study on Physic nut (*Jatropha curcas* L.) in Thailand. *Pak. J of Biological Science* 7(9): 1620-1623.

Chapter - 10

A STUDY ON *JATROPHA CURCAS* L. CUTTING BASED NURSERY

Kapil Sekhri, Kamendra Singh and Ghanashyam Bhanja

Tinna Overseas Ltd., Plantation, R & D Division, Sultanpur, Mehrauli, New Delhi-110030 E-mail: g.bhanja@yahoo.com

ABSTRACT

Jatropha curcas L. (Physic nut) is a wonder plant that produces biodiesel, a smoke-free renewable fuel for regular diesel engines. The word 'Jatropha' is derived from two Greek words jatros meaning a 'doctor' and trophe meaning "nutrition". It could be served as panacea for energy shortage in the future world i.e. main source of biodiesel. The seeds contain 35-45% oil. For raising betters plantation, good quality planting material is necessary. *Jatropha* can be propagated from seeds as well as cuttings. Propagation through seed is unable to maintain the purity of germplasm due to possibility of germplasm intermixing during pollination. Cutting is best to maintain the purity of germplasm, so it is necessary to study the best period to raise cutting and effect of metrological and climatic parameters on rooting. Jatropha cuttings are highly sensitive to climatic factors. Cuttings from three year old plants planted from 0-4 day interval indicate that December to February is the best period to develop cutting rised nursery. That means low temperature and dormancy or defoliation is suitable for vegetative propagation. Furthermore PGRs enhance the rooting ability and shoot growth of Jatropha cuttings. Number of leaves and length of shoots were significantly higher in BA treatment whereas, the number of roots and length of roots were higher at NAA treatment. Although the survivability and days to bud break was higher under control condition, growth regulators can be used for enhancing the shoot and root growth and development. The plantlets developed under different treatment combinations were transferred to the field subsequently.

Keywords: *Jatropha curcas,* germplasm, cutting, temperature, PGRs, bud break

Abbreviations: NAA-α-Naphthalene acetic acid; BAP-6-Benzylaminopurine; GA_3-Gibberellic acid; PGR- Plant growth regulator; CRD- Completely randomized design.

INTRODUCTION

J. curcas L. belonging to the family Euphorbiaceae is a multipurpose shrub or small tree of about 6-8 meter height with grey or reddish bark, mark by white patches. It is a deciduous woody glabrous shrub or under tree with spreading branches and twigs with soppy or milky latex exudates. The peculiarity of the plant is it defoliate during winter. The green leaves are simple, apically crowded, ovate and acute with a length of 6-40 cm, 6-35 cm broad, 3-5 lobed outlines. The leaves are arranged alternately with 2.5- 7.5 cm long petiole. Normally, it has tap root system. Five roots are formed during germination; one is tap root and other four lateral roots. Plant developed from cuttings does not have tap root. The plant flowers in March-April and June-July with several to many greenish cymose inflorescences. The flowers are bell shaped actino-morphic, yellowish green or creamy in colour. Male flowers are smaller than the female one and female flower born singly with elliptic, 3 celled, triovulate ovary with 3 spreading bifurcate stigma. Generally male flowers are more and the sex ratio is 1:8 to 1:30 (F:M). Male flower with 10 stamens, 5 united at the base only and 5 united into a column.5 nectar gland present at the base of the each stamen. Fruit is a capsule 2.5-4 cm long and finally split in to 3 valves, all or two of which commonly have an oblong black seed of 2 x 1 cm size. The fruits mature during August to September. At maturity it changes its colour from green to yellow than dark brown.

Commercial cultivation of *J. curcas* is being carried out all over world, mainly in tropical and sub tropical countries. The present study on Jatropha cutting attempted with a view to develop quality planting material for plantation. Response of Jatropha cutting under different day interval from 0 day to 4th day in different month indicates the percentage of rooting. Other study of plant growth regulator indicates that BA and NAA enhance both rooting and shooting.

MATERIALS AND METHOD

Plant materials

Stem cuttings from three year old plant of germplasm GL-3 were taken as plant material for this study, which have 2.5-4 cm diameter and 25 cm length. These cuttings were cut (100 nos) by pruning secateur, only 20 nos planted in poly bags filled with nursery mixture (soil, sand and Farm Yard Manure (3 : 1 :1) These cuttings were kept in poly house at 30°C and 60-70% humidity for improving the rooting percentage. Rema-ining cuttings were also planted in poly bag filled with nursery mixture at 1-4 day interval, each day 20 cuttings. These poly bags were watered weekly twice to maintain optimum soil moisture. In the same way every month from October-2005 to October-2006 cuttings were planted to study the effect of day interval and climatic parameters on rooting of *J. curcas*.

Effect of plant growth regulators on Jatropha cuttings

Cuttings from three year old germplasm BLS local were taken for this study. At first, three cuttings of 2.5-4 cm diameter and 25 cm height were selected for each treatment. The desired cuttings were immersed for 30-45 minutes in PGRs solution of each treatment and were then transferred to a shaded place. After 2-3 hrs these cuttings were planted in polybags filled with nursery mixture (soil, sand and FYM (3:1:1)). The cuttings were kept in polyhouse at 30°C and 60-70% humidity for improving the rooting percentage.

RESULTS

After planting, the aerial end of cutting got sealed by dry soppy latex. The first bud break started after 5-7 days and continued till 15 day for rooting (20-25 days). After bud break it was observed that callus like tissue (undifferentiated cells) were developed on tip of rooting end, that tissue later differentiated into lateral root. The average percentage of rooting increases with decrease in the temperature and does not get affected by day interval. Highest average rooting percentage was observed in the month of February (97%) when maximum and minimum day temperature is 29.5 and 17.8°C, where as lowest rooting was observed in the month of August, when maximum and minimum day temperature is 29.6 and 25.4°C (Table 1)

Table 1. Monthly performance of Jatropha cutting under different day intervals

S. No.	Month	Day interval	% rooting	Germplasm	Average % of rooting	Average temperature (°C)	
						Maximum	Minimum
1.	October 2005	0	60	GL-3	64	30.5	24.8
		1	70	GL-3			
		2	70	GL-3			
		3	75	GL-3			
		4	45	GL-3			
2.	November 2005	0	90	GL-3	77	29.7	18.6
		1	95	GL-3			
		2	60	GL-3			
		3	70	GL-3			

		4	70	GL-3			
3	December 2005	0	100	GL-3	96	27.3	15.8
		1	95	GL-3			
		2	95	GL-3			
		3	95	GL-3			
		4	95	GL-3			
4	January 2006	0	95	GL-3	92	27.7	15.3
		1	85	GL-3			
		2	95	GL-3			
		3	85	GL-3			
		4	100	GL-3			
5	February 2006	0	100	GL-3	97	29.5	17.8
		1	100	GL-3			
		2	100	GL-3			
		3	95	GL-3			
		4	90	GL-3			
6	March 2006	0	100	GL-3	73	31.5	22.8
		1	75	GL-3			
		2	55	GL-3			
		3	70	GL-3			
		4	65	GL-3			
7	April 2006	0	65	GL-3	25	32.8	24.3
		1	25	GL-3			
		2	15	GL-3			
		3	10	GL-3			
		4	10	GL-3			
8	May 2006	0	30	GL-3	25	33.0	26.7
		1	25	GL-3			
		2	30	GL-3			
		3	20	GL-3			
		4	20	GL-3			
9	June 2006	0	25	GL-3	18.75	33.8	27.2
		1	25	GL-3			
		2	15	GL-3			
		3	10	GL-3			

		4	0	GL-3			
10	July 2006	0	0	GL-3	25	32.3	24.4
		1	25	GL-3			
		2	25	GL-3			
		3	0	GL-3			
		4	0	GL-3			
11	August 2006	0	0	GL-3	10	29.6	25.4
		1	0	GL-3			
		2	10	GL-3			
		3	10	GL-3			
		4	10	GL-3			
12	September 2006	0	40	GL-3	29	29.7	24.8
		1	35	GL-3			
		2	30	GL-3			
		3	20	GL-3			
		4	20	GL-3			
13.	October 2006	0	60	GL-3	39	29.8	25.0
		1	65	GL-3			
		2	30	GL-3			
		3	30	GL-3			
		4	10	GL-3			

The result of the present study was statistically significant for number of leaves, length of shoot, number of roots and length of roots under different treatments (Table 2). Number of leaves and length of shoot were significantly higher in T_7 whereas, the number of roots and length of roots were higher at T_4. Although the survivability and days to bud break was higher under control condition, growth regulators can be used for enhancing the shoot and root growth and development. The plantlets developed under different treatment combinations were transferred to the field subsequently.

Table 2. Treatment details (A) and effect of PGRs on cuttings (B)

(A)

No. of accessions	:	1 (BLS Local)
No. of treatments	:	T_1= Control
		T_2= NAA 50 ppm
		T_3= NAA 100 ppm

T_4= NAA 150 ppm
T_5= BA 50 ppm
T_6= BA 100 ppm
T_7= BA 150 ppm
T_8= GA_3 50 ppm
T_9= GA_3 100 ppm
T_{10}= GA_3 150 ppm
T_{11}= NAA 50 ppm + BA 50 ppm
T_{12}= NAA 50 ppm + GA_3 50 ppm
T_{13}= BA 50 ppm + GA_3 50 ppm
T_{14}= NAA 50 ppm + BA 50 ppm + GA_3 50 ppm

No. of replications : 3 (7 x 3)
Design : CRD
Month of planting : November

(B)

Treatment	Days to bud break	No. of leaves	Length of shoot (cm)	No. of root	Length of root (cm)
T_1	8.46	2.31	23.17	1.67	3.63
T_2	9.20	1.82	25.02	4.68	6.46
T_3	11.20	1.35	25.43	7.13	7.18
T_4	9.42	1.64	24.04	17.13	9.07
T_5	12.25	2.27	27.83	3.94	3.81
T_6	13.31	2.24	27.18	3.66	3.27
T_7	11.80	3.29	30.15	2.49	3.46
T_8	12.18	1.55	22.04	0.78	2.79
T_9	12.25	1.81	24.36	0.92	2.86
T_{10}	10.59	1.55	22.24	0.50	1.92
T_{11}	11.44	2.24	27.49	8.14	7.95
T_{12}	12.58	1.42	23.21	6.42	7.50
T_{13}	12.33	1.42	27.09	1.72	4.16
T_{14}	11.94	1.81	21.33	1.81	4.43
Mean	11.35	1.91	25.04	4.36	4.89
SEm	NS	0.15	0.52	0.71	0.55
CD (0.05)	NS	0.47	1.60	2.18	1.69
CD (0.01)	NS	0.66	2.25	3.05	2.37

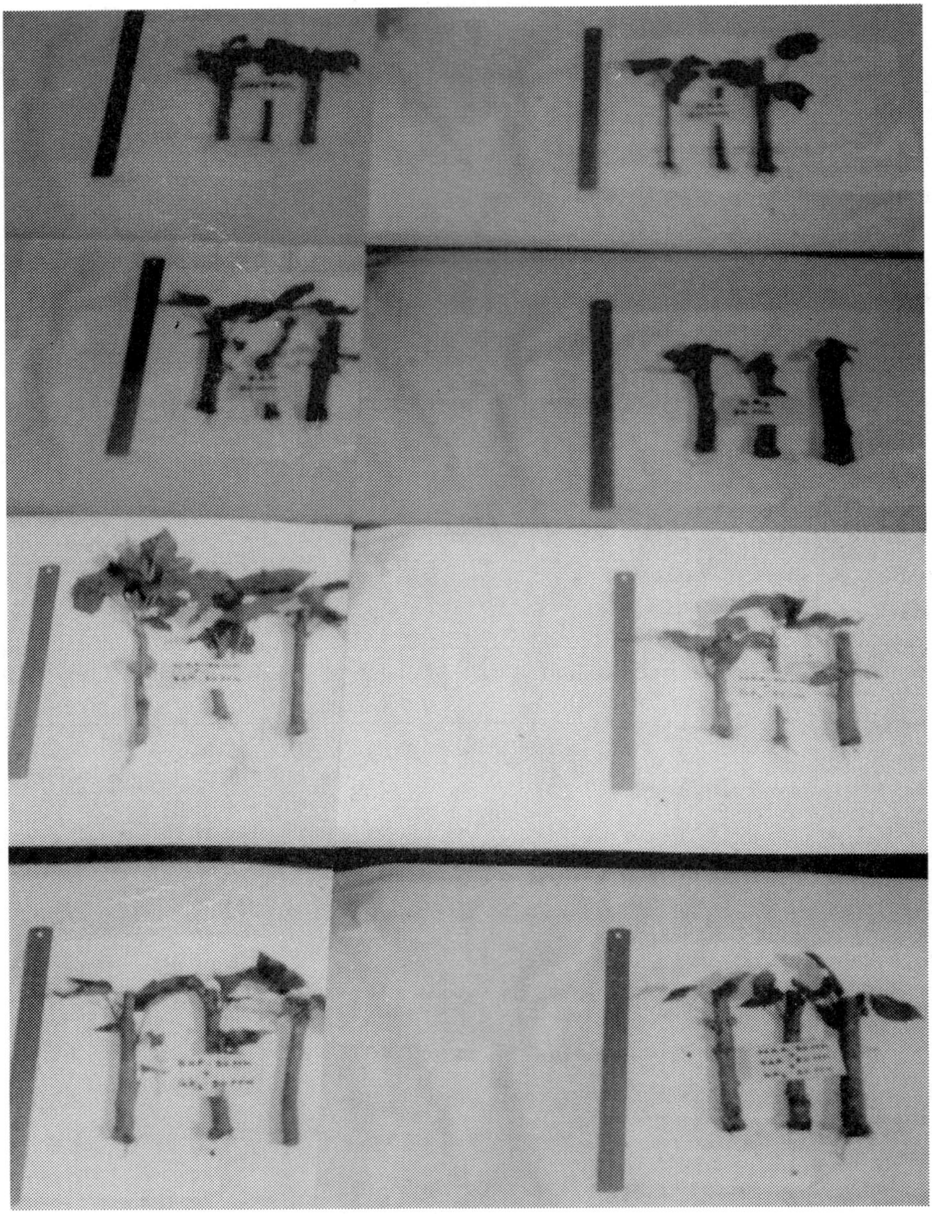

Fig. 1. Effect of plant growth regulators on *Jatropha* cuttings

DISCUSSION

Though the main aim of Jatropha plantation is for biodiesel but other medicinal and valuable uses also expected. Further suitable planting materials are not always available. Therefore, for the establishment of cutting raised nursery it was necessary to standardize the concentration of PGRs for the growth of root and shoot in Jatropha cuttings. Plant growth regulators enhance both rooting and shooting in Jatropha cuttings. Climatic and metrological parameters also play a vital role in cutting development. Temperature and humidity are the major climatic parameters that affect rooting and shooting pattern of cutting nursery in Jatropha. It was observed that before root initiation callus like tissue developed cutting end. It means micropropagation through plant tissue culture can be possible in Jatropha. The physiology of bud break indicates that low temperature and humidity enhance both rooting and shooting. Further plant growth regulators also play a vital role on rooting and shooting of Jatropha cuttings. During dormancy and defoliation period in winter season, plants of deciduous tree have the property to store starchy food material within plant body and it is the region of profuse shooting and rooting. Auxin and other plant growth regulator enhance the physiology of rooting and shooting, which need further study.

REFERENCES

Anonymous 1976. Wealth of India, Vol X. Publications and Information Directorate, CSIR, New Delhi

Brown, R.G. and Hull, F.E. 1951. Taxus (Yew) poisoning of cattle. *J. Am. Vet. Med. Asoc.*

Chauhan, K.S. and Reddy, T.S. 1974. Effect of growth regulators and mist on rooting in stem cuttings of plum. *Ind. J. Hort.* 32: 229-231.

Chen, H.H., Shen, Z.Y. and Li, P.H. 1982. Adaptability of crop plants to high temperature stress. *Crop Sci.* 22: 719-725.

Duncan, D.R. and Widholm, J.M. 1990. Measurements of viability suitable for plant tissue. In: J.W. Pollard and J.M. Walker (Eds.) *Plant Cell and Tissue Culture*. Humana Press, New Jersey, pp. 29-36.

Dubey, K.P. 1997. Himalayan yew (Taxus *baccata*) conservation: A vegetative approach. *Ind. For.* 123:1150-1154.

Gomez, K.A. and Gomez, A.A. 1984. Statistical Procedures for Agricultural Research, 2nd Ed. John Willey, New York.

Greenberg, J.T. 1996. Programmed cell death: A way of life for plants, *Proc. Natl. Acad. Sci.* USAVol. 93: 12094–12097.

Guinand, Y., Hitimana, L., Franzel, S. and Akyeampog, E. 1992. Complement de diagnostic agroforestier dans la region naturelle du Kirimiro. *AERENA* Report No. 66 ICRAF, Nairobi.

Gressel, J. 1980. Uses and drawbacks of cell cultures in pesticide research. In: F. Sal *et al.* (Eds.) Plant Cell Cultures- Results and Perspectives. Elsevier, Amsterdam, pp. 379-388.

Nagasawa, A. and Finer, J. 1988. Induction of morphogenic callus cultures from leaf tissue of garlic. *Hort. Sci.* 23: 1068- 1070.

Novak, F.J., Havel, L. and Dolezel, J. 1986. Onion, garlic and leek (*Allium* species). In: Bajaj, Y.P.S. (Eds.). Biotechnology in Agriculture and Forestry, Vol. 2. Springer- Verlag, Berlin 2:387- 404.

Orme, K. 1983. Vegetative propagation of Tasmanian Eucalypts. Colloque International sur les Eucalyptus Résistants au Froid". Bordeaux, 1983 (IUFRO-AFOCEL): 559-569.

Saxena, H.O. and Brahmam. M. 1995. The Flora of Orissa, Vol 3. Orissa Forest Development Corporation Ltd., Bhubaneswar, Orissa.

Singh, R.R. and Chander 2001. Effect of auxin on rooting behaviour of Neem (*Azadirachta indica* L) *Ind. For*. 9: 1019-1024.

Troup, R,S. 1921. The Silviculture of Indian Trees, Vol III. Clarenden press, Oxford, pp. 785-1195.

Chapter - 11

PONGAMIA PINNATA - VARIABILITY IN SEED CHARACTERS AND OIL CONTENT AMONG THE PLUS TREES OF KARNATAKA

R. Nainamalai, Geeta Joshi, A.N.Arunkumar and Bhimi Ram

Forest Research Centre, P.O. Hakimpet, Dullapally, Hyderabad – 500 014
E- mail: rnainamalai@icfre.org

Abstract

Energy demand in India is increasing at the rate of 6% annually compared to 2% for many other countries. Currently, imported petroleum crude supplies about 70% of the energy requirement. Vegetable oils, fats and their derivatives have been proposed as an alternate renewable and eco-friendly energy source. As India imports more than 40% of its edible oil requirement, it has to depend on non-edible oils for biodiesel. Various non-edible tree borne oils such as pongamia, mahua, jatropha and neem, are available in the country. To take care of the increasing demand of the biofuel requirement, *Pongamia pinnata* is expected to play a very important role. The significant variation observed with reference to two commercially important traits viz; kernel weight and oil content shows the potential genetic variability, this species harbours. With reference to kernel weight and oil content, the correlation studies showed a strong positive relationship, while the high heritability values confirming the possibilities of achieving higher gain during selection process. This important relationship information can be utilized in developing further improvement and future breeding strategies in *P. pinnata,* a prized oil yielding species.

Keywords: *Pongamia pinnata*, seed variability, oil content, heritability, biofuel

INTRODUCTION

With petroleum product prices rising steadily, diesel alone has become 25 percent costlier over the last year. Apart from the search for alternatives, it is the need to achieve energy independence that is directing so much focus on biofuel and the crops that will help yield these oils. The enthusiasm for biofuel must also be viewed against the backdrop of the country's thirst for oil which is about 114 million tonnes every year and 75 percent of which is imported at a cost of Rs 1,20,000 crore. About

112 million tonnes of oil is consumed just by the transportation sector. Biofuel is any kind of fuel produced from dry organic matter or combustible oils produced by plants. It is a source of renewable energy, unlike other natural resources, such as petroleum, coal and nuclear fuels. One of the important indigenous species that has been indentified and accepted as a suitable source for biofuel is *Pongamia pinnata*.

Pongamia pinnata also popularly known as Karanj tree (referred to as the Indian beech) is extensively grown in India. It is considered to be native to the Western Ghats. It is common along the banks of rivers, streams and in coastal areas. It flourishes in dry areas where agriculture is unproductive and in poor or saline soils. The tree can be identified by its greying bark, imparipinnate shining leaves and pinkish white flowers. The tree has multipurpose uses, wood for fuel, leaves as fodder and also as manure. The seed oil is used for making soaps, treatment of skin disease and rheumatism. With its potential utility as a source for biofuel, presently Karanj is being extensively planted on large areas and there is a huge demand for identifcation of superior genotypes, which can yield higher oil content. This calls for a systematic approach in improvement of this species. Karnataka Forest Department has already initiated the tree improvement work in Karanj by identifying candidate plus trees in different silivicultural zones of the state. This study was initiated to evaluate these identified candidate plus trees with reference to seed traits and oil content.

MATERIAL AND METHODS

For the present study, from each silivicultural zone (Central, East, North and South) in Karnataka, five plus trees of *P. pinnata* totalling to 20 plus trees were selected (Table 1). The geographical range of these plus trees extended from 12°25' to 15°28' N latitude and from 74°01' to 77°44' E longitude with the average rainfall ranging from 55 to 325 cm. From the individual plus trees, pods were collected and were marked separately to maintain the identity. The pods were dried in shade and stored in air tight plastic containers. The pods were then used for measuring following observations: pod length, pod width, kernel length, kernel width, kernel weight and oil content.

To measure pod length and pod width, 50 pods each in four replications were used. Pod length (cm) and width (cm) was measured using a scale upto one decimal place. Pods were manually broken and the kernels were removed carefully. Fifty kernels each in four replications were used to measure individual kernel length and width using digital vernier caliper (Mitutoyo, model CD-6” CS), and kernel weight (grams) was recorded on a top pan electronic balance (Adair Dutt MJ, 300). For estimating the oil content, 25 grams of kernel was used in four replications. The oil extraction process was carried out in Sox-hlet extractor according to AOAC (1970). Hexane was used as a solvent for the oil extraction. Oil content was calculated by dividing oil weight with sample weight and expressed in percentage.

Table 1. Geographical and rainfall data of the locations of 20 plus trees

Plus tree	Locality	Division	Silvicultural Zone	Latitude (°N)	Longitude (°E)	Mean annual Rainfall (cm)
P1	Kengeri	Bangalore Urban	Central			
P2	Kengeri	Bangalore Urban	Central			
P3	Malleswaram	Bangalore Urban	Central	12°59`	77°35`	90
P4	Malleswaram	Bangalore Urban	Central			
P5	IISC campus	Bangalore Urban	Central			
P6	Bellary	Bellary	Eastern	15°09`	76°56`	55
P7	Bellary	Bellary	Eastern			
P8	Bellary	Bellary	Eastern			
P9	Challekere	Bellary	Eastern			
P10	Challekere	Bellary	Eastern			
P11	Dharwad	Dharwad	Northern	15°28`	74°01`	80
P12	Sirsi Road	Dharwad	Northern			
P13	Sirsi Road	Dharwad	Northern			
P14	Konankeri	Dharwad	Northern			
P15	Sirsi Road	Dharwad	Northern			
P16	Somwarpet	Madikeri	Southern	12°25`	77°44`	325
P17	Somwarpet	Madikeri	Southern			
P18	Somwarpet	Madikeri	Southern			
P19	Madikeri	Madikeri	Southern			
P20	Madikeri	Madikeri	Southern			

The data obtained was subjected to statistical analysis. Estimates of mean, standard error and critical difference were worked. Variance analysis was estimated for each character separately, Karl Pearson's correlation was also worked out (Panse and Sukhatme, 1961). The genetic parameters such as genotypic variance (GV), phenotypic variance (PV), were calculated by using respective mean sum of square from variance table (Johnson *et al.*, 1955).

RESULTS

Observation pertaining to pod, kernel traits and oil content of all the plus trees were recorded. It is evident from the Table 2 that all the characters under study revealed significant differences ($p < 0.01$) among the plus trees. Though, it is not mentioned here, there was no significant difference among the plus trees within respective zones. Pods collected from Southern silvicultural zone recorded the highest length as compared to other three zones. Plus tree number P18 and P20 recorded the highest length (6.72cm) and (6.70cm) respectively. The lowest pod length was recorded by P6 (4.11cm), which represented the Eastern silvicultural zone. The general mean value for this trait was 5.53cm and the coefficient of variation was 13.39%. For pod width, it was found that P3 recorded the highest width (2.78 cm), followed by P10 (2.58 cm). All the plus trees from Southern silvicultural zone showed the lower pod width values. The CV for this trait was 15.91%, exhibiting considerable variability. Plus trees from Central silvicultural zone recorded the highest kernel length. Plus tree P1 (20.81 mm) was followed by P3 (20.26 mm) and the lowest was observed by plus tree P6 (16.98 mm). The general mean for this trait was 18.44mm (Table 3) and the CV was 6.28%. With reference to kernel width, compared to other zones, kernels from the plus trees of Northern zone recorded the highest width of 13.34 mm in P15 followed by P11 (13.16mm). The lowest kernel width was recorded by P4 (11.43mm), which represented Central zone. The general mean for this trait was 12.25 mm with CV value of 4.68%. Among all the zones, plus trees from Northern silvicultural zone recorded consistently higher kernel weight. Plus tree P14 and P15 recorded highest weight (2.09 g) and the lowest was observed in P17 (1.12 g). In case of oil content, plus trees of Northern zone registered highest oil content compared to other zones. Plus tree P14 (29.70%), followed by P12 (29.21g) recorded the highest oil, while all the plus trees from Southern silvicultural zone recorded the lowest. Correlation studies (Table 5) revealed that there was significant positive relationship (p = 0.01) between kernel weight and oil content (r = 0.859). Significant positive relationship was also observed between kernel length and kernel weight (r = 0.645), kernel width and oil content (r = 0.618) and between kernel width and weight (r = 0.476, p = 0.05).

Table 2. Mean sum of squares for pod, kernel traits and oil content

Source	d.f	Mean sum of squares					
		Pod length	Pod width	Kernal length	Kernal width	Kernal weight	Oil content
Plus trees	19	2.199**	0.048**	5.371**	1.315**	0.697**	10.10**
Error	60	0.273	0.004	0.503	0.139	0.074	0.007

**Significant at p=0.01

Table 3. Summary of statistics for pod, kernel traits and oil content

	Pod length (cm)	Pod width (cm)	Kernel weight (g)	Kernel length (mm)	Kernel width (mm)	Oil content (%)
Mean	5.53	2.18	1.629	18.44	12.248	27.02
Median	5.42	2.25	1.648	18.04	12.00	26.56

Range	4.11-6.72	1.64-2.78	1.11-2.08	16.97-20.80	11.43- 13.34	24.00-29.70
Variance	0.550	0.121	0.174	1.342	0.328	2.53
SD	0.741	0.347	0.417	1.158	0.573	1.59
CV(%)	13.39	15.91	25.59	6.28	4.68	5.88

Table 4. Mean value for pod length, width, kernal length, width, weight and oil content for different plus trees

S.N.	Pod length (cm)	Pod width (cm)	Kernel length (mm)	Kernel width (mm)	Kernel weight (g)	Oil content (%)
P1	5.28	2.26	20.81	11.82	2.01	26.71
P2	2.08	2.13	20.22	11.93	2.01	28.58
P3	5.51	2.78	20.26	11.99	1.97	27.73
P4	5.39	2.44	19.97	11.43	1.99	26.61
P5	5.61	2.25	20.04	12.01	1.97	28.68
P6	4.11	2.23	16.98	11.52	1.19	26.47
P7	5.46	2.38	17.65	11.73	1.30	26.36
P8	6.08	2.53	18.11	12.34	1.33	25.70
P9	4.63	2.43	17.97	11.97	1.28	25.96
P10	5.06	2.58	18.62	12.18	1.30	26.00
P11	4.78	2.36	17.88	13.16	2.08	28.38
P12	5.38	2.31	18.33	13.09	2.05	29.21
P13	5.73	2.36	18.49	12.89	2.07	29.12
P14	5.16	2.20	18.04	13.13	2.09	29.70
P15	4.64	2.20	18.06	13.34	2.09	28.38
P16	6.38	1.64	17.53	12.16	1.25	25.98
P17	6.39	1.64	17.32	11.98	1.12	25.38
P18	6.72	1.66	17.92	12.48	1.17	25.60
P19	6.53	1.65	17.17	11.82	1.18	24.99
P20	6.70	1.68	17.54	12.00	1.15	24.00
Mean	5.53	2.18	18.44	12.25	1.63	27.02
Sem (±)	0.37	0.048	0.501	0.264	0.03	0.06
CD 0.05	0.73	0.094	0.983	0.517	0.07	0.12

Table 5. Correlation matrix for pod, kernel traits and oil content

	Pod length	Pod width	Kernel length	Kernel width	Kernel weight	Oil content
Pod length	1					
Pod width	-0.646**	1				
Kernal length	-0.182	0.481*	1			
Kernal width	-0.120	0.065	-0.190	1		
Kernal weight	-0.432*	0.492*	0.645**	0.476*	1	
Oil content	0.470*	0.414*	0.380	0.618**	0.859	1

DISCUSSION

The most successful tree improvement programmes are those in which proper seed sources/provenances are identified. Documenting variation within a species is necessary for effective tree improvement programme. In order to evolve superior material, the tree breeder has to asses the traits to be improved, and their variation both with reference to morphological and biochemical characters. Therefore, success in establishment and productivity of forest tree plantations is dertemined largely by the use of proper improved plant material either in terms of seed or through clonal methods (Zobel and Talbert, 1984).

P. pinnata is one of the species in which preliminary improvement work has already been initiated through identification of candidate plus trees by the Karnataka Forest Department. Evaluating these plus trees with reference to pod, kernel and oil content strengthens further breeding programme. In the present investigation, it is evident that there was significant variation (p=0.01) for all the traits studied, *i.e.*, pod length and width, kernel length, width and weight and oil content. This significant variation can be attributed to the differences in the population, which could have been as a result of differences in the environmental conditions, viz; nutrients, light or water, to which the mother plants were subjected during the growing season (Gutterman, 1992). Similar observations with reference to variation in pod length and pod width and seed weight was also observed in case of *Acacia mangium* (Salazar, 1989), *Pongamia pinnata* (Kumaran, 1991), *Acacia catechu* (Ramachandra, 1997), *Azadirachta indica* (Philomina, 2000) and *Madhuca latifolia* (Khanna, 2001). In case of seed oil, significant variation was also observed from the seeds collected from different sources and the value ranged from 24.00 to 29.70%. Philomina (2000), while working on 73 one parent families of *Azadirachta indica* (neem) reported significant variation for seed oil content ranged from 26.83 to 32.5%. Jenner (2003) studied one parent families of *Madhuca latifolia* and reported that seed oil content ranged from 43.25% to 50.45% and showed significant difference for this trait.

Correlation studies revealed a very interesting result. There is a strong positive correlation betweeen kernel weight and oil content (r = 0.859). This relationship can be used as an indicator for further selection of superior genotypes.

CONCLUSION

To take care of the increasing demand of the biofuel requirement, *Pongamia pinnata* is expected to play a very important role. The significant variation observed with reference to two commercially important traits viz; kernel weight and oil content shows the potential genetic variability this species harbours. With reference to kernel weight and oil content, the correlation studies showed a strong positive relationship. This important relationship information can be utilised in developing further improvement and future breeding strategies in *P. pinnata,* a prized oil yielding species.

ACKNOWLEDGEMENT

We thank the Director and Co-ordinator (Research) of the Institute for their encouragement. We sincerely thank all the Silviculturists of the respecive silviculture zone of Karnataka State Forest Department for providing necessary help during seed collection process.

REFERENCES

AOAC 1970. Official Methods of Analysis of the Association of Official Agricultural Chemists. 11th ed., Washington, D.C.

Gutterman, Y. 1992. Maternal effects of seed during development. In: Fenner, M. Seeds: The Ecology of Regeneration in Plant Communities. CAB International, Wallingford, UK . 27–59.

Johnson, H.W., Robinson, H.F. and Constock R.E. 1995. Estimation of genetic and environmental variability in soybean. *Agron. J.* 47: 314-318.

Kumaran, K. 1991. Genetic analysis of seed and juvenile seedling attributes in neem (*Azadirachta indica* A. Juss) and (*Pongamia pinnata* Linn. Pierre). M.Sc. Thesis. Tamil Nadu Agrl. Univ., Coimbatore, 153 pp.

Panse, V.G. and Sukhatme, P.V. 1961. Statistical Methods for Agricultural Workers. New Delhi: ICAR. 381p.

Philomina, D. 2000. Genetic analysis of one parent families for variability, diversity, stability and propagation techniques in neem (*Azadirachta indica* A. Juss). Ph.D. (Agri) Thesis. Tamil Nadu Agricultural University, Coimbatore.

Ramachandra, N.G., Nautiyal, S., Negi, D.S. and Thapliyal, R.C. 1997. Seed source variation in chlorophyll contents of leaves of *Acacia catechu* under different water stress conditions. *Ann. For.* 5 (1):88-96.

Salazar, R. 1989. Genetic variation in 16 provenances of *Acacia mangium* at nursery level in Turrialba, Costa Rica. CATIE, Costa Rica. 17

Umesh Khanna, S. 2001. Genetic analysis, biochemical and molecular characterization of *Madhuca latifolia* Macb. Ph.D. Thesis. Tamil Nadu Agricultural University, Coimbatore.

Zobel, B. and Talbert, J. 1984. Applied Forest Tree Improvement. John Wiley and Sons, New York.

Chapter - 12

PERFORMANCE EVALUATION OF TWENTY TWO PROVENANCES OF *PONGAMIA PINNATA* (KARANJA) IN MADHYA PRADESH

A.K. Pandey, Dhan Shri Sharma and A.K. Mandal

Tropical Forest Research Institute, P.O. RFRC, Mandla Road
Jabalpur 482021 (M.P.) E-mail: akpandey10@rediffmail.com

ABSTRACT

Pongamia pinnata (Karanja) belonging to family Fabaceae is a large tree that grows in arid and semi arid regions of India. It is one of the nitrogen fixing tree species and produce seeds containing 30-40% oil. *P. pinnata* has been recognized as one of the potential species for biodiesel production as it can be grown under different land use situations. Lot of variation has been recorded in various growth characteristics including seed and oil yield. Therefore, it is of immense need to identify promising Karanja provenances with better growth and higher seed yield. With this objective, 22 provenances of *P. pinnata* received from various network partners were evaluated for their performance under tropical climate of Jabalpur, Madhya Pradesh. Provenance trials (national and zonal trial) were established in August 2005 and 2006 in randomized block design with three replications at institute campus. In national trial, 5 provenances received from 2 participating institutes/centers (TNMP-Tamilnadu and RAK-Rahuri) were included. However, zonal trial comprises of 17 provenances received from 6 participating institutes/centers (IGAU-Raipur, TFRI Jabalpur, NRCAF-Jhansi, CSAUAT-Kanpur, CCSHAU-Bawal and JNKVV Jabalpur). In order to study existing natural variation among all provenances, regular growth data for growth parameters i.e. height, collar diameter, number of branches was recorded and statistically analyzed. Significant variation was found in all studied growth parameters. The best performing provenance in national trial is TNMP-6 (height-127.83cm, collar diameter- 2.47cm, number of branches 4.46) followed by TNMP-3 (height-108.13 cm, collar diameter-2.05 cm, number of branches-4.33) and RAK-5 Rahuri (height-84.56 cm, collar diameter 2.21cm, number of branches 4.73) while in zonal trial best performance has been shown by TFRI-2 Jabalpur (height-159.12cm, collar diameter-2.68 cm, number of branches-9.33) followed by IGAU-1-Raipur (height-134.48 collar dia.-2.33, number of branches 13.0) and IGAU 2-Raipur (height- 131.56 cm, collar diameter -2.71 cm, number of branches-7.54). This information can be utilized in developing further improvement and future breeding programmes of *P. pinnata* to increase productivity and oil yield.

Keywords: *Pongamia pinnata*, performance evaluation, provenance.

INTRODUCTION

Pongamia pinnata commonly known as Pongam and Karanj, is a member of family Fabaceae subfamily Papilionaceae. It is a medium sized tree that grows in arid and semi arid regions of India. It produces seeds containing 30-40% oil. Due to its higher seed oil content and suitability to grow under different land use situations it has been recognized as one of the potential species for biodiesel production. It is found mainly in Western Ghats in India, northern Australia, Fiji and in some regions of Eastern Asia and has been successfully introduced to humid tropical regions of the world as well as parts of Australia, New Zealand, China and the USA (Scott *et al.,* 2008).The natural distribution of Karanja is along coasts and river banks in India and Burma. It is a hardy tree that mines water for its needs from 10 meter depths without competing with other crops. It grows all over the country, from the coastline to the hill slopes. It require very little care and cattle do not browse it. It has rich leathery evergreen foliage that can be used as green manure. This medium sized tree attains a height of about 8 m and a trunk diameter of more than 50 cm. However, Troup (1921) reports trees attaining heights of 18 m. The trunk is generally short with thick branches spreading into a dense hemispherical crown of dark green leaves. It is grown for shade and as wind break in tea plantation (Duke, 1983). The bark is thin grey to grayish- brown, and yellow inside. The leaves are a good source of green manure and being leguminous, they enrich the soil with nitrogen. The seed yield a thick yellow-orange to brown oil having many uses. The seed oil is mostly used as a fuel for cooking and illuminating lamps; as a lubricant, water-paint binder, pesticide, and in soap making and tanning industries. The wood is used for cabinet making, cart wheels, posts (NAS 1980), agricultural implements, tool handles and combs. The oil is also known to have value in folk medicine for the treatment of rheumatism, as well as human and animal skin diseases. The seed oil is a potential source of biofuel and now it is mainly promoted as an alternative of fossil fuel.

With the declining reserves of fossil fuels it has become apparent that biofuels (biodiesel) are destined to make a substantial contribution to the future energy demands of the domestic and commercial economies. To meet the future energy demands through biodiesel obtained from karanja oil, it will be important to establish commercial-scale karanja plantations. There is an immense need to identify promising karanja provenances with better growth and higher seed yield. Source variation tests are necessary to screen the naturally available variation to select the quality planting material for higher productivity (Gera *et al.,* 1996; Bhat and Chouhan, 2002). To evaluate the performance of karanja provenances, provenance trials (national and zonal trial) were established in August 2005 and 2006 at Tropical Forest Research Institute (TFRI) campus.

MATERIALS AND METHOD

For the study 22 provenances were selected for evaluating their performance through provenance trials. The provenances were received from various network partners for evaluating their performance in tropical climate of Jabalpur, Madhya Pradesh. Provenances collected by this institute have been exchanged among the other members of network for evaluation of their performance. The provenances were selected on the basis of overall growth, seed yield and oil content.

Establishment of provenance trial

Provenance trial in the form of national and zonal trial was established in 2005-2006 at TFRI campus. The plantation was laid down in randomized block design in three replications.

National trial

It was established in the year 2005 comprising of 5 provenances received from 2 participating institutes/centres. The details of provenances are given in Table 1.

Table 1. Details of 5 provenances of national trial

Provenance	Participating institute/centre
TNMP-1	TNAU, Mettupalayam
TNMP-3	TNAU, Mettupalayam
TNMP-6	TNAU, Mettupalayam
TNMP-14	TNAU, Mettupalayam
RAK	RAK, Rahuri

Zonal trial

Zonal trial I was established in 2005 comprising of 12 provenances received from 5 participating institutes/centres and Zonal trial II was established in 2006 comprising of 5 provenances recieved from participating institutes/centres. The details of provenance are given in Table 2 and 3.

Table 2. Details of 12 provenances of Zonal trial I

Provenance	Participating institute/centre
IGAU-1	IGAU, Raipur
TFRI-1	TFRI, Jabalpur
IGAU-2	IGAU, Raipur
CCSHAU-1	CSAUAT, Kanpur
IGAU-3	IGAU, Raipur
NRCAF-1	NRCAF, Jhansi
IGAU-4	CCSHAU-Bawal
NRCAF-2	NRCAF, Jhansi
TFRI-2	TFRI, Jabalpur
CCSHAU-2	CSAUAT, Kanpur
IGAU-5	IGAU, Raipur
TFRI-3	TFRI, Jabalpur

Table 3. Details of 5 provenances of Zonal trial II

Provenance	Participating institute/centre
JNKVV-12	JNKVV, Jabalpur
JNKVV-20	JNKVV, Jabalpur
JNKVV-15	JNKVV, Jabalpur
JNKVV-29	JNKVV, Jabalpur
JNKVV-32	JNKVV, Jabalpur

Nursery

Pongamia pinnata is easily established by seedling or by stump cutting. Seeds require no treatment before sowing. Seeds received from different pastures were sown in the nursery to raise seedlings. The collected seeds were tested for their germination. Standard procedures to evaluate germination percentage were followed. The seeds were sown in the nursery in polythene bags of size 15 x 23 cm containing soil, sand and FYM (1:1:1 ratio) with regular irrigation. The seedlings become transplantable on reaching height of 60 cm.

Layout and designing

The plantation was laid out in randomized block design in three replications. The experimental field was divided into 15 equal size blocks for national trial, 36 blocks for zonal trial I and 15 blocks for zonal trial II and 9 plants were planted in each block at a distance of 5x5 m. The pits were dug before one month of plantation and organic manure *viz*. FYM was applied @ 5 kg/pit two weeks earlier to planting. Nitrogen, phosphorus and potassium were applied in the form of urea, single super phosphate and muriate of potash as a basal dose (20 g urea+ 20 g SSP +20 g MOP) at the time of planting.

Plantation

One year old seedlings were found better for the establishment of plantation. Plantation was done at the onset of rainy season (July-Aug.) Normal cultural practices were applied for the management of plantation. Plants were watered once in winter month and twice in summer months.

Observation

Regular growth data was recorded to find out variation among different provenances on the basis of various growth parameters like height, collar diameter, and no. of branches per plant. (National Trial I: Tables 4, 4A, 4B; Zonal Trial I: Tables 5, 5A, 5B and Zonal Trial II: Tables 6, 6A, 6B and Figures 1, 2, 3).

Table 4. Growth performance of different provenances of national trial

Treatment	Provenance	Height	Collar diameter	No. of branches
T1	TNMP-1	94.53	33.68	4.06
T2	TNMP-3	108.13	20.53	4.33
T3	TNMP-6	127.83	24.73	4.46
T4	TNMP-14	105.53	20.16	5.23
T5	RAK-5	84.56	22.16	4.73

Table 4A. Analysis of variance for different characters in *Pongamia pinnata* (national trial)

Variable	Range	Mean	SD	CD	SE	F Value
Height	127.83-84.58	104.12	16.23	13.052	5.66	16.46
Collar diameter	33.68-20.16	24.25	5.57	19.478	8.4466	NS
No. of branches	5.23-4.06	4.56	0.44	NS	NS	NS

*Significant at 0.05 percent level. NS=Non significant.

Table 4B. Simple correlations between different characters (national trial)

Variable	Treat-ment	Replication	Height	Collar diameter	No. of branches
Treatment	1.0000				
Replication	0.0000	1.0000			
Height	-0.1985	0.2696	1.0000		
Collar diameter	-0.3416	0.3198	0.1371	1.0000	
No. of branches	0.3835	-0.3073	-0.0670	0.2662	1.0000

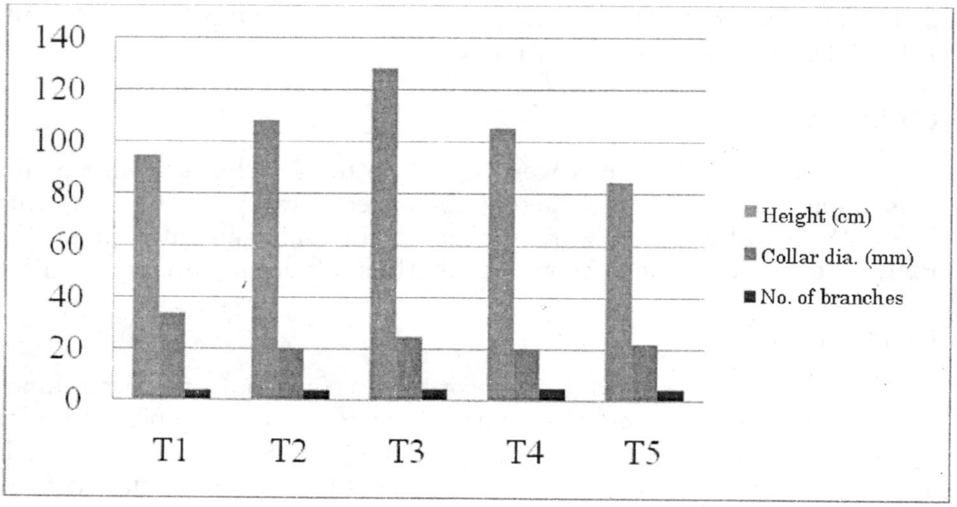

Fig. 1. Growth performance of provenances of national trial

Table 5. Growth performance of provenances of zonal trial I

Treatment	Provenance	Height	Collar diameter	No. of branches
T1	IGAU-1	134.56	23.34	13.00
T2	TFRI-1	110.07	25.91	10.00
T3	IGAU-2	131.56	`27.16	7.54
T4	CCSHAU-1	110.07	27.81	8.20
T5	IGAU-3	115.02	23.58	9.22
T6	NRCAF-1	116.03	23.20	10.11
T7	IGAU-4	110.26	22.30	8.88
T8	NRCAF-2	111.22	20.45	8.88
T9	TFRI-2	159.10	26.80	9.51
T10	CCSHAU-2	118.98	22.54	11.22
T11	TFRI-3	129.33	23.54	10.96
T12	IGAU-5	120.37	24.91	9.85

Table 5A. Analysis of variance for different characters in *Pongamia pinnata* (zonal trial I)

Variable	Range	Mean	CD	SE	F Value
Height	159.10-110.07	122.21	4.0207	1.9387	108.40
Collar diameter	27.81-20.45	24.03	4.2334	2.0413	3.86
No. of branches	13.00-7.54	9.78	2.0205	0.9742	4.50

*Significant at 0.05 percent level

Table 5B. Simple correlations between different characters

Variables	Treatment	Replication	Height	Collar diameter	No. of branches
Treatment	1.0000				
Replication	0.0000	1.0000			
Height	0.1118	0.0580	1.0000		
Collar diameter	-0.2435	-0.0785	0.1334	1.0000	
No. of branches	0.0282	0.1775	0.1794	-0.3105	1.0000

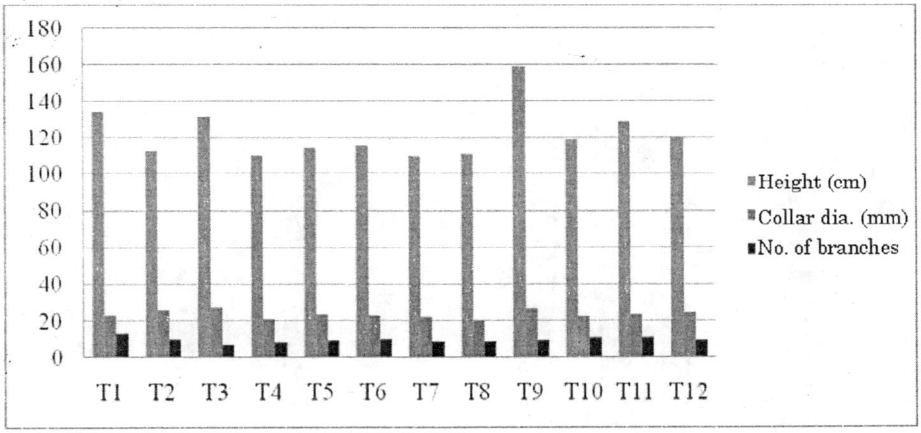

Fig. 2. Growth performance of provenances of zonal trial I

Table 6. Growth performance of provenances of zonal trial II

Treatments	Provenance	Height (cm)	Collar diameter(mm)	No. of branches
T1	JNKVV-12	50.46	12.06	3.16
T2	JNKVV-20	40.83	13.13	3.00
T3	JNKVV-15	31.46	09.13	2.80
T4	JNKVV-29	28.46	09.53	3.03
T5	JNKVV-32	37.60	13.33	3.26

Table 6A. Analysis of variance for different characters in *Pongamia pinnata* (zonal trial II)

Variable	Range	Mean	CD	SE	F Value
Height	50.46-28.46	36.742	5.4216	2.3511	26.74
Collar diameter	13.33-09.13	11.096	2.3371	1.0135	7.69
No. of branches	3.26-2.80	3.05	NS	NS	NS

Table 6B. Simple correlations between different characters

Variables	Treatment	Replication	Height	Collar diameter	No. of branches
Treatment	1.0000				
Replication	0.0000	1.0000			
Height	-0.6389	0.2434	1.0000		
Collar diameter	-0.0750	0.0812	0.5710	1.0000	
No. of branches	0.0904	0.3804	0.2488	0.6455	1.0000

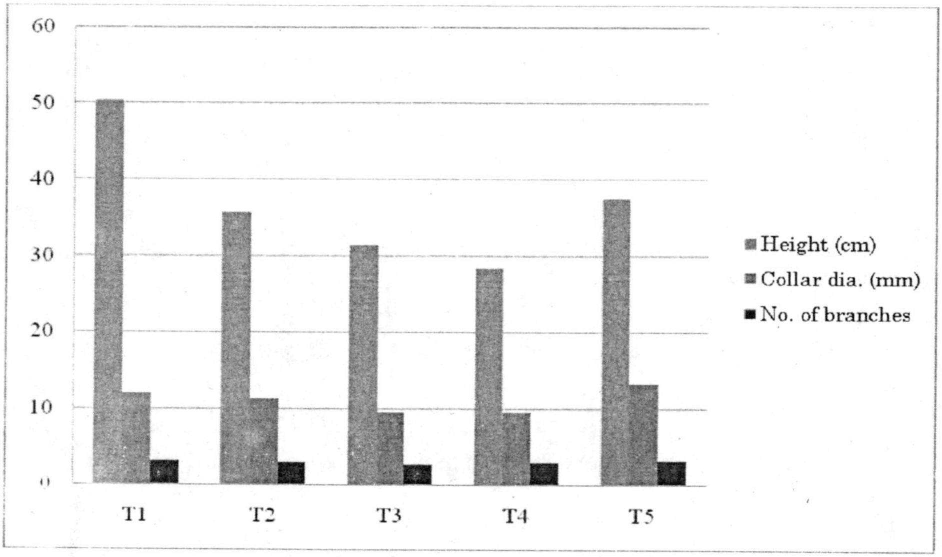

Fig. 3. Growth performance of provenances of zonal trial II

RESULTS AND DISCUSSION

Variation studies for growth performance among 22 provenances (National 5, Zonal I-12 and Zonal II-5) showed significant differences in height and collar diameter except number of branches as it varied season to season. Seed size influenced the germi-nation and vigour of Karanja seeds. The same findings were reported by Manonmani *et al.,* 1996.

Plant height

In national trial the plant height was recorded maximum (127.83 cm) in TNMP-6 and minimum in RAK-5 Rahuri (84.56 cm). However, in Zonal trial I, maximum plant heigh (159.12 cm) was found in TFRI-2 followed by IGAU-1 Raipur (134.48 cm) and IGAU-2 Raipur (131.56 cm) and in Zonal Trial II highest value (50.46 cm) was recorded in JNKVV-12 Jabalpur followed by JNKV-32 (37.6 cm).

Collar diameter

Collar diameter value was maximum in TNMP-6 (2.47mm) and minimum in RAK-5 Rahuri (22.1mm) in national trial. In zonal trial I, highest value of collar diameter recorded in TFRI-2 (26.8 mm) followed by IGAU-1 Raipur (23.3) and IGAU-2 Raipur (27.1 mm) and in Zonal Trial II highest value was recorded in JNKVV-12 (12.06 mm) Jabalpur followed by JNKV-32 (13.33 mm) and JNKVV-20 (11.43 mm). Significant variation was found in collar diameter among provenances.

Number of branches

In national trial no. of branches was found maximum in RAK-5 Rahuri (4.73) followed by TNMP-6 (4.46). However, in zonal trial I, maximum number of branches recorded in IGAU-1 Raipur (13.00) followed by TFRI-2 (9.33) and IGAU -2 Raipur (7.54) and in Zonal Trial II, number of branches were maximum in JNKV-32 (3.26) followed by JNKVV-12 (3.16) and JNKVV-20-(3.00). The provenances did not differ significantly with respect to number of branches.

CONCLUSION

Significant variation was found in respect of height and collar diameter growth in national trial and zonal trial II. The best performing provenance in national trial is TNMP-6 and in zonal trial, TFRI-2 Jabalpur followed by IGAU-1 Raipur and IGAU-2 Raipur. This information can be utilized in developing future breeding programs of *P. pinnata* to increase productivity and oil yield.

REFERENCES

Bhat, G.S. and Chouhan, P.S. 2002. Provenance variation in seed and seedling traits of *Albizzia lebbek* Benth. *Journal of Tree Scientists* 21:52-57.

Duke, J.A. 1983. Handbook of energy crops, unpublished, excerpt from the internet, http://www.hort.purdue.edu/newcrop/duke_energy/Pongamia_pinnata.html.

Gera, M. and Ginwal, H. S. 1996. Performance of seventeen different multipurpose tree species under semi arid region of central India. *Indian Forester* 122(3): 250-257.

Manonmani, V.K. and Vanangamudi 1996. Effect of seed size on seed germination and vigour in *Pongamia pinnata*. *Journal of Tropical Forest Science* 9(1): 1-5.

NAS 1980. Firewood crops: shrub and tree species for energy production, Volume 1. National Academy of Sciences, Washington, D.C. pp 237.

Scott, P.T., Pregelj, L., Chen, N., Hadler, J.S., Djordjeevic. M.A. and Gresshoff, P.M. 2008. *Pongamia pinnata*: An untapped resource for the biofuels industry of the future. *Bioener. Res.* 1: 2-11.

Singh, R.V. 1982. Fodder Trees of India. Oxford and IBH Publishing Co. New Delhi, India. pp 663.

Troup 1921. The Silviculture of Indian Trees, Leguminosae. Government of India Press, Nasik, India. IV: 345.

Chapter - 13

INITIATIVES BY NABARD IN PROMOTING AND FUNDING OF BIODIESEL PLANTATIONS ON WASTELANDS

M.S. Haque and K.G. Karmakar

National Bank for Agriculture and Rural Development (NABARD), Bandra-Kurla Complex, Bandra East, Mumbai-400 051
E-mail: haque.ms@rediffmail.com

ABSTRACT

Realizing the potential of *Jatropha curcas* and *Pongamia pinnata* in meeting national objectives of wasteland development, substituting costly fossil fuel and rural employment generation, NABARD formulated model bankable schemes for channeling bank credit to farmers. For *J. curcus,* it recommends spacing of 2 m x 2 m accommodating 2500 plants per ha. The unit cost of cultivation is Rs. 30,000 per ha over a period of 3 years. The plantation will be raised through seeds and commercial yield will start from 3rd year with production of 0.5 kg seed per plant, increasing to 2.0 kg by 7th year, which will continue to produce at least for 30 years. Application of fertilizers, weeding, soil working and pruning have been recommended in the scheme. The IRR of the scheme is 30% thus making it bankable. One ha Jatropha plantation will create employment at 138 man days in first year, 65 man days in second year and 34 man days in third year. *Pongamia pinnata* is a medium sized tree with long life, hence the spacing recommended for raising commercial plantations is 5 m x 4 m accommodating 500 plant per ha. Because of wide spacing, the farmers can practice intercropping for many years. Besides, the plant is a legume and fixes atmospheric nitrogen which improves soil fertility. The estimated unit cost for cultivation is Rs. 19,000 per ha over a period of 5 years. The plantation will start commercial yield in the 6th year when 6 kg seeds will be available per tree and which will stabilize at 20 kg per tree by 11th year. Considering the sale price of fruits at Rs.5 per kg and based on the estimated unit cost of cultivation and projected yield, the IRR of the scheme is 35.06%. One ha Pongamia plantation will create employment at the rate of 105, 51 and 51 man days in the first, second and third years. Based on these models, NABARD is making vigorous efforts to promote bankable bio-diesel projects in the potential Indian states, but facing difficulties in absence of authentic data on planting stocks, agrotechnology, yield and prices, which needs to be solved soon by research organizations.

Keywords: Bio-diesel plantations, NABARD, economics, wastelands, bank credit

INTRODUCTION

National Bank for Agriculture and Rural Development (NABARD) is an apex developmental bank providing credit for promotion of agriculture, village and cottage industries and other allied economic activities in rural areas for bringing about integrated rural development. The Planning Commission has identified *Jatropha curcas* and *Pongamia pinnata* as the most suitable and promising species for biodiesel production. Realizing the potential of these two species in biofuel production, NABARD took initiatives in promoting and funding them. In Tripura state, D1 Williamson Magor Bio-fuel Ltd. has proposed to raise Jatropha plantations under contract farming system in 2 lakh ha of privately owned wastelands in a time span of 3-4 years and requested NABARD for clearing the proposal. Tripura Gramin Bank is likely to channelize credit to the project. The company has a proposal to set up oil expelling facilities in every block and a central refinery to produce bio-diesel from Jatropha oil seed in Assam State. NABARD undertook studies and recommended the unit costs, cultural operations, estimated yield, returns etc, and sent to the Bank. Biofuel Missions, an Australian company with an office in Mumbai has raised more than 50000 hectare Jatropha plantations in various states of India and approached NABARD for funding. Our team visited Orissa state and we may consider to fund around 5000 hectare under UPNRM (Umbrella Project for Natural Resource Management) fund on pilot basis. North Eastern Council (NEC), Shillong which is a regional Planning Commission for North East, approached NABARD for formulating models for Jatropha plantations on wastelands of Tea Estates. We prepared 3 models viz. two rows border plantings, three rows border plantings and block plantings which were sent to NEC for implementation in Tea Estates.

Project sanctioned: NABARD has already sanctioned a big biofuel project under RIDF (Rural Infrastructural Development Fund) with loan of Rs. 70.00 crores to Andhra Pradesh State Forest Department for raising Pongamia plantation in 30,600 ha of degraded forest lands in Adilabad, Ananthapur, Chittoor, Kadapa, East Godavari, Guntur, Karimnagar, Khammam, Kurnool, Mahaboobnagar, Medak, Nalgonda, Nellore, Nizamabad, Prakasam, Ranga Reddy, Visakhapatnam, Vizianagaram, Warangal and West Godavari districts. It has also sanctioned an agroforestry based *Jatropha* plantation project to Uma Tariang Estates in Meghalaya with financial outlay of Rs. 216 lakhs in 500 ha area of jhumlands. D1 Williamson Magor Biofuel Ltd. has supplied the technology and seedlings for the project which is being financed by Indian

Bank. It has also proposed for Co-financing with NABARD. Besides, NABARD has sanctioned under Co-financing agreement, 2 projects for processing of TBOs (Tree Borne Oilseeds) in Andhra Pradesh and Maharashtra states by the tribals after collecting from the wild trees.

R & D Project: Prior to realize by others the importance of initiating systematic research on oil yielding trees like *P. pinnata* and *Madhuca indica*, NABARD took the lead and sanctioned a project to the Institute of Forest Productivity, (IFP) Ranchi, under Indian Council of Forestry Research and Education, Dehradun in 2004 for "Identification, collection and genetic evaluation of important oil-yielding trees of Jharkhand". The major objectives of the project were: survey and collection of candidate plus trees (CPTs), collection and evaluation of clones, genetic evaluation of CPTs based on progeny trials, assessment of variability, estimation of genetic parameter, assessment of interrelationship among characters, estimation of genetic diversity of CPTs through biometric and enzyme approaches and establishment of germplasm bank of selected oil yielding trees. NABARD approved the project to IFP to undertake research on two most important TBOs viz. *Pongamia* and *Madhuca,* so that these could be promoted under agroforestry systems. The institute has been able to successfully clone the trees and raise trial plantations. On a visit to the project site in 2007, the senior author was highly satisfied with the progress and excellent performance of the project. It has been proposed to take the project to second phase where clonal trial plantations will be raised on farmer's fields. This will have a demonstration effect to the farmers and will be able to produce enough biodiesel feedstock for the industries.

RESULTS AND DISCUSSION

India has vast tracts of degraded lands in most villages, mostly in areas with adverse agro-climatic condition, where hardy, tree borne-oilseeds species like *Jatropha, and Pongamia* can be grown easily. It has been estimated that if 30 million hectare wastelands can be planted under bio-diesel, then it can completely replace the current use of fossil fuels. India has nearly 100 varieties of tree borne oilseeds popularly known as non-edible oil or minor oilseeds. At present, above two species have the potential to change India's oil scenario. One of the major constraints faced by the farmers in raising biodiesel plantations are the non-availability of quality planting stocks. Hence, evolving high yielding varieties for commercial cultivation of these species are the need of the hour. These important aspects were brought out by the present authors in their earlier publication (Karmakar and Haque, 2004). Since this was the first published paper dealing with economics and bankability aspects of *J.curcas*, it has been extensively quoted by many authors especially in relation to CDM (Clean Development Mechanism) projects (Singh and

Kalha, 2006). Regarding seed yield, detailed studies were undertaken by Agroforestry and Wasteland Development Foundation over a period from 1986 to 2003 in *Jatropha* which indicated maximum yield of 3000 kg per ha (Patil *et al.*, 2003). There are unconfirmed reports that claim yields of 10 tons per ha. These are unrealistic and farmers should not be carried away by these tall claims. The studies by CRIDA, Hyderabad (Rao and Korwar, 2003) and BAIF, Pune (Kermane *et al.,* 2003) on *Jatropha* seed yield was also not very encouraging. These studies indicated that research to increase per ha seed yield should get top priority. The good news is that private sector companies like Jain Irrigation Systems, D1 oils U.K., Biofuel Missions, Australia etc. have also started research on this plant. Regarding *Pongamia,* no systematic plantations were undertaken by Research Institutes; however, authentic yield data from forestry plantations are available which indicates high yield.

NABARD and other financial institutions will be in a position to fund biodiesel projects if these are raised on wastelands that are presently not being under cultivation in villages. NABARD will however evaluate such projects to see the technical and financial viability. It will however adopt proactive approach in popularizing the *Jatropha* and *Pongamia* plants. In fact it has already included both the plants as components in some projects sanctioned under RIDF and TDF (Tribal Development Fund) including inserting chapters on *Jatropha* in the annual PLPs (Potential Linked Credit Plans). It has organized seminars and workshops both in Mumbai and Regional Offices to create awareness. Many states have ambitious plans for bio-diesel plantations and NABARD is closely associated to these programmes. Besides, NABARD and Banks will also fund nursery development schemes, seed collection centers, oil extraction and esterification units when submitted to them. Govt. of Andhra Pradesh (GoAP) has created a new department for the development of the rain shadow areas of the state. This department is responsible for promoting use of bio-diesel in state to provide long term income security to the farmers. It has been decided by the state government to take up *Jatropha* and *Pongamia* plantations in 17.5 lakh acres in 22 districts of AP. The state government has already constituted state level and district level task forces in the identified districts for implementation of the programme. GoAP has announced 90% subsidy for installing drip system for bio-diesel plantation in the above districts under Andhra Pradesh Micro Irrigation Project (APMIP). Besides, the state government also has plans to promote contract farming and introduce an income guarantee scheme to farmers taking up bio-diesel plantation. Already three corporates viz. Southern Biofuels, Natural Bio energy Ltd. and Tree Oils India Ltd. has decided to establish bio-diesel plants in the state. In Tamil Nadu state also the Govt. has given a lot of importance to bio-diesel projects and has proposed to provide subsidy to the farmers, however quality farm lands need not be diverted for these plantations. Companies like Bannary Amman Sugar have started Jatropha plantations in a big way. D1 oils are planning a refinery in the state.

Credit requirements in the bio-diesel sector are mainly for raising of nurseries, plantation establishment, setting up of seed collection and oil expelling centres and bio-diesel manufacturing units. Already commercial banks are extending credit support to some of these activities. NABARD, as an apex institution with regard to policy, planning and operations in the field of rural and agricultural credit is actively involved in extending credit support for bio-diesel programme in rural areas along with other financial institutions. Raising bio-diesel crops / TBOs in wastelands has been identified as a thrust area by NABARD. Considering the importance of this sector, NABARD is extending 100% refinance to the banks at a concessional rate of interest for wasteland development. Besides, refinance support is also available for bio-diesel expeller units. NABARD has also provision to co-finance bio-diesel manufacturing plants along with other banks. To popularize *J. curcas and P. pinnata* in wastelands for bio-diesel production, NABARD has circulated the model plantation schemes to all the banks for utilization of wastelands; however no good agriculture land should be diverted for this. Ambitious bio-diesel plantation programmes envisaged by GoAP can become a reality only if the farmers are able to access quality planting material in adequate quantities. As both *J. curcas* and *P. pinnata* are less researched crops, there is urgent need for evolving high yielding strains of these species along with technology for their mass multiplication through Agricultural Universities and ICAR Institutes in various states. A visit to TNAU, Coimbatore in 2008 by the first author indicated that it is possible to obtain Jatropha seed yield of 2.1 kg per plant from 5th year onwards, when the plantations are properly maintained by providing adequate irrigation, fertilization and proper plant protection measures (TNAU, 2008). The university, however, cautioned that with irrigation lot of pest attacks were noticed. Another visit undertaken by the author in Khurda District of Orissa in 2008 indicated that the plantations were neither well maintained nor fertilized, hence the very survival of the plantations were questioned.

Tax relief by central Govt. for industries involved in biofuel related activities would also help in development of this sector. Banks, wherever needed, for taking up such plantations on wastelands can extend credit support. Necessary circulars have been issued by NABARD to the banks for raising biodiesel plantations. The Self Help Groups (SHGs) can be a good launching platform for biofuel plantations. Vana Samrakshana Samithis organized under the Joint Forest Management Programme can also play a useful role in promoting *Jatropha* and *Pongamia* cultivation in degraded forest lands. Public-Private partnerships between bio-diesel manufacturing plants and SHGs for taking up bio-diesel plantations also need to be tried.

CONCLUSION

India's energy scenario calls for the effective management of all available natural resources. Fuel security is as important as food security. Hence it is high time for all related institutions including banks to join hands in development of the biofuel sector. A recent news article published in the Times of India on 31st December, 2008 brought a new hope for *J. curcas* (TOI, 2008). A passenger jet of Air New Zealand powered by 50:50 blend of *Jatropha* oil and standard A1 jet fuel successfully completed a two hour flight that could lower Airplane Green House Gas (GHG) emission and cut costs. Biofuels were once regarded as impractical for aviation because most freeze at low temperatures encountered at cruising altitudes. But recent tests showed that biodiesel produced from *Jatropha* seeds has even lower freezing point than jet fuel. This experiment by Air New Zealand has opened new vistas not only for *Jatropha* but also for sustainable development of biofuel sector. Air New Zealand chief executive officer Mr. Rob Fyfe called the flight "a milestone for the airline and commercial aviation". Should we continue to doubt about the potential of *Jatropha* biodiesel as a substitute for fossil fuel? We must make up our mind that the biofuel sector must play a major role in reducing GHGs, thus saving the earth from climate change disasters. NABARD is progressing in this direction and is thinking to claim carbon credits from its 30,600 ha *Pongamia pinnata* plantation raised in forest wastelands of Andhra Pradesh Government.

REFERENCES

Anonymous 2003. Report of the committee on development of bio-fuel. Planning Commission, Government of India.

Jain Irrigation 2004. JISL in the service of *Jatropha* farmers.

Karmakar, K.G. and Haque, M.S. 2004. Economics of cultivation of bio-diesel tree, *Jatropha curcas* in wastelands. Paper presented at International Conference on Biofuels held in New Delhi from Sept. 16-17 2004.

Karmakar, K.G. and Haque, M.S. 2004. Economics of cultivating Jatropha. *Biofuels India*. II(III): 8-12.

Keremane, B.G., Hegde, G.V. and Sheshachar, V.S. 2003. *Jatropha curcas*: Production systems and uses. Proceedings of the National Workshop held at Pune from 5-8 August, organized by BAIF Development Research Foundation, Pune.

Patil, V., Singh, K. and Bhandare, T.N. 2003. Experience and observations on *Jatropha* plantations from 1986-2003. Agroforestry and Wasteland Development Foundation, Nasik, Maharashtra.

Rao, G.R. and Korwar, G.R. 2003. Research priorities to popularize *Jatropha* among the farming community. *Ibid*.

Singh, M.P. and Kalha, G. 2006. Village empowered: Rural bio-energy production as a bundled CDM Project. Paper presented at the 3rd International Biofuels Conference held at New Delhi from 18-20 January, 2006.

TNAU 2008. Brochure on package and practices for *Jatropha curcus*. Centre of Excellence in Biofuels, Agricultural Engineering College and Research Institute, Tamil Nadu Agricultural University, Coimbatore.

TOI 2008. Now, Jetliner flies on vegetable oil. The Times of India, Mumbai edition, 31st December, 2008.

Chapter - 14

ECONOMIC EVALUATION OF LAB-SCALE BIODIESEL PRODUCTION FROM *PONGAMIA PINNATA*

Swati Dindorkar[1] and S.J. Kakde[2]

[1]Department of Unconventional Energy Sources & Electrical Engineering, College of Agricultural Engineering & Technology,
[2]Deptt. of Agricultural Economics, PDKV, Akola.
E-mail: swateek@rediff.com

ABSTRACT

The study, entailed a detailed analysis of the energy input-yield ratios of producing biodiesel from *Pongamia pinnata* (Karanja) and cottonseed oils; and on the basis of energy balance study, economic feasibility of producing these fuels was estimated. The factors considered for the study were those involved in the energy use in producing the crop (production of pesticides and fertilizer, running farm machinery and irrigating, crushing and transporting the crop) and in actual production of biodiesel from oil in the small scale processor developed specially for the purpose. It was noticed that the cost of the edible/ non edible oil feedstock is the single largest factor in biodiesel production costs. And that increasing the area under the species would decrease the principal cost of feedstock, which would consequently reduce the cost of biodiesel. Factors like sophistication in the production process, recycling of alcohol used and use of catalyst free process would add to curtail the process cost to some extent. The cost of biodiesel production in 2006 in the small scale processor was estimated to be Rs. 36 per litre, which was comparable with the cost of petrodiesel and the process was found feasible on socio-economic viewpoint.

Keywords: Transesterification, energy balance, biofuel, cottonseed oils, lab-scale, economic evaluation

INTRODUCTION

Increase in fuel prices since mid-2004 has stimulated interest in the technical and economic feasibility of (lab-scale) fuel production. Local and international evidence have revealed that it is technically feasible to produce biofuels from a variety of agricultural substrates such as fats and oilseeds. However, less clear is the economic wisdom of small-scale investment in biofuel production.

Farmers are concerned about the price increases for the reasons: (i) the current generation of farmers, or their parents, remember the economic hardships in agriculture in the mid to late 1980s, largely attributable to unanticipated huge rises in fuel and credit prices. Admittedly farmers currently are not exposed to historically high credit prices but fuel costs form a significant component of farm operating costs and so farmers are concerned about the impact on their businesses of high prices of fuel, (ii) There is evidence that the natural environment for agricultural production in India is worsening, further weakening farmers' abilities to respond to other adverse impacts such as rising fuel prices. If the recent rise in fuel prices is prolonged then, in combination with any worsening of the natural environment, farming in this country will face some serious commercial challenges.

In responding to those challenges farmers will need access to innovation and R&D services to deliver the needed productivity improvement (Miller and Paul, 2003, Kingwell and Plunkett, 2003). Support for such innovation will be one way that farmers can restructure the likely adverse impacts of a sustained period of high fuel prices and a worsening trend in the natural environment.

Many studies have been conducted on the potential macroeconomic benefits of large scale biodiesel production around the country. These studies also give some indication of the potential economic impacts across the nation. The major economic benefits which would accrue to the country from the increased use of biodiesel are:

1. Biodiesel production will expand demand for feedstock oil, which makes processors to pay the farmers high cost for these seeds or oil.
2. Farmers near the biodiesel plant may receive slightly higher prices for their raw feedstock like Jatropha and Karanj seed.
3. The presence of a facility that creates energy from these non edible oils adds value to the nations industrial and income base.

Planning Commission of India (2004) has calculated the cost of Jatropha seed at Rs. 6 per kg and biodiesel at Rs. 20 per liter. As the prices of petrol and diesel have increased in recent years, attention has been drawn to the merits of farmers producing biofuels. A popularly quoted option for farmers is biodiesel. The technical feasibility of producing biodiesel is widely acknowledged so far and portable plants displaying the production process are available. Nevertheless, potential studies on economic evaluation of small scale (farm yard based) have not been done to a large extent. So, in accordance with the land holding of farmers of Vidarbha region, a small scale biodiesel processor was developed to cater the needs of small marginal farmers, and economic feasibility of this processor was studied.

Production process

Biodiesel is an alternative fuel for diesel engines that can be made straight from vegetable oil, animal fat such as tallow (a byproduct of the slaughter houses) and used oils.

There are three basic routes to biodiesel production from oils and fats:

- Base catalysed transesterification of the oil.
- Direct acid catalyzed transesterification of the oil.
- Conversion of the oil to fatty acids and then to biodiesel.

Biodiesel is made by way of a simple chemical process known as transesterification (similar to saponification- the chemical reaction used to make soap out of vegetable oil/animal fat). Transesterification is the general term used to describe the important class of organic reactions where an ester is transformed into another through interchange of the alkoxy moiety, a triglyceride reacts with an alcohol in the presence of strong base (NaOH), producing a mixture of fatty acid alkyl esters and glycerol.

Process of small scale biodiesel production begins with the collection of seeds. After collection, they are crushed to get the oil, which is filtered to remove any particles in suspension and is then heated in an appropriate reactor vessel. Once the oil has reached 60° C, mixture of methanol and sodium hydroxide (~20% methanol/volume of oil and anywhere from 3.5 to 9 g/lt of oil depending on the quality of oil) are added to the heated oil once a homogenous solution is obtained. The mixture is then stirred for ~1 hour and allowed to settle into a vessel containing biodiesel and glycerin as a by-product. The finished product is then cooled down to room temperature before using as fuel. Small capacity batch type "PKV-Biodiesel Processor" was developed in the project and was used for production of biodiesel from *Pongamia pinnata* oil and cottonseed oil. The base catalysed transesterfication process was followed for the purpose.

The most important variable that influence the transesterification reaction time and conversion efficiency are temperature, catalyst type and its concentration, alcohol to ester ratio, and stirring rate. Purity of reactants, presence of water, free fatty acids, and other contaminants found in unrefined oils (or other feedstocks) are also very important.

Used cooking oil offers the potential of a cheaper feedstock but only small quantities are available. Crushing or pressing the seed for oil consists of several stages. Solvent extraction is used to maximize extraction of oil. The present scale of Pongamia and Jatropha production in India do not justify a solvent extraction plant and therefore mechanical extraction methods are used.

Economic evaluation of cottonseed methyl ester on the basis of energy balance

For calculating the value of energy inputs and outputs, the predefined energy values of sources and materials are taken from energy tables and cost of the inputs are computed in parallel. The analysis was made using the costs of production of biodiesel from 1 ha of cotton grown in the experimental area of the university.

RESULTS AND DISCUSSION

Key assumptions of the analysis are that the 15 kg cotton seed sown on one ha area, gave, yield of 13 quintals, out of which 78 % seeds are obtained after separation of fiber i.e. 1015 kg of seeds. Cottonseed has approximately 18% oil content. The oil is extracted using a small scale crusher and after chemical catalytic transformation with methanol sample of biodiesel is produced (oil obtained from 1015 kg seeds = 182.7 kg from 182.7 kg oil 96 per cent ester i.e. 175.39 kg of biodiesel ≈ 159.8 litre biodiesel) and used in a standalone CI engine for testing. This figure (160 litre/ha) will be used henceforth to simplify the calculations. Key parameter of assumptions and calculated values are listed in Table 1.

Table 1. Agricultural inputs for cotton seed production

S. No.	Particular (Input)	Kg or L per ha	Rate / unit (Rs.)	Cost (Rs.)
1	Nitrogen	15 + 15	Urea (per kg)	150
2	Phosphorus	15	5	
3	Fuel (diesel)	45	36	1620
4	Seeds (Sowing)	15	500	7500
5	FYM	one bullock cart	500 /bullock cart	500
	Total = 5166.22 MJ /ha			8,770 for 160 kg = 61.06 per litre

Table 2. Cost input for oil extraction and refining

S. No.	Input source	Unit/ litre	Cost (Rs.)
Extraction			
1	Electricity (kWh)	0.14	0.336
Total for extraction			0.336
Refining			
1	Electricity (kWh)	0.002	0.0048
Total for Extraction + Refining			0.36

Cottonseed oil

Cottonseed Biodiesel

The cost of production for the processor was around Rs. 5000/- and life of plant is considered approximately 7 years. And the plant is supposed to run for 8 hr/day i.e. 208 per month and 1456 hr for 7 years. Depreciation on machinery is 10 per cent of cost i.e. Rs. 500/1456 is Rs. 3.43 per lite. and 1% maintenance cost. So, Rs. 0.343 and interest of 10%, 5000 x 7 = 35000 / 1456 is Rs. 24.0 per hr. Capacity of processor is 10 litre per hr. so, in 8 hrs, 80 ltrs/day.

Fixed cost of processor	
Cost of machine	Rs. 5000/-
Depreciation @ 10 per cent	Rs. 3.43 per hr
Interest @ 10 per cent	Rs. 0.343 per hr
Maintenance @ 1 per cent	Rs. 24 per hr
Labour cost @ Rs. 47 /day	Rs. 5.87 per hr
Total	33.64/hr
Produces 10 litre per hr	3.36 /litre

Cotton plant

Cotton ball

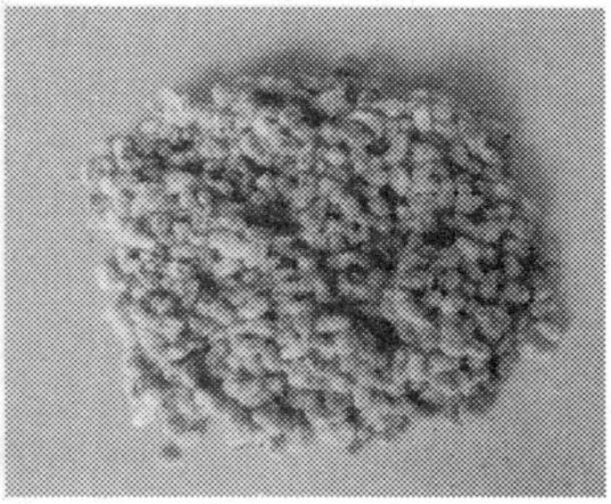

Cotton seeds

Glycerin

Seed cake

Table 3. Expences for esterification of cottonseed oil

S. No.	Energy source	Unit / litre	Rate per unit	Cost (Rs.)
1	Electricity (Kwh)	0.72	Rs. 2.40	1.728
	(For stirring heating coil and washing of ester)	water -one litre	0.50	0.50
2	Methanol (litre)	0.20 (200 ml)	Rs. 23 /litre	4.6
3	Catalyst (Kg)	0.005 (5 gms)	Rs. 32 /kg	0.16

Total for transesterification 6.988 ≈ 7.0

Fixed cost of processor per litre– Rs. 3.36

Total = Rs. 10.36 per litre

Table 4. Byproducts energy value (on replacement credit weight basis)

Cottonseed meal		Cost (Rs.)	Glycerin		Cost (Rs.)
Weight	value		Weight	value	
81 percent	822.15 Kg*	6577.2 per ha	14 percent	22.73kg**	909.2 /ha
Total		41.10 per litre			5.68 per litre

Cottonseed meal has 41% protein and is used as feed for milking animals. One kg of this meal replaces the cereal/grain meal equivalent to 2.5 kg of cereal grain meals due to its high protein value. The amount of meal produced is approx. 81% of the cottonseed per ha. So 1015 x 0.81, which is equal to 822 kg. One hectare yields 160 litre of biodiesel and one kg meal is equivalent to 2.5 wt credit .The meal is sold at Rs. 8/kg hence cost of meal obtained is Rs. 30.82.

Also, 14% weight basis glycerin is obtained from biodiesel production. One ha yields (160 x 0.14) 22.73 kg of glycerin which is sold at Rs. 40/kg, so total cost obtained from glycerin is Rs. 5.68 per litre.

Table 5. Input and output costs involved in cottonseed biodiesel production

S. No.	Input type	Economic value (Rs.)
1	Agriculture	61.06
2	Expelling & refining	2.8
3	Esterification	10.36
	Total	74.22
4	Less by-products	46.78
	Cost of biodiesel produced per ltr.	27.44

The cost involved in overall lifecycle production of one litre biodiesel is Rs. 74.22; less by-product value (i.e. Rs. 46.78 per litre) gives actual cost of biodiesel as Rs. 27.44 per litre which is comparable to the present cost of petro-diesel, thus the biodiesel production from cottonseed oil will be economically feasible. Hence small-scale biodiesel production by using "PKV-Biodiesel Processor" was found to be affordable and comparable to the existing diesel prices.

To make any crop commercially more attractive as a feedstock for biodiesel production, requires the development of crop that is relatively cheap to grow and which has high yield of oil per hectare and has a comparative advantage over traditional oilseed production. For example, if the non edible oilseed crops could be profitably grown in environments or locations not suited to current oilseed varieties or if it had agronomic or economic characteristics that ensured its comparative advantage over other possible crops grown at that location then a case could be made for its use in biodiesel production.

Cost benefit scenario

Cost benefits scenario will depend on various factors such as seed yield, area of production, its gestation period, and raw oil yield during various stages of bio-diesel production-plantation, extraction, and type of transesterification process adopted.

CONCLUSION

In summary, it is likely that biofuels, especially fuel blends, eventually will become attractive as a fuel source, in the face of further rises in petroleum and mineral diesel prices. At current costs and prices, on-farm production of biodiesel is likely to be commercially doubtful in many setup. However, if cheaper or more suitable feedstocks become available or significant reductions in processing costs are achieved through economies of size and capital sharing or if fuel prices continue to escalate then in some situations on-farm or localized production of biodiesel is likely to become justified on large commercial grounds also.

REFERENCES

Miller, P. 2003. Value-added tree in a biobased vision for Montana and the Pacific Northwest. Presentation at the Greening Under the *Big Sky conference*. Big Sky, MT. June.

Planning Commission. 2004. Report of Committee on Development on Biofuels, Govt. of India, New Delhi.

Chapter - 15

SCREENING OF SOME OIL BEARING SEED PLANTS OF SUBTROPICAL REGION OF MADHYA PRADESH FOR BIODIESEL PRODUCTION

Gulab Singh Thakur and P.S. Bisen

Research and Development Centre, Bisen Biotech and Biopharma Pvt. Ltd., M-7 Laxmipuram Transport Nagar, Gwalior- 474010

E-mail- gsingh@bisenbiotech.com

ABSTRACT

The ever increasing demand of diesel, continuous rise in prices and the spread of the fossil energy have forced us to search for ecologically sustainable alternative energy source. In this search biodiesel from vegetable oil has emerged as a viable alternative source, particularly non-edible vegetable oils. There are many plant species which bear seeds rich in oil. In the present study some promising plant species have been evaluated for biodiesel production such as *Jatropha curcas*, *Ricinus communis, Pongamia pinnata* and *Argemone mexicana* and found them most suitable for the production of biodiesel. In this study plants have been screened for biodiesel production on the basis of their oil percentage and biodiesel conversion rate. The oil percentage has been found 37%, 34%, 32%, 30% from *Jatropha*, *Ricinus*, *Pongamia* and *Argemone* respectively. For the biodiesel production from obtained oil, transesterification process has also been standardized for parameters such as concentration of NaOH, reaction time and reaction temperature. The biodiesel conversion rate was obtained 76%, 72%, 66% and 78% in *Jatropha*, *Ricinus*, *Pongamia* and *Argemone* seed oils respectively. Biodiesel (methyl ester part) was characterized compared to biodiesel standard (DIN EN 14214) to checkout viscosity, acid value and iodine value of methyl ester layer. This is the 1st report on screening of some oil bearing seed plants of subtropical region of Madhya Pradesh for biodiesel production on the basis of oil percentage and biodiesel conversion rate. *Argemone mexicana,* a weed of subtropical region is also 1st time reported suitable for biodiesel production. Present study offer a good opportunity to use these plant species of subtropical region of Madhya Pradesh for biodiesel production particularly *Argemone mexicana*.

Keywords: Biodiesel production, oil bearing plants, *Jatropha curcas*

INTRODUCTION

Today's transportation services in industrialized countries are primarily based on fossil fuels especially crude oil derivatives. The spread of this fossil energy intensive approach to developing countries and economies in transitions with large population may be constrained by limited resource availability and concerns about environment and human health. One of the most important questions that are haunting us today is the global warming, climate change and the soil and environmental pollution. Increasing industrialization in the developing world is leading to spiraling increase in the demand of fossil fuel. Therefore, search for an alternative to fossil fuel is very essential. Biodiesel is one such option and many countries are taking initiatives in this direction (Francis *et al.*, 2005).

Biodiesel has drawn attention because it is environmentally safe, can be made from renewable sources and prepared locally. Biodiesel can replace petroleum diesel or can be applied in various blends with conventional diesel. Biodiesel is obtained in a chemical process called transesterification (Devansen *et al.*, 2007), where organically derived oils are combined with alcohol and chemically altered to produce fatty acid alkyl esters, methyl and ethyl esters.

While the country is short of petroleum reserve, it has large arable land as well as good climatic condition with adequate rainfall in large part of the area to account for large biomass production every year. For the reason of edible oil demand being higher than its domestic production, there is no possibility of diverting this oil for production of biodiesel. Fortunately there is a large junk of degraded forest land and unutilized public land, field boundaries and fallow lands of farmers where non-edible oil-seeds can be grown. Many plant species have been reported to bear seeds rich in oil. Of these some promising plant species have been evaluated and it has been found that there are a number of them which would be very suitable in our conditions. Selection of the crop for biodiesel production is perhaps the most important and also the most neglected issue. As mentioned earlier, for India it appears that non-edible oils are the choice feed stock for biodiesel production. There is need to collect scientific data to get the realistic figures by critical analysis of the seed yield, oil percentage, biodiesel conversion rate and its quality.

In the present study, four plant species viz. *Jatropha curcas* (Euphorbiaceae), *Ricinus communis* (Euphorbiaceae), *Pongamia pinnata* (Fabaceae) and *Argemone mexicana* (Papaveraceae) were taken up for biodiesel production on the basis of oil percentage and biodiesel conversion rate from obtained oil. All these plant species were found easily in the sub-tropical region of Madhya Pradesh. Transesterification process for biodiesel production from all those plants was also standardized. Finally, present study offers a good opportunity to use these data for biodiesel production.

Plate-1. Plant of *Jatropha curcas*

Plate-4. Plant of *Ricinus communis*

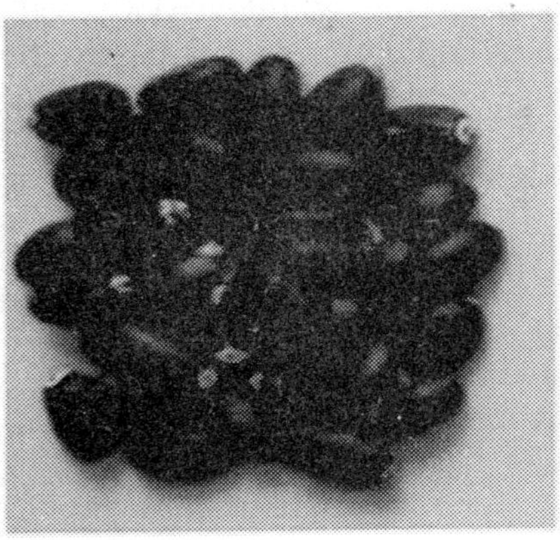

Plate-2. Seeds of *Jatropha curcas*

Plate-5 Seeds of *Ricinus communis*

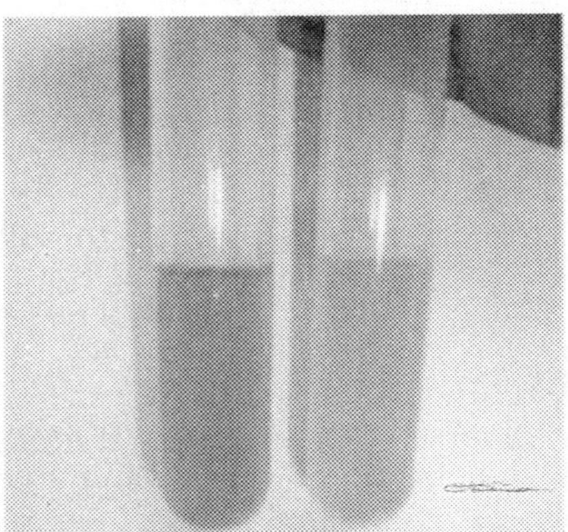

Plate-3. Oil and Biodiesel sample of *Jatropha curcas*

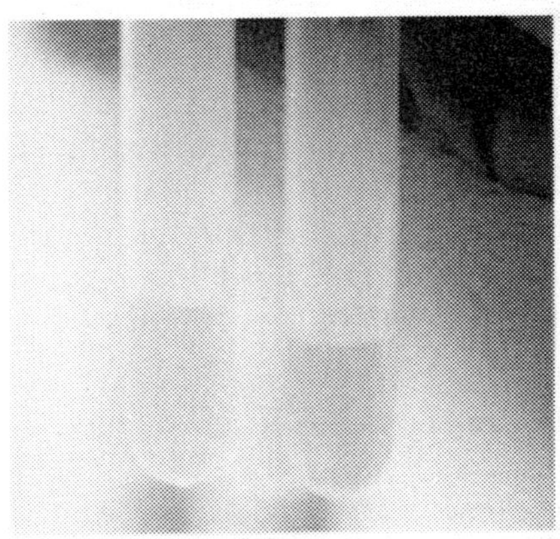

Plate-6. Oil and Biodiesel sample of *Ricinus communis*

Fig. 1. *Jatropha curcas* and *Ricinus communis* -plants, seeds, oil and biodiesel.

Plate-7 Plant of *Pongamia pinnata*

Plate-10 Plant of *Argemone mexicana*

Plate- 8 Seeds of *Pongamia pinnata*

Plate-11. Seeds of *Argemone mexicana*

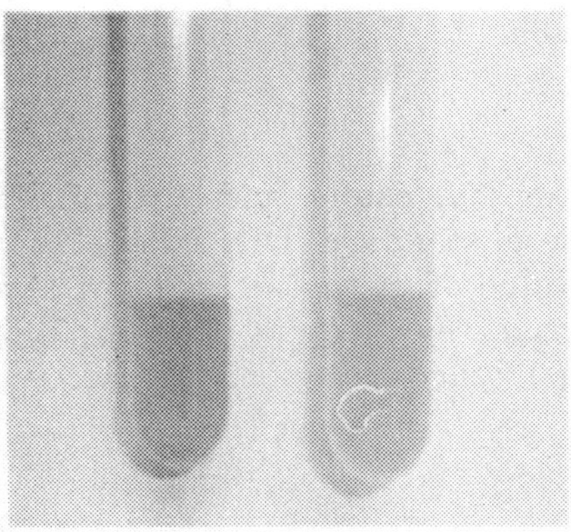

Plate-9 Oil and Biodiesel sample of *Pongamia pinnata*

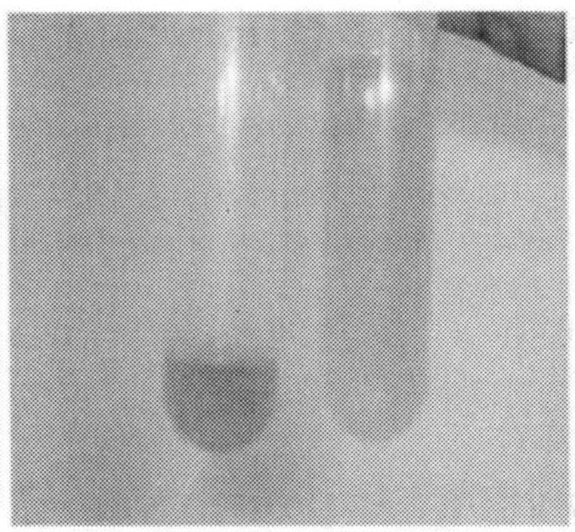

Plate-12. Oil and Biodiesel sample of Argemone mexicana

Fig. 2. *Pongamia pinnata* and *Argemone mexicana* -plants, seeds, oil and biodiesel.

MATERIALS AND METHODS

Collection of seeds

The fruits of *J. curcas, R. communis, P. pinnata* and *A. mexicana* were collected from Gwalior-Chambal region by spreading polythene under the plant species. The collected fruits were washed and dried under the sun light before decortication. The kernels were removed before extraction of oil. The photographs of the plants, fruits, seeds, seed oil and biodiesel obtained from the selected species are depicted in Fig. 1 and Fig. 2.

Extraction and estimation of oil by solvent extraction method

The dried and cleaned seeds were cut into pieces and crushed using a mortar and pestle. The aforesaid crushed material was used for oil estimation with the help of Soxhlet apparatus (ASGI) using petroleum ether as solvent at 40°C. The oil percentage was estimated as ml/100 g seed.

Tranesterification of oil

Extracted oils were converted into biodiesel using one step-alkali (base-catalyst) transesterification process. In the alkali (base catalyst) transesterification, NaOH was used as catalyst and methanol was used as solvent. The concentration of NaOH used for transesterification was standardized by titration of Jatropha oil with NaOH using phenol red indicator. In the entire process of base catalyst transesterification, the reaction temperature and reaction time was taken upto 30°C-70°C and 10-50 minutes respectively. After transesterification process two layers were formed and separated with the help of separating funnel, methyl ester layer was measured with the help of measuring cylinder for biodiesel conversion rate of the Jatropha oil.

Characterization of biodiesel

After separation of two layers of methyl esters and glycerol, the methyl ester part was compared to biodiesel standard (DIN EN 14214) to check viscosity, acid value and iodine value of methyl ester layer. Acid value was calculated using method described by Cox and Pearson (1962) and iodine value was calculated using method described by Horowitz (1975).

RESULT AND DISCUSSION

The percentage of oil collected from *J. curcas, R. communis, P. pinnata* and *A. mexicana* seeds were found to be 37%, 34%, 32%, and 30% respectively (Table 1) and their biodiesel conversion rate were achieved 76%, 72%, 66% and 78% (Table 1) in alkali-based catalyzed transesterification. The results showed that *J. curcas* have higher oil percentage than *R. communis, P. pinnata* and *A. Mexicana.*

Table 1. Analysis of oil percentage and biodiesel conversion rate of collected seeds

Plant Species	Oil%	Biodiesel conversion rate
Jatropha curcas	37%	76%
Ricinus communis	34%	72%
Pongamia pinnata	32%	66%
Argemone mexicana	30%	78%

In the transesterification process alkali catalyst like sodium hydroxide, potassium hydroxide and potassium methoxide are more effective other than acidic (Ma *et al.*, 1999) for neutralization of free fatty acids. NaOH was more effective alkali catalyst other than sodium methoxide (Ma *et al.,* 1998). 4.0, 3.5, 3.5 and 4.0 g conc. of NaOH was found suitable for higher coversion rate of biodiesel in transesterification process in *R. communis, P. pinnata* and *A. mexicana* respectively (Table 2).

Table 2. Standardization of NaOH conc. for transesterification of biodiesel

Plant Species	Reaction time	Reaction temperature	Biodiesel conversion rate					
			NaOH Conc.					
			2%	2.5%	3%	3.5%	4%	4.5%
Jatropha curcas	4 Hrs.	55°C	0%	42%	58%	67%	76%	68%
Ricinus communis	4 Hrs.	55°C	0%	32%	62%	72%	66%	48%
Pongamia pinnata	4 Hrs.	55°C	0%	28%	48%	66%	60%	51%%
Argemone mexicana	4 Hrs.	55°C	0%	35%	41%	62%	78%	71%

Reaction time and reaction temperature is also important factor for transesterification of oils (Freedman *et al.,* 1984). 4 hrs reaction time and 50°C reaction temperature for alkali base catalysis were observed for higher biodiesel conversion rate. Ma *et al.* (1998) reported that the reaction time increases with conversion rate in beef tallow. Freedman (1984) found ester yield to be 94%, 87% and 64% at different reaction temperature i.e. 60°C, 45°C, and 32°C. In the present study, it was recorded that the reaction time and temperature played a vital role, these initially increases the conversion rate and on attaining the optimum time and temperature, the conversion rate decreases due to evaporation of the solvent (methanol) used in transesterification process. Insufficient quantity of methanol for transesterification showed reversible reaction and resulted in soap formation.

The viscosity of methyl ester layer was found to be 4.8, 4.0, 4.3 and 4.2 in *J. curcas, R. communis, P. pinnata* and *A. Mexicana,* respectively. The iodine value was found to be 114.8, 108.2, 110.4, and 106.6 respectively and the acid value was found to be 0.40, 0.38, 0.36 and 0.37 (Table 3), respectively in alkali based transesterification processes which were compared to biodiesel standards DIN EN 14214.

Table 3. Comparison of biodiesel sample with biodiesel standards

Biodiesel Standards	Viscosity (mm^2s^{-1}) In 15°C	Iodine value	Acid value
DIN EN 14214*	3.5-5.0	Maximum 120	Maximum 0.50
Jatropha curcas	4.8±0.56	114.8	0.40±0.06
Ricinus communis	4.0±0.52	108.2	0.38±0.04
Pongamia pinnata	4.3±0.48	110.4	0.36±0.08
Argemone mexicana	4.2±0.58	106.6	0.37±0.08

Source: Gerhard Knothe, JAOCS, Vol. 83(10) 2006.

CONCLUSION

In the present study, it is concluded that *J. curcas, R. communis, P. pinnata* and *A. mexicana* are found to be the most suitable plants for the production of biodiesel. *J. curcas* is having higher oil percentage than *R. communis, P. pinnata* and *A. Mexicana.* Above study offers a good opportunity to use all these plants which were found easily in the subtropical region of Madhya Pradesh for biodiesel production.

REFERENCES

Cox, H.E. and Pearson, D. 1962. The chemical analysis of food chemicals. Publishing Co Inc New York, pp 420.

Devansen, M.G., Viruthagiri, T. and Sugumar, N. 2007. Transesterification of Jatropha oil using immobilized *Pseudomonas fluorescens*. *African Journal of Biotechnology* 6 (21): 2497-2501.

Francis, G., Edinger, R. and Becker K. 2005. A concept for simultaneous wasteland reclamation, fuel production and socio economic development in degraded areas in India need potential and perspectives of Jatropha plantations. *Natural Resources Forum* 29:12-24.

Freedman, B., Pryde, E.H. and Kwolek, W.F. 1984. Thin Layer Chromatography/ Flame Ionization Analysis of Transesterified Vegetable Oils. *J. Am. Oil Chem. Soc.* 61(7): 1215-1220.

Freedman, B., Pryde, E.H. and Mounts, T.B. 1984. Variables affecting the yield of fatty esters from transesterified vegetable oils. *J. Am. Oil Chem. Soc.* 61 (10): 1638-43.

Gerhard, Knothe. 2006. Analyzing Biodiesel: Standards and Other methods. *J. Am. Oil Chem. Soc.* 83(10): 823-833

Ma, F., Clements, L.D. and Hanna, M.A. 1998. Biodiesel fuel from animal fat: Ancilliary studies on transesterification of beef tallow. *Ind. Eng. Chem. Res.* 37: 3768-3771.

Ma, F., Clements, L.D. and Hanna, M.A. 1999. The effect of mixing on transesterification of beef tallow. *Bioresour. Technol.* 69: 289-93.

Horowitz, W. 1975. Official methods of Analysis of Association of Official Analytical Chemists, Washington (12th Ed.) pp 488.

Chapter - 16

BIODIESEL PRODUCTION TECHNOLOGY-LOOKING AHEAD

Naveen Kumar

Mechanical Engineering Department, Delhi Technological University, (Formerly Delhi College of Engineering), Delhi-110 042

E-mail: naveenkumardce@gmail.com

ABSTRACT

Biodiesel has become an attractive alternative fuel for diesel engines because it is renewable and provide environmental benefits. Amongst the different routes of biodiesel production, transesterification of lipid is most preferred. The transesterification has always found to be the best route as it not only reduced the viscosity but also addressed the problem of un-saturation up to a great extent. As removal of catalysts from raw biodiesel is always a problem, use of enzymes and hetrogenous catalysts is going to increase in a big way to alleviate this problem. Ultrasonic and microwave assisted biodiesel synthesis have a major role to play in coming time as these technologies shall reduce the reaction time drastically and reduce the processing cost.

Keywords: Vegetable oil, transesterification, biodiesel, technology.

INTRODUCTION

Vegetable oils have been considered as the potential fuels of diesel engines ever since diesel engines were developed. In fact the inventor of diesel engine, Rudolph diesel was very optimistic about these fuels. However, due to inherent drawback of vegetable oils having high viscosity and un-saturation, the straight vegetable oils could not be considered as a viable substitute to petroleum diesel. The problem of high viscosity can be addressed using dilution, heating, microemulsification, pyrolysis, and transesterification techniques (Chauhan *et al.,* 2010). However, the transesterification has always found to be the best route as it not only reduced the viscosity but also addressed the problem of un-saturation up to a great extent. Biodiesel is transesterfied vegetable oils or animal fats and considered a potential alternative diesel fuel (Kumar and Sharma, 2005).

Transesterifcation

Vegetable oils contains fatty acids with long methylene group (CH_2) chains, similar in structure and carbon to hydrogen ratio [C:H] as in conventional diesel fuel (Crookes *et al.*, 1997).

Vegetable oils consist of triglycerides to about 97%, the other 3% distribute among di- and monoglycerides and structurally, a triglyceride is a reaction product of one molecule of glycerol with three fatty acid molecules to yield three molecules of water and one molecule of triglyceride as shown in Figure 1 (Chauhan et al., 2010).

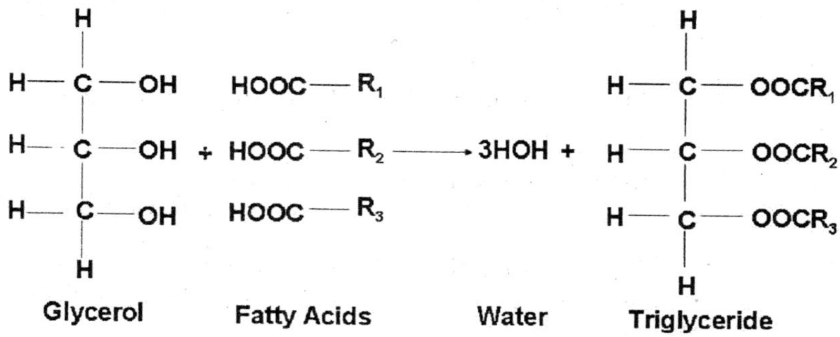

Fig. 1 Chemical structure of vegetable oils

Table 1: Chemical structure of some of the common fatty acids

Fatty acid	Common name	Formula	Chain length
4:0	butyric	$CH_3(CH_2)_2CO_2H$	short
6:0	caproic	$CH_3(CH_2)_4CO_2H$	short
8:0	caprylic	$CH_3(CH_2)_6CO_2H$	short/medium
10:0	capric	$CH_3(CH_2)_8CO_2H$	medium
12:0	lauric	$CH_3(CH_2)_{10}CO_2H$	medium
14:0	myristic	$CH_3(CH_2)_{12}CO_2H$	medium
16:0	palmitic	$CH_3(CH_2)_{14}CO_2H$	
18:0	stearic	$CH_3(CH_2)_{16}CO_2H$	
18:1 9*c*	oleic	$CH_3(CH_2)_7CH{=}CH(CH_2)_7CO_2H$	
18:2 9*c*12*c*	linoleic	$CH_3(CH_2)_4(CH{=}CHCH_2)_2(CH_2)_6CO_2H$	
18:3 9*c*12*c*15*c*	α-linolenic	$CH_3CH_2(CH{=}CHCH_2)_3(CH_2)_6CO_2H$	
22:1 13*c*	erucic	$CH_3(CH_2)_7CH{=}CH(CH_2)_{11}CO_2H$	long
20:5 5*c*8*c*11*c*14*c*17*c*	EPA*	$CH_3CH_2(CH{=}CHCH_2)_5(CH_2)_2CO_2H$	long
22:6 4*c*7*c*10*c*13*c*16*c*19*c*	DHA*	$CH_3CH_2(CH{=}CHCH_2)_6CH_2CO_2H$	long

The chemical formula of any triglyceride is given in Figure 1 where R_1, R_2, and R_3 are longer alkyl chains. The three fatty acids R_1COOH, R_2COOH and R_3COOH can be all different, all the same, or only two the same. Chain lengths of the fatty acids in naturally occurring triglycerides can be of varying lengths but 16, 18 and 20 carbons are the most common. Natural fatty acids found in plants and animals are typically composed only of even numbers of carbon atoms. Table 1 gives chemical structure of some of the common fatty acids.

However, there are five types of common chains that are found in vegetable oil and animal fats (others are present in small amounts). Table 2 represents common fatty acid chain normally found in vegetable oils and animal fats (Gerpen et al., 2004).

Table 2: Common chains present in vegetable oil and animal fats

Palmitic:	$R = -(CH_2)_{14}-CH_3$	16 carbons, (including the one that R is attached to.) (16:0)
Stearic:	$R = -(CH_2)_{16}-CH_3$	18 carbons, 0 double bonds (18:0)
Oleic:	$R = -(CH_2)_7\ CH{=}CH(CH_2)_7CH_3$	18 carbons, 1 double bond (18:1)
Linoleic:	$R = -(CH_2)_7\ CH{=}CH\text{-}CH_2\text{-}CH{=}CH(CH_2)_4CH_3$	18 carbons, 2 double bonds (18:2)
Linolenic:	$R = -(CH_2)_7\ CH{=}CH\text{-}CH_2\text{-}CH{=}CH\text{-}CH_2\text{-}CH{=}CH\text{-}CH_2\text{-}CH_3$	18 carbons, 3 double bonds (18:3)

Transesterification was known to be as early as 1864 when Rochleder described glycerol preparation through ethanolysis of castor oil. It refers to conversion of an organic acid ester into another ester of the same acid (Ali and Hanna, 1994).

Transesterification also known as alcoholysis, is the reaction of fat or oil with an alcohol to form esters and glycerin. A catalyst is used to improve the reaction rate and yield. The transesterification reaction is represented by the general equation:

$$RCOOR' + R''OH \xrightleftharpoons{\text{Catalyst}} RCOOR'' + R'OH$$

If methanol is used in the above reaction, it is termed methanolysis. The reaction of triglyceride with methanol is represented by the general equation (Shrivastava and Prasad, 2000)

$CH_2O\text{–}R1$			$CH_2\text{-}OH$		$CH_3O\text{–}R1$
$CHO\text{–}R2$	+ $3\ CH_3OH$	$\longrightarrow$	$CH\text{–}OH$	+	$CH_3O\text{–}R2$
$CH_2O\text{–}R3$			$CH_2\text{-}OH$		$CH_3O\text{–}R3$
Fat/Oil	Methanol		Glycerol		Fatty Esters

During transesterification, triglycerides (TG) react with methanol to produce diglycerides (DG) which further reacts with methanol to produce monoglyceride (MG). Finally, monoglyceride reacts with methanol to give glycerol. At each reaction step, one molecule of methyl ester is produced for each molecule of methanol consumed. This kinetic mechanism can be represented by the following reactions (Diasakou *et al.*, 1998):

$$TG + MeOH \xrightarrow{k_1} DG + ME \quad (1)$$

$$DG + MeOH \xrightarrow{k_2} MG + ME \quad (2)$$

$$MG + MeOH \xrightarrow{k_3} GL + ME \quad (3)$$

Among the alcohols, methanol, ethanol, propanol and butanol (Lang *et al.*, 2001) are used for transesterification process but methanol and ethanol are preferred because of their low cost, easy availability and physico-chemical advantages. Methanol is further preferred to ethanol as it reacts faster in the medium and can be easily dehydrated as compared to ethanol. In case of methyl esters, separation of glycerol is also faster with low stoichiometric ratio.

During transesterification process, 3:1 molar ratio of alcohol to triglycerides is required theoretically; however, higher molar ratio is always used for higher yield. Alkalis, acids enzymes, can catalyze the reaction; however, alkali transesterification is faster and used commercially (Ramadhas *et al.*, 2004).

During the base catalyzed transesterification process, a mixture of vegetable oil and mono alcohol mixed with a base catalyst such as sodium hydroxide or potassium hydroxide are heated and maintained at 65°C for 1 h, while the solution is continuously stirred. Two distinct layers are formed, the lower layer is glycerin and the upper layer is ester. The upper layer (ester) is separated and can be purified either using wet wash methods or dry wash method. Transesterification is the change of the trivalent glycerine molecules against three molecules of mono-valent alcohols i.e. methanol or ethanol. Each alcohol molecule then form with the fatty acid residue, one molecule of monoester. Three consecutive and reversible reactions do occur. The Forward reaction can be pseudo first order or second order depending upon the conditions used. The Reverse reactions are of second order and at the molar ratio of 6:1 of methanol to vegetable oil; a shunt reaction is usually observed (Freedman *et al.*, 1986).

During this process glycerol from triglycerides is removed and replaced with radicals from the alcohols used for the conversion process. Furthermore, the ester fuel also reduces the tendency for polymerization during storage and combustion (Vellguth, 1983; Geyer *et al.*, 1984). The process has the effect of increasing the volatility and decreasing the viscosity of the oil, making it similar to the diesel fuel in these characteristics (Aggarwal, 1998). A schematic diagram of transesterification process is shown in Figure 2.

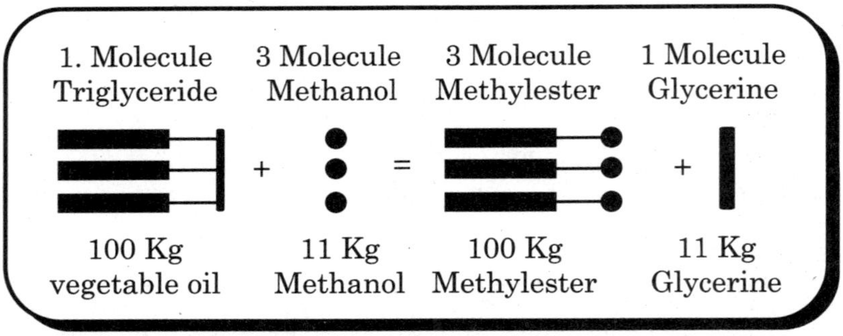

Fig. 2. Transesterification process (Laforgia and Ardito, 1995)

The yield of transesterification reaction is affected by parameters such as alcohol to triglyceride molar ratio, catalyst type and concentration, mixing intensity, reaction temperature, free fatty acids in oil and water content of oils.

FACTORS AFFECTING THE YIELD OF AN ESTER

Effect of molar ratios

The molar ratio of alcohol to vegetable oil is one of the important factors that affect the conversion efficiency as well as production cost of biodiesel. The conversion efficiency is defined as the yield of the process represented in terms of percentage. Molar ratio is the ratio of number of moles of alcohol to number of moles of glycerides in the oil. Theoretically, transesterification reaction requires three moles of alcohol for each mole of oil. However, in practice, the molar ratio should be higher than that of stoichiometric ratio in order to drive the reaction towards completion (Ramadhas *et al.,* 2005).

The alcohol–oil molar ratio that should be used varies from 1:1–6:1. However, 6:1 is the most used ratio giving an important conversion for the alkali catalyst without using a great amount of alcohol. The types of alcohol are usually methanol and ethanol. The last one has fewer safety problems because it is less toxic (Marchetti *et al.,* 2007). The glycerol formed is removed by density separation; however, a very high molar ratio can result in difficulties associated with glycerol removal.

Type and concentration of catalyst

A catalyst is required to speed up the transesterification reaction. Transesterification may be either an acidolysis, where an acid component of an ester is replaced by another acid, or an alcoholysis, where an alcohol component of an ester is replaced with a different alcohol (Peterson, 1991).

The acid catalysis is considerably slower than alkali catalysis for alcoholysis of esters. With acid catalysis, reaction times ranges from 18-69 hours at normal temperature and pressure (Mariinkovic and Tomasaic, 1998). The acid catalysts used in transesterification process are bronsted acids, preferably sulfonic and sulfuric acids. These catalysts give very high yields in alkyl esters, but the reactions are slow, requiring typically temperatures above 100°C and more than 3 h to reach complete conversion. The methanolysis of soybean oil, in the presence of 1 mol% of H_2SO_4, with an alcohol/oil molar ratio of 30:1 at 65°C, took 50 h to reach complete conversion of the vegetable oil (> 99%), while the butanolysis (at 117°C) and ethanolysis (at 78°C), using the same quantities of catalyst and alcohol, took 3 and 18 h, respectively (Schuchardt *et al.,* 1998).

Base catalysts are always considered to be faster than the acid catalysts. The common base catalysts include sodium hydroxide, potassium hydroxide, sodium hydride, potassium hydride, sodium amide, potassium amide and sodium methoxide. Alkali metal alkoxides are the most effective transesterification catalyst compared to the acidic catalyst. Sodium alkoxides are among the most efficient catalysts used for this purpose, although KOH and NaOH can also be used. CH_3ONa for the methanolysis are the most active catalysts, since they give very high yields (> 98%) in short reaction times (30 min) even if they are applied at low molar concentrations (0.5 mol%).

Transmethylation occurs approximately 4000 times faster in presence of an alkaline catalyst than the same amount of acidic catalyst. As they are less corrosive to industrial equipments, alkaline catalysts are preferred in industrial processes. The concentration in the range of 0.5–1.0% (w/w) has found to yield 94–99% conversion of vegetable oils into ester and further increase of catalyst concentration does not affect the conversion but add to extra cost, as the catalyst needs to be removed from the reaction mixture after completion of reaction (Murugesan, 2009).

Alkaline metal hydroxides (KOH and NaOH) are cheaper than metal alkoxides, but lesser active. Nevertheless, they are a good alternative since they can give the same high conversions of vegetable oils just by increasing the catalyst concentration to 1 or 2 mol%. Either KOH or NaOH can be used. NaOH is cheaper but requires a high temperature (70°C) and produces a worthless by-product. The KOH reacts at room temperature and by-product potassium sulphate, can be sold as a fertilizer. However, even if a water-free alcohol/oil mixture is used, some water is produced in the system by the reaction of the hydroxide with the alcohol. The presence of water gives rise to hydrolysis of some of the produced ester, with consequent soap formation. This undesirable saponification reaction reduces the ester yields and considerably difficults the recovery of the glycerol due to the formation of emulsions. Potassium carbonate, used in a concentration of 2 or 3 mol% gives high yields of fatty acid alkyl esters and reduces the soap formation (Phan and Phan, 2008; Ferella *et al.,* 2010).

Catalyst concentration is also believed to affect the biodiesel yield to a great extent. Meher *et al.* (2006) investigated effect of catalyst concentration (KOH) during transesterification of *Pongamia pinnata* oil and found 1% KOH was optimal in the reaction with a yield of 96% in 3 h and the reaction was completed 86.4% in 10min. With the increase in the concentration of catalyst, there was decrease in the yield of methyl esters. This was due to the formation of soap in presence of high amount of catalysts, which increased the viscosity of the reactants and lowered the yield (Meher *et al.,* 2006).

Mixing intensity

A factor of particular importance in the transesterification process is the degree of mixing between the alcohol and triglyceride phases. Triglyceride and alcohol phases are not miscible and form two liquid layers upon their initial introduction into the reactor. Mechanical mixing is normally applied to increase the contact between the reactants, resulting in an increased mass transfer rate. Therefore the variations in mixing intensity are expected to alter the kinetics of transesterification reaction (Noureddini and Zhu, 1997). The transesterification reaction is diffusion-controlled and poor diffusion between the phases results in a slow rate. As methyl esters are formed, they act as a mutual solvent for the reactants and a single-phase system is formed. The mixing effect is most significant during the slow rate region of the reaction. As the single phase is established, mixing becomes insignificant. The understanding of the mixing effects on the kinetics of the transesterification process is a valuable tool in the process scale-up and design (Srivastava and Prasad, 2000).

Temperature

The rate of reaction is strongly influenced by the reaction temperature. However, given enough time, the reaction will proceed to near completion even at room temperature. Generally, the reaction is conducted close to the boiling point of methanol (60 to 70°C) at atmospheric pressure. These mild reaction conditions, however, require the removal of free fatty acids from the oil by refining or pre- esterification. Therefore, the degummed and deacidified feedstock is used at these conditions. The pretreatment is not required if the reaction is carried out under high pressure (9 MPa) and high temperature (240°C). Under these conditions, simultaneous esterification and transesterification take place. The maximum yield of esters occurs at temperatures ranging from 60 to 80°C at a molar ratio (alcohol to oil) of 6:1 (Srivastava and Prasad, 2000). Further increase in temperature is reported to have a negative effect on the conversion.

At higher reaction temperatures, there is a chance of loss of methanol and increase in darkness of the product. High reaction temperature increase the production cost of biodiesel also (Ramadhas *et al.,* 2005).

The butanolysis and methanolysis of soyabean oil were studied at different temperatures in the presence of acidic and alkaline catalysts. These studies indicated that given enough time, transesterification can proceed satisfactorily at ambient temperatures in the case of the alkaline catalyst (Sensoz *et al.,* 2000). Freedman *et al.* (1984) conducted some experiments with refined soybean oil at 60°C, 45°C and 32°C with 6:1 molar ratio and 1% NaOH as catalyst and observed that after 0.1 hr of time, the percentage conversion into ester was 94%, 87% and 64% for temperatures of 60°C, 45°C and 32°C respectively, showing the influence of temperature on ester conversion (Freedman *et al.,* 1984).

Presence of impurities such as free fatty acids and moisture

The triglycerides used for alkali catalysed alcoholysis should meet certain specifications. The glyceride should have an acid value less than 1, and all materials should be substantially anhydrous (Freedman *et al.,* 1984). Transesterification reaction is very sensitive to the feedstock purity that directly affects the reaction performance. Contrarily to refined vegetable oils, waste oils and animal fats usually have a lot of impurities, such as FFA and water that negatively affect the reaction yield, reducing the reaction rate by several orders of magnitude, even in small amounts. The addition of more alkaline catalyst compensates for higher acidity, but the resulting soap causes an increase in viscosity or formation of gels that interferes in the reaction as well as with separation of glycerol (Meher *et al.,* 2006).

IMPROVED METHODS OF BIODIESEL PRODUCTION

Enzymatic Transesterification

Although chemical methanolysis using an alkali catalyst gives higher conversion levels of triglycerides to their corresponding methyl esters in short reaction times, the reaction has several drawbacks: it is energy intensive, recovery of glycerol is difficult, the alkaline catalyst has to be removed from the product, alkaline wastewater requires treatment, and free fatty acids and water interfere with the reaction (Fukuda *et al.,* 2001).

Recently, enzymatic methanolysis using lipases has become more attractive for biodiesel fuel production, since it can overcome the above problems. In particular, it should be noted that the byproduct glycerol, can be easily recovered without complex processing, and also that free fatty acids contained in waste oils and fats can be completely converted to methyl esters. Enzymes are generally effective biocatalyst for having substrate specificity, functional group specificity and stereo specificity in aqueous media. Besides, chemical reactions can also be raised directly using lipase in an organic media. The esterification reaction of alcohol and carboxylic acid is difficult in aqueous media, whereas this reaction takes place easily in organic solvent. Several extracellular lipases from *Candida* sp., *Pseudomonas* sp. and *Rhizopus* sp. have also resulted in effective methanolysis reactions. With these lipases, methyl ester content in the reaction mixture of more than 90% is obtained using either low- or high-water content systems. However, the use of extracellular lipase as a catalyst requires complicated recovery, purification, and immobilization processes for industrial application. However, the production of biodiesel fuel by enzymatic method has not been adopted industrially, because of the high cost of enzyme catalyst. In order to use the enzyme catalyst repeatedly, the process of immobilization must be carried out using appropriate method (Iso *et al.,* 2002; Ban *et al.,* 2002).

The lipase catalyst could not be reused in spite of using immobilized enzyme. This problem was basically due to the irreversible inactivation of the lipase by contact with insoluble methanol (Shimada *et al.,* 2002). Also their reaction systems were insufficient to apply for industrial biodiesel fuel production from the following reasons:

(i) Reaction efficiency was poor and a large amount of enzyme was necessary.

(ii) Methanolysis of triacylglycerols (TAGs) efficiently proceeded in the presence of organic solvent, but not in the absence.

(iii) Though immobilized lipase was used as a catalyst, the enzyme preparation could not be reused (Watanble *et al.,* 2002).

There are two major limitations of lipase-catalyzed biodiesel synthesis. One is higher cost (which can be reduced up to a certain extent by immobilization) and another is its inactivation by methanol and glycerol (Kumari *et al.,* 2009).

Heterogenous catalysts

Heterogeneous catalysts can provide new routes for the environmentally benign production of biodiesel. Particulate heterogeneous catalysts can be readily separated from products following reaction allowing the catalyst to be reused, generating less waste, and consuming less energy. A considerable amount of energy is required for the purification of products and catalyst separation during chemical catalyst assisted biodiesel production, and furthermore these catalysts are not reusable. This process results in substantial energy wastage and the production of large amounts of chemical waste. Particulate heterogeneous catalysts can be readily separated from products following reaction, allowing the catalyst to be reused and consuming less energy. Heterogeneous catalyst can also use high FFA feedstock so that biodiesel can be made competitive with diesel from a economical point of view (Hara, 2009).

Based on the above reasons, developing a new solid catalyst seems to be an appropriate solution to overcome problems associated with homogeneous catalysts. Metal hydroxides , metal complexes, metal oxides such as calcium oxide, magnesium oxide, zirconium oxide and supported catalysts have been investigated as solid catalysts. The catalysts are not consumed or dissolved in the reaction and therefore, can be easily separated from the products. As a result, the products do not contain impurities of the catalyst and the cost of final separation could be reduced. The catalysts can also be readily regenerated and reused and it is more environmentally benign because there is no need for acid or water treatment in the separation step (Zabeti, 2009). A number of researchers have used different kind of heterogeneous catalysts in their study and the acceptability of these catalysts is increasing day by day (Sercheli *et al.*, 1999; Leclercq, 2001; Suppes *et al.*, 2004; Wang *et al.*, 2006; Xie and Chen, 2006).

Co-solvent Method

Catalyzed transesterification suffers from two main drawbacks; the process is relatively time consuming and purification of the product for catalyst and saponified products are necessary. The first problem due to the two phase nature of vegetable oil/methanol mixture requires vigorous stirring to proceed in the transesterification reaction. Boocock *et al.* (1996) have suggested that the use of simple ether such as tetrahydrofuran can make this two phase nature into one phase of its mixture and that methyl esters can be produced in less than 15 min depending on the catalyst concentration. They found that at the 6:1 methanol: oil molar ratio the addition of a cosolvent, such as 1.25 volumes of tetrahydrofuran (THF) per volume of methanol, produces an oil-dominant one-phase system in which methanolysis speeds up dramatically. The critical separation of the glycerol-rich phase still occurs and does so faster than in the cosolvent-free system. For THF, recycle of solvent is simplified because of the similar boiling points of THF (67°C) and methanol (65°C). Tetrahydrofuran, when used as a co-solvent during transesterification of *Pongamia pinnata* oil increased the conver-sion to 95% from 92% when no co-solvent was used (Karmee and Chadha, 2005). Use of dimethyl ether has also been reported during esterification in presence of acid catalysts (Guan *et al.,* 2009). Yet, the catalyst problem cannot be solved for purification. Therefore, this conventional process still requires a high production cost and energy. The overall process, thus, includes transesterification reaction, recovery of un-reacted methanol, and purification of methyl esters from catalyst and separation of glycerin as a co-product from saponified products.

Supercritical alcoholysis

Acid or base catalyzed transesterification suffers from main drawback of purification of end products. One method to overcome these difficulties has been reported as catalyzed transesterification of vegetable oil in supercritical methanol. The supercritical state of methanol is believed to solve the two phase nature of oil/methanol mixture to form a single phase due to a decrease in dielectric constant of methanol in supercritical state resulting the reaction to be complete in a very short time within 2-4 min. In addition, because of non-catalytic process, the purification of products after transesterification reaction is much simpler and environmentally friendly, compared with the conventional commercial method in which all the catalyst and saponified products have to be removed for biodiesel fuel (Kusdiana and Saka, 2001a). High temperatures and pressures are required, but energy costs of production are similar or less than catalytic production routes (Kusdiana and Saka, 2001b). The process can tolerate water in the feedstock; free fatty acids are converted to methyl esters instead of soap, so a wide variety of feedstocks can be used (Vera *et al.,* 2005).

Demirbas (2002) carried transesterification of six vegetable oils (cotton seed, hazelnut kernel, poppy seed, rape seed, safflower seed and sun-flower seed by supercritical methanol in a set up capable of operating at 100Mpa and 850K. The transesterification of all the vegetable oil were completed within 15 minutes.

The synthesis of biodiesel from edible oils like palm oil and groundnut oil and from crude non-edible oils like *Pongamia pinnata* and *Jatropha curcas* was investigated in supercritical methanol and ethanol without using any catalyst from 200 to 400°C at 200 bar. The variables affecting the conversion during transesterification, such as molar ratio of alcohol to oil, temperature and time were investigated in supercritical methanol and ethanol. Very high conversions (>80%) were obtained within 10 min and nearly complete conversions were obtained at within 40 min for the synthesis of biodiesel in supercritical alcohols (Rathore and Madras, 2007).

Ultrasonic-reactor method

In the last two decades, ultrasonic cavitation has seen diverse application in a number of fields, such as synthesis of nanostructured materials, processing of biomass, sonofusion, sonodynamic therapy, and sonochemical degradation of pollutants and hazardous chemicals. The ultrasonic field is known to produce unique chemical and physical effects. In the ultrasonic biodiesel synthesis, the ultrasonic waves cause the reaction mixture to produce and collapse bubbles constantly. Normally, low frequency ultrasounds (28 and 40 kHz) are used. This cavitation provides simultaneously the mixing and heating required to carry out the transesterification process. Experiments show that ultrasonic wave facilitates ester exchange reaction, reducing the reaction time to only $5 \sim 30$ minutes. The molar ratio of alcohol/oil used is only 6:1. Normal chain alcohols react fast, while secondary and tertiary alcohols show some or no conversion after 60 min of reaction. The ultrasonic process uses only $30\% \sim 50\%$ of the catalysts used in the liquid alkali catalysis technology and its methyl ester conversion rate can reach as high as 99%. 40 kHz ultrasounds are much more effective in the reduction of the reaction time (10–20 min). 28 kHz give slightly better yields (98–99%), but longer reaction time, while higher frequencies are not useful at all for the transesterification of fatty acids.

However, commercializing the ultrasonic biodiesel technology requires a series of processes such as feedstock pre-processing, methanol recovery, methyl ester cleaning and methyl ester desiccation (Kaidong *et al.*, 2005; Stavarache *et al.*, 2005; Hanh and Maeda, 2005). The schematic of ultrasonic transesterification is shown in Figure 3.

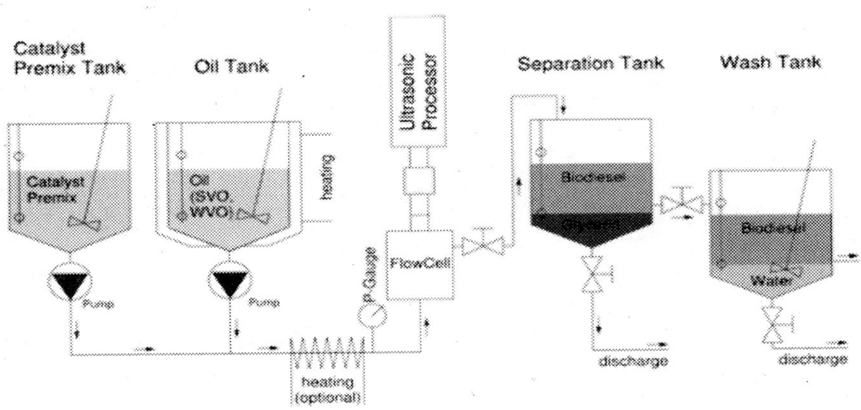

Fig. 3. The Schematic of Ultrasonic Transesterification
(Source- www.greencongress.com)

The effect of ultrasonics on biodiesel production from soybean oil was studied by Chand *et al.* (2010). Esterification of palm fatty acid distillate (PFAD) with methanol in the presence of concentrated H_2SO_4 as a catalyst has been investigated in the presence of ultrasonic irradiations at

22 kHz frequency and 120 W as supplied power dissipation (Deshane *et al.*, 2009).

Microwave method

During conventional heating in transesterification process, heat energy is transferred to the reaction through convection, conduction and radiation from the surfaces of the reactor, which is an inefficient heat transfer. Large amount of energy is used to heat the media. Thus, long reaction time (usually from 30 min to 8h) is required to achieve a satisfactory conversion of oil to biodiesel. Microwave radiation, on the other hand, delivers energy directly to the reactants. Therefore, preheating step is eliminated. Heat transfer is more effective than conventional heat. Consequently, transesterification with microwave radiation can be completed in much shorter time (Lertsathapornsuk *et al.*, 2008).

In the electromagnetic radiation spectrum, the microwave radiation region is located between infrared radiation and radio waves. Microwaves have wavelengths of 1 mm - 1 m, corresponding to frequencies between 0.3 - 300 GHz. In general, in order to avoid interference, industrial and domestic microwave apparatus are regulated to 12.2 cm, corresponding to a frequency of 2.45 GHz, but other frequency allocations do exist. Microwaves, a nonionizing radiation incapable of breaking bonds, are a form of energy and not heat and are manifested as heat through their interaction with the medium or materials wherein they can be reflected (metals), transmitted (good insulators that will not heat) or absorbed (decreasing the available microwave energy and rapidly heating the sample). Microwaves, representing a non-ionizing radiation, influence molecular motions such as ion migration or dipole rotations, but not altering the molecular structure. Polar solvents of low molecular weight and high dielectric constant irradiated by microwaves increase their temperature very rapidly, reaching boiling point in a short time: in this class of solvents the rate of a given reaction is more enhanced. Typical solvents widely employed in microwave chemistry are water, methanol, ethanol, and acetone. In the presence of microwaves, common solvents are found to boil at higher temperatures: for water the difference is about 5°C; 19°C for methanol, up to 36°C was the difference measured for tetrahydrofuran and acetonitrile. Microwave irradiation, has been used for a variety of applications including organic synthesis, wherein chemical reactions are accelerated because of selective absorption of MW energy by polar molecules, non-polar molecules being inert to the MW dielectric loss. The application of microwave irradiation in conjunction with the use of catalysts or mineral-supported reagents enables organic reactions to occur expeditiously at ambient pressure, thus providing unique chemical processes with special attributes such as enhanced reaction rates, higher yields, and the associated ease of manipulation (Refaat and Sheltawy, 2008).

The methodology allows for the reaction to be run under atmospheric conditions; it is complete in a matter of a few minutes and can be performed on batch scales up to 3 kg of oil at a time. It can be utilized with new or used vegetable oil and with methanol or ethanol with a 1:6 ratio of oil/alcohol (Leadbeater and Stencel, 2006). Rapeseed oil was converted to biodiesel by transesterification using microwave heating. The results indicated that microwave heating has effectively increased the biodiesel yield and decreased the reaction time (Azcan and Danisman, 2008).

Although it is still in the lab-scale development stage, the microwave method holds great potential to be an efficient and cost-competitive method for commercial-scale biodiesel production.

Purification of biodiesel

After separation from the glycerol phase, crude biodiesel is mainly contaminated with residual catalyst, water, unreacted alcohol, free glycerol, and soaps that were generated during the transesterification reaction. Normally, crude biodiesel enters a neutralization step and then passes through an alcohol stripper before the washing step. In some cases, acid is added to crude biodiesel to neutralize any remaining catalyst and to split any soap. Soaps react with the acid to form water soluble salts and free fatty acids. Neutralization before the washing step reduces the materials required for the washing step and minimizes the potential for emulsions being formed during the washing step. Unreacted alcohol should be removed with distillation equipment before the washing step to prevent excess alcohol from entering the wastewater effluent. Three main approaches are adopted for purifying biodiesel: water washing, dry washing, and membrane extraction (Dennis *et al.*, 2010). However, water washing is becoming obsolete due to large requirement of water and the environmental legislations which mandates treatment of effluent water. In this context dry washing is becoming more relevant for purification of biodiesel.

During dry washing water is replaced with an ion exchange resin or a magnesium silicate powder in order to remove impurities. These two dry washing methods can bring the free glycerol level down and is reasonably effective for removing soaps. Both the ion exchange process and the magnesol process have the advantage of being waterless and thus eliminate many of the problems outlined above. The magnesol process has a better effect on the removal of methanol than the ion resins. Sabudak *et al.* (2010) used magnesol and ion exchange resin (Purolite-PD 206) for purification of biodiesel produced from waste cooking oil and found that methyl ester yield was higher in case of dry washing as compared to wet washing. Table 3 summarizes relative features of the three purification techniques (Dennis *et al.*, 2010).

Table 3. Relative features of different purification techniques

Approaches	Primary material used	Function	Phases separation	Advantage	Disadvantage
Water washing	Distilled warm water	Prevents precipitation of saturated fatty acid esters Retards the emulsion formation	Separation funnel, centrifuge, molecular sieves, silica gels, etc.	Very effective in removing contaminants	Increased cost and production time, liquid effluent, product loss, emulsions formation
	Softened water	Eliminates calcium and magnesium contamination			
Dry washing	Ion exchange resin Magnesium silicate powder	Brings the free glycerol level down and removing soaps	–	Waterless	Overruns the limit in the EN Standard
Membrane extraction	Polysulfone	Remove the contaminants	–	Avoids the emulsion formation and decreases the refining loss	Probably high cost and low throughput due to contaminants existed

CONCLUSION

Biodiesel has become an attractive alternative fuel for diesel engines because it is renewable and provide environmental benefits. Amongst the different routes of biodiesel production, transesterification of lipid is most preferred. An alkali catalyzed transesterifcation process is mostly adopted for biodiesel production and the yield is dependent upon number of operating variables such as molar ratio, catalyst type and concentration, temperature, level of agitation and impurities such as FFA. However, there are short comings in existing process and there is an urgent need to develop and adapt new technologies. As removal of catalysts from raw biodiesel is always a problem, use of enzymes and heterogenous catalysts is going to increase in a big way to alleviate this problem. Ultrasonic and microwave assisted biodiesel synthesis have a major role to play in coming time as these technologies shall reduce the reaction time drastically and reduce the processing cost. With the stringent environmental legislation, the dry washing of biodiesel is the future technology which shall rule the industry in future.

REFERENCES

Aggarwal, A.K. 1998. Vegetable Oil Versus Diesel Fuel, *TERI Information Digest on Energy,* 8:191-205.

Ali, Y. and Hanna, M.A. 1994. Alternative Diesel Fuels from Vegetable Oils, *Bioresource Technology,* 50:153-163.

Azcan, N. and Danisman, A. 2008. Microwave assisted transesterification of rapeseed oil. *Fuel*, 87:1781-1788.

Ban, K., Hama, S. and Nishizuka, K. 2002. Repeated use of whole-cell biocatalyst immobilized within biomass support particles for biodiesel production. Journal of Molecular Catalysis, 17: 157-165.

Boocock, D.G.B., Konar, S.K., Mao, V. and Sidi, H. 1996. Fast one-phase oil-rich processes for the preparation of vegetable oil methyl esters. Biomass and Bioenergy, 11(1): 43-50.

Chand, P., Chintareddy, V.R., Verkade, J.G. and Grewell, D. 2010. Enhancing biodiesel production from soybean oil using ultrasonics. *Energy Fuels,* 24(3): 2010-2015.

Chauhan, B.S., Kumar, N. and Cho, H.M. 2010. Performance and emission studies on an agriculture engine on neat Jatropha oil. *Journal of Mechanical Science and Technology,* 24(2):529-535.

Chauhan, B.S., Kumar, N., Jun, Y.D. and Lee, K.B. 2010. Performance and emission study of preheated Jatropha oil on medium capacity diesel engine. *Energy,* 35: 2484-2492.

Crookes, R.J., Kiannejad, F. and Nazha, M.A.A. 1997. Systematic Assessment of Combustion Characteristics of Biofuels and Emulsions with Water for Use as Diesel Engine Fuel. *Energy Conversion and Management,* 38:1785-1795.

Demirbas, A. 2002. Biodiesel from vegetable oil via transesterification in supercritical methanol. *Energy Conversion and Management*, 43: 2349-2356.

Deshmane, V.G., Gogate, P.R. and Pandit, A.B. 2009. Ultrasound-assisted synthesis of biodiesel from palm fatty acid distillate. *Industrial and Engineering Chemistry Research,* 48(17):7923-7927.

Ferella, F., Mazziotti Di Celso, G., De Michelis, I., Stanisci, V. and Veglio, F. 2010. Optimization of the transesterification reaction in biodiesel production. *Fuel,* 89:36-42.

Freedman, B. Pryde, E.H. and Mounts, T.L. 1984.Variables affecting the yields of fatty esters from transesterified vegetables oils. *Journal of American Oil Chemists Society,* 61(10):1638-1643.

Freedman, B., Butterfield, R.O. and Pryde, E.H. 1986. Transesterified Kinetics of Soybean Oil. *Journal of American Oil Chemists Society,* 63(10):1375-1380.

Fukuda, H., Kondo, A. and Noda, H. 2001. Biodiesel fuel production by transesterification of oils. *Journal of Bioscience and Bioengineering*, 92(5): 405-416.

Gerpen, J.V., Shanks, B. and Pruszko, R. 2004. Biodiesel Production Technology, National Renewable Energy Laboratory subcontractor report NREL/SR-510-36244.

Geyer, S.M., Jacobus, M.J. and Lestz, S.S. 1984. Comparison of Diesel Engine Performance and Emissions from Neat and Transesterified Vegetable Oils, *ASAE* 27(2):375-381.

Guan, G., Kusakabe, K., Sakurai, N. and Moriyama, K. 2009. Transesterification of vegetable oil to biodiesel fuel using acid catalysts in the presence of dimethyl ether. *Fuel*, 88(1):81-86.

Hanh, H.D., Dong, N.T., Okitsu, K., Nishimura, R. and Maeda, Y. 2009. Biodiesel production through transesterification of triolein with various alcohols in an ultrasonic field. *Renewable Energy*, 34(3):766-768.

Hara, M. 2009. Environmentally benign production of biodiesel using heterogeneous catalysts. *Chem. Sus. Chem.*, 2(2):129-35.

http://www.enpromer2005.eq.ufrj.br/nukleo/pdfs/0818_paper_818.pdf

Iso, M., Chen, B. Eguchi, M. and Kudo, T. 2001. Production of biodiesel fuel from triglycerides and alcohol using immobilized lipase. *Journal of Molecular Catalysis, 16:53-58.*

Kaidong, Y., Ke, G., Peng, Z. and Ming, G. 2009. A Study into ultrasonic biodiesel production technology. *Journal of Sino-Global Energy,* 5.

Karmee, S.K. and Chadha, A. 2005. Preparation of biodiesel from crude oil of *Pongamia pinnata*. *Bioresource Technology*, 96(13):1425-1429.

Kumar, N. and Sharma, P.B. 2005. *Jatropha curcas*: A sustainable source for production of biodiesel. *Journal of Scientific and Industrial Research,* 64:883-889.

Kumari, A., Mahapatra, P., Garlapati, V.K. and Banerjee, R. 2009. Enzymatic transesterification of Jatropha oil. *Biotechnology for Biofuels*, 2:1.

Kusdiana, D. and Saka, S. 2001. Kinetics of transesterification in rapeseed oil to biodiesel fuel as treated in supercritical methanol. *Fuel*, 80:693-698.

Laforgia, D. and Ardito, V. 1995. Biodiesel fueled IDI engines: performances, emissions and heat release investigation. *Bioresource Technology,* 51(1):53-59.

Lang, X., Dalai, A.K., Bakshi, N.N., Reaney, M.J. and Hertz, P.B. 2001. Preparation and characterization of bio-diesel from various bio-oils. *Bioresource Technology,* 80:53-62.

Leadbeater, N.E. and Stencel, L.M. 2006. Fast, easy preparation of biodiesel using microwave heating. *Energy Fuels*, 20(5):2281–2283.

Leclercq, E. Finiels, A. and Moreau, C. 2001. Transesterification of rapeseed oil in the presence of basic zeolites and related solid catalysts. *Journal of American Oil Chemists Society*, 78(11):1161.

Lertsathapornsuk, V., Pairintra, R., Aryusuk, K. and Krisnangkura, K. 2008. Microwave assisted in continuous biodiesel production from waste frying palm oil and its performance in a 100 kW diesel generator. *Fuel Processing Technology*, 89:1330-336.

Leung, D.Y.C., Wu, X. and Leung, M.K.H. 2010. A review on biodiesel production using catalyzed transesterification. *Applied Energy*, 87:1083–1095.

Louloudi, M.D.A. and Papayannakos, N. 1998. Kinetics of the non-catalytic transesterification of soybean oil, *Fuel,* 77 (12):1297-1302.

Marchetti, J.M., Miguel, V.U. and Errazu, A.F. 2007. Possible methods for biodiesel production. *Renewable and Sustainable Energy Reviews,* 11:1300-1311.

Mariinkovic, S.S. and Tomasaic, A. 1998. Transesterification of sunflower oil *in situ*. *Fuel,* 77(12):1389-1391.

Meher, L.C., Dharmagadda, V.S.S, Naik, S.N. 2006. Optimization of alkali-catalyzed transesterification of *Pongamia pinnata* oil for production of biodiesel. *Bioresource Technology,* 97:1392:1397.

Murugesan, A., Umarani, C., Chinnusamy, T.R., Krishnan, M., Subramanian, R. and Neduzchezhain, N. 2009. Production and analysis of bio-diesel from non-edible oils: A review. *Renewable and Sustainable Energy Reviews,* 13:825-834.

Noureddini, H. and Zhu, D. 1997. Kinetics of transesterification of soybean oil. *Journal of American Oil Chemists Society,* 74(11):1457-1463.

Peterson, C.L., Fieldman, M., Korus, R. and Auld, D.L. 1991. Batch size transesterification process for winter rape oil. *ASAE,* 7(6):711-716.

Phan, A.N. and Phan, T.M. 2008. Biodiesel production from waste cooking oils. *Fuel,* 87(17-18):3490-3496.

Ramadhas, A.S., Jayaraj, S. and Muraleedharan, C. 2004. Use of Vegetable Oils as an I.C Engine Fuels: A review, *Renewable Energy,* 29:727-742.

Ramadhas, A.S., Jayaraj, S. and Muraleedharan, C. 2005. Biodiesel production from high FFA rubber seed oil. *Fuel,* 84:335-340.

Rathore, V. and Madras, G. 2007. Synthesis of biodiesel from edible and non-edible oils in supercritical alcohols and enzymatic synthesis in supercritical carbon dioxide. *Fuel*, 86(17-18):2650-2659.

Refaat, A.A. and Sheltawy, S.T. El. 2008. Time Factor in microwave-enhanced biodiesel production. *WSEAS Transactions on Environment and Development*, 4(4):279-288.

Sabudak, T. and Yildiz, M. 2010. Biodiesel production from waste frying oils and its quality control. *Waste Management*, 30(5):799-803.

Schuchardt, U., Sercheli, R. and Vargas, R.M. 1998. Transesterification of Vegetable Oils: a Review. *Journal of Brazilian Chemists Society,* 9(1):199-210.

Sensoz, S., Angin, D. and Yorgun, S. 2000. Influence of particle size on the pyrolysis of rapeseed: fuel properties of bio-oil. *Biomass and Bioenergy,* (19):271-279.

Sercheli, R., Vargas, R.M. and Schuchardt, U. 1999. Alkyl guanidines-Catalysed heterogeneous transesterification of soybean oil. *Journal of American Oil Chemists Society*, 76(10):1207.

Shimada, Y., Watanble, Y., Sugihra, A. and Tominaga, Y. 2002. Enzymatic alcholysis for biodiesel fuel production and application of the reaction to oil processing. *Journal of Molecular Catalysis*, 17:133-141.

Srivastava, A. and Prasad, R. 2000.Triglycerides-based diesel fuels. *Renewable and Sustainable Energy Reviews,* 4:111-133.

Stavarache, C., Vinatoru, M., Nishimura, R. and Maeda, Y. 2005. Fatty acids methyl esters from vegetable oil by means of ultrasonic energy. *Ultrasonics Sonochemistry*, 12:367–372.

Suppes, G.J., Dasari, M.A., Doskocil, E.J., Mankidy, P.J. and Goff, M.J. 2004. "Transesterification of soybean oil with zeolite and metal catalysts. *Applied Catalysts A: General*, 257(2):213.

Vellguth, G. 1986. Performance of Vegetable Oils and Their Monoesters as Fuels for Diesel Engines. *SAE- 831358.*

Vera, C.R., D'Ippolito, S.A., Pieck, C.L. and Parera, J.M. 2005. Production of biodiesel by a two-step supercritical reaction process with adsorption refining, 2nd Mercosur Congress on Chemical Engineering.

Wang, Y., Ou, S., Liu, P., Xue, F. and Tang, S. 2006. Comparison of two different processes to synthesize biodiesel by waste cooking oil. *Journal of Molecular Catalysis A: Chemical*, 252:107–112.

Watanble, Y., Shimada, Y., Sugihra, A. and Tominaga, Y. 2002. Conversion of degummed soyabean oil to biodiesel fuel with immoblized *Candida antarctica* lipase. *Journal of Molecular Catalysis*, 17:151-155.

Xie, W. and Chen, H.P.L. 2006. Calcined Mg–Al hydrotalcites as solid base catalysts for methanolysis of soybean oil. *Journal of Molecular Catalysis A: Chemical*, 246(1-2):24-32.

Zabeti, M., Daud, W.M.A.W. and Aroua, M.K. 2009. Activity of solid catalysts for biodiesel production: A review. *Fuel Processing Technology*, 90:770–777.